3 0388 00121883 6

D0072743

The Facts On File
DICTIONARY
of
ORGANIC CHEMISTRY

The Facts On File

DICTIONARY of ORGANIC CHEMISTRY

Edited by
John Daintith

The Facts On File Dictionary of Organic Chemistry

Facts On File, Inc.
132 West 31st Street
New York NY 10001

Library of Congress Cataloging-in-Publication Data

The Facts on File dictionary of organic chemistry / edited by John Daintith.
p. cm.
Includes bibliographical references.
ISBN 0-8160-4928-9 (alk. paper).
1. Chemistry, Organic—Dictionaries. I. Title: Dictionary of organic chemistry. II. Daintith, John.

QD246.F33 2004
547'.003—dc22 2003061227

Facts On File books are available at special discounts when purchased in bulk quantities for businesses, associations, institutions, or sales promotions. Please call our Special Sales Department in New York at (212) 967-8800 or (800) 322-8755.

You can find Facts On File on the World Wide Web at http://www.factsonfile.com

Compiled and typeset by Market House Books Ltd, Aylesbury, UK

Printed in the United States of America

MP 10 9 8 7 6 5 4 3 2 1

This book is printed on acid-free paper

CONTENTS

PREFACE

This dictionary is one of a series covering the terminology and concepts used in important branches of science. *The Facts On File Dictionary of Organic Chemistry* has been designed as an additional source of information for students taking Advanced Placement (AP) Science courses in high schools. It will also be helpful to older students taking introductory college courses.

This volume covers organic chemistry and includes basic concepts, classes of compound, reaction mechanisms, and important named organic compounds. In addition, we have included a number of compounds that are important in biochemistry, as well as information on certain key biochemical pathways. The definitions are intended to be clear and informative and, where possible, we have illustrations of chemical structures. The book also has a selection of short biographical entries for people who have made important contributions to the field. There are a number of appendixes, including structural information on carboxylic acids, amino acids, sugars, and nitrogenous bases and nucleosides. There is also a list of all the chemical elements and a periodic table. The appendixes also include a short list of useful webpages and a bibliography.

The book will be a helpful additional source of information for anyone studying the AP Chemistry course, especially the section on Descriptive Chemistry. It will also be useful to students of AP Biology.

ACKNOWLEDGMENTS

Contributors

John O. E. Clark B.Sc.
Richard Rennie B.Sc., Ph.D.

ABA *See* abscisic acid.

abscisic acid (ABA) A PLANT HORMONE once thought to be responsible for the shedding (abscission) of flowers and fruit and for the onset of dormancy in buds (hence its early name, *dormin*). The compound is associated with the closing of pores (stoma) in the leaves of plants deprived of water.

absolute alcohol Pure alcohol (ethanol).

absolute configuration A particular molecular configuration of a CHIRAL molecule, as denoted by comparison with a reference molecule or by some sequence rule. There are two systems for expressing absolute configuration in common use: the *D–L convention* and the *R–S convention*. *See* optical activity.

absolute temperature Symbol: T A temperature defined by the relationship:

$$T = \theta + 273.15$$

where θ is the Celsius temperature. The absolute scale of temperature was a fundamental scale based on Charles' law, which applies to an ideal gas:

$$V = V_0(1 + \alpha\theta)$$

where V is the volume at temperature θ, V_0 the volume at 0, and α the thermal expansivity of the gas. At low pressures (where real gases show ideal behavior) α has the value 1/273.15. Therefore, at $\theta = -273.15$ the volume of the gas theoretically becomes zero. In practice substances become solids at these temperatures; however, the extrapolation can be used for a scale of temperature on which –273.15°C corresponds to 0° (*absolute zero*). The scale is also known as the *ideal-gas scale*; on it temperature intervals were called *degrees absolute* (°A) or *degrees Kelvin* (°K), and were equal to the Celsius degree. It can be shown that the absolute temperature scale is identical to the currently used thermodynamic temperature scale (on which the unit is the KELVIN).

absolute zero The zero value of thermodynamic temperature; 0 kelvin or –273.15°C. *See* absolute temperature.

absorption **1.** A process in which a gas is taken up by a liquid or solid, or in which a liquid is taken up by a solid. In absorption, the substance absorbed goes into the bulk of the material. Solids that absorb gases or liquids often have a porous structure. The absorption of gases in solids is sometimes called *sorption*. There is a distinction between *absorption* (in which one substance is assimilated into the bulk of another) and ADSORPTION (which involves attachment to the surface). Sometimes it is not obvious which process is occurring. For example, a porous solid, such as activated CHARCOAL may be said to *absorb* a large volume of gas, but the process may actually be *adsorption* on the high surface area of internal pores in the material.
2. The process in which electromagnetic radiation, particles, or sound waves lose energy in passing through a medium. Absorption involves conversion of one form of energy into another.

absorption spectrum *See* spectrum.

accelerator A substance that increases the rate of a chemical reaction. In this sense the term is synonymous with CATALYST. It is common to refer to catalysts as 'accelerators' in certain industrial applications. For example, accelerators are used in the VULCANIZATION of rubber and in the polymerization of adhesives. Also, in the production of composite materials using polyester resins a distinction is sometimes made between the catalyst (which initiates the polymerization reaction) and the accelerator (which is an additional substance making the catalyst more effective). The terms *promoter* and *activator* are used in a similar way.

acceptor The atom or group to which a pair of electrons is donated in forming a COORDINATE BOND.

accessory pigment *See* photosynthetic pigments.

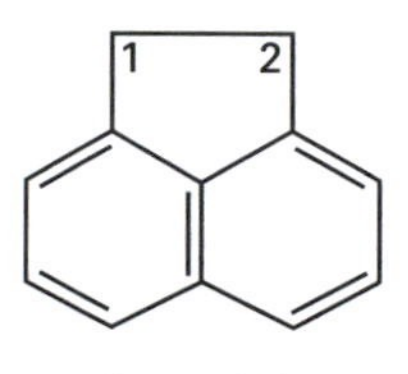

Acenaphthene

acenaphthene ($C_{12}H_{10}$) A colorless crystalline derivative of naphthalene, used in producing some dyes.

acetal A type of compound formed by reaction of an alcohol with either an aldehyde or a ketone. The first step in formation of an acetal is the formation of an intermediate, known as a *hemiacetal*. For example, ethanal (acetaldehyde; CH_3CHO) reacts with ethanol (C_2H_5OH) as follows:

$$CH_3CHO + C_2H_5OH \rightleftharpoons CH(OH)(CH_3)(C_2H_5O)$$

The hemiacetal has a central carbon atom (from the aldehyde) attached to a hydrogen, a hydroxyl group, a hydrocarbon group (CH_3), and an alkoxy group (C_2H_5O). If a ketone is used rather than an aldehyde, the resulting hemiacetal contains two hydrocarbon groups. For example, reaction of the ketone R^1COR^2 with the alcohol R^3OH is:

$$R^1COR^2 + R^3OH \rightleftharpoons CR^1R^2(OH)(OR^3)$$

The formation of a hemiacetal is an example of NUCLEOPHILIC ADDITION to the carbonyl group of the aldehyde or ketone. The first step is attack of the lone pair on the O of the alcohol on the (positively charged) C of the carbonyl group. This is catalyzed by both acids and bases. Acid catalysis occurs by protonation of the O on the carbonyl, making the C more negative and more susceptible to nucleophilic attack. In base catalysis the OH^- ions from the base affect the –OH group of the alcohol, making it a more effective nucleophile.

In general, hemiacetals exist only in solution and cannot be isolated because they easily decompose back to the component alcohol and aldehyde or ketone. However, some cyclic hemiacetals are more stable. For example, cyclic forms of SUGAR molecules are hemiacetals.

Further reaction of hemiactals with another molecule of alcohol leads to a full acetal. For example:

$$CH(OH)(CH_3)(OC_2H_5) + C_2H_5OH \rightleftharpoons CH(CH_3)(OC_2H_5)_2$$

The overall reaction of an aldehyde or ketone with an alcohol to give an acetal can be written:

$$R^1COR^2 + R^3OH \rightleftharpoons CR^1R^2(OR^3)_2$$

It is also possible to have 'mixed' acetals with the general formula $CR^1R^2(OR^3)(OR^4)$. Note that if the acetal is derived from an aldehyde, then R^1 and/or R^2 may be a hydrogen atom. The mechanism of formation of an acetal from a hemiacetal is acid catalyzed. It involves protonation of the –OH group of the hemiacetal followed by loss of water to form an oxonium ion, which is attacked by the alcohol molecule.

Formerly it was conventional to use the terms 'hemiacetal' and 'acetal' for compounds formed by reaction between aldehydes and alcohols. Similar reactions between ketones and alcohols gave rise to compounds called *hemiketals* and *ketals*. Current nomenclature uses 'hemiacetal' and 'acetal' for compounds derived from either an aldehyde or a ketone, but reserves 'hemiketal' and 'ketal' for those derived from ketones. In other words, the ketals are a subclass of the acetals and the hemiketals are a subclass of the hemiacetals.

acetaldehyde *See* ethanal.

acetamide *See* ethanamide.

acetate *See* ethanoate.

acetic acid *See* ethanoic acid.

acetone *See* propanone.

acetonitrile *See* methyl cyanide.

acetophenone *See* phenyl methyl ketone.

acetylation *See* acylation.

acetyl chloride *See* ethanoyl chloride.

acetylcholine (ACh) A neurotransmitter found at the majority of synapses, which occur where one nerve cell meets another.

acetylene *See* ethyne.

acetyl group *See* ethanoyl group.

acetylide *See* carbide.

acetyl CoA (acetyl coenzyme A) An important intermediate in cell metabolism, particularly in the oxidation of sugars, fatty acids, and amino acids, and in certain biosynthetic pathways. It is formed by the reaction between pyruvate (from GLYCOLYSIS) and COENZYME A, catalyzed by the enzyme pyruvate dehydrogenase. The acetyl group of acetyl CoA is subsequently oxidized in the KREBS CYCLE, to yield reduced coenzymes and carbon dioxide. Acetyl CoA is also produced in the initial oxidation of fatty acids and some amino acids. Other key roles for acetyl CoA include the provision of acetyl groups in biosynthesis of fatty acids, terpenoids, and other substances.

acetyl coenzyme A *See* acetyl CoA.

acetylsalicylic acid *See* aspirin.

ACh *See* acetylcholine.

achiral Describing a molecule that does not have chiral properties; i.e. one that does not exhibit OPTICAL ACTIVITY.

acid A substance than contains hydrogen and dissociates in solution to give hydrogen ions:

$$HA \rightleftharpoons H^+ + A^-$$

More accurately, the hydrogen ion is solvated (a hydroxonium ion):

$$HA + H_2O \rightleftharpoons H_3O^+ + A^-$$

Strong acids are completely dissociated in water. Examples are sulfuric acid and tri-

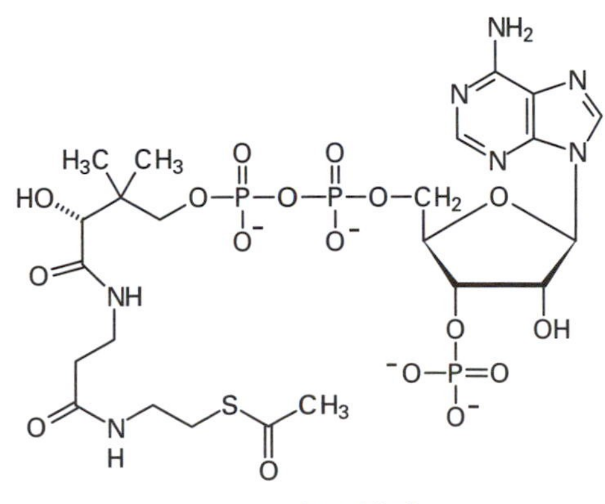

Acetyl CoA

choloroethanoic acid. *Weak acids* are only partially dissociated. Most organic carboxylic acids are weak acids. In distinction to an acid, a *base* is a compound that produces hydroxide ions in water. Bases are either ionic hydroxides (e.g. NaOH) or compounds that form hydroxide ions in water. These may be metal oxides, for example:

$$Na_2O + H_2O \rightarrow 2Na^+ + 2OH^-$$

Ammonia, amines, and other nitrogenous compounds can also form OH^- ions in water:

$$NH_3 + H_2O \rightleftharpoons NH_4^+ + OH^-$$

As with acids, *strong bases* are completely dissociated; *weak bases* are partially dissociated.

This idea of acids and bases is known as the *Arrhenius theory* (named for the Swedish physical chemist Svante August Arrhenius (1859–1927)).

In 1923 the Arrhenius idea of acids and bases was extended by the British chemist Thomas Martin Lowry (1874–1936) and, independently, by the Danish physical chemist Johannes Nicolaus Brønsted (1879–1947). In the *Lowry–Brønsted theory* an acid is a compound that can donate a proton and a base is a compound that can accept a proton. Proton donators are called *Brønsted acids* (or *protic acids*) and proton acceptors are called *Brønsted bases*. For example, in the reaction:

$$CH_3COOH + H_2O \rightleftharpoons CH_3COO^- + H_3O^+$$

the CH_3COOH is the acid, donating a proton H^+ to the water molecule. The water is the base because it accepts the proton. In the reverse reaction, the H_3O^+ ion is the acid, donating a proton to the base CH_3COO^-. If two species are related by loss or gain or a proton they are described as *conjugate*. So, in this example, CH_3COO^- is the *conjugate base* of the acid CH_3COOH and CH_3COOH is the *conjugate acid* of the base CH_3COO^-.

In a reaction of an amine in water, for example:

$$R_3N + H_2O \rightleftharpoons R_3NH^+ + OH^-$$

The amine R_3N accepts a proton from water and is therefore acting as a base. R_3NH^+ is its conjugate acid. Water donates the proton to the R_3N and, in this case, water is acting as an acid (H_3O^+ is its conjugate base). Note that water can act as both an acid and a base depending on the circumstances. It can accept a proton (from CH_3COOH) and donate a proton (to R_3N). Compounds of this type are described as *amphiprotic*.

One important aspect of the Lowry–Brønsted theory is that, because it involves proton transfers, it does not necessarily have to involve water. It is possible to describe reactions in nonaqueous solvents, such as liquid ammonia, in terms of acid–base reactions.

A further generalization of the idea of acids and bases was the *Lewis theory* put forward, also in 1923, by the US physical chemist Gilbert Newton Lewis (1875–1946). In this, an acid (a *Lewis acid*) is a compound that can accept a pair of electrons and a base (a *Lewis base*) is one that donates a pair of electrons. In a traditional acid–base reaction, such as:

$$HCl + NaOH \rightarrow NaCl + H_2O$$

the effective reaction is

$$H^+ + OH^- \rightarrow H_2O$$

The OH^- (base) donates an electron pair to the H^+ (acid). However, in the Lewis theory acids and bases need not involve protons at all. For example, ammonia (NH_3) adds to boron trichloride (BCl_3) to form an adduct:

$$NH_3 + BCl_3 \rightarrow H_3NBCl_3$$

Here, ammonia is the Lewis base donating a LONE PAIR of electrons to boron trichloride (the Lewis acid).

The concept of acid–base reactions is an important generalization in chemistry, and the Lewis theory connects it to two other general ideas. One is oxidation–reduction: oxidation involves loss of electrons and reduction involves gain of electrons. Also, in organic chemistry, it is connected with the idea of electrophile–nucleophile reactions. Acids are ELECTROPHILES and bases are NUCLEOPHILES. In organic chemistry a number of inorganic halides, such as $AlCl_3$ and $TiCl_4$, are important Lewis acids, forming intermediates in such processes as the FRIEDEL–CRAFTS REACTION.

acid anhydride A type of organic compound containing the group –CO.O.CO–.

Simple acid anhydrides have the general formula RCOOCOR′, where R and R′ are alkyl or aryl groups. They can be regarded as formed by removing a molecule of water from two molecules of carboxylic acid. For example, ethanoic anhydride comes from ethanoic acid:

$$2CH_3COOH - H_2O \rightarrow CH_3CO.O.COCH_3$$

A long-chain dicarboxylic acid may also form a cyclic acid anhydride, in which the –CO.O.CO– group forms part of a ring. Acid anhydrides can be prepared by reaction of an acyl halide with the sodium salt of a carboxylic acid, e.g.:

$$RCOCl + R'COO^-Na^+ \rightarrow RCOOCOR' + NaCl$$

Like the acyl halides, they are very reactive acylating agents. They hydrolyze readily to carboxylic acids:

$$RCOOCOR' + H_2O \rightarrow RCOOH + R'COOH$$

See also acylation; anhydride.

acid dyes The sodium salts of organic acids used in the dyeing of silk and wool. They are so called because they are applied from a bath acidified with dilute sulfuric or ethanoic acid.

acid halide *See* acyl halide.

acidic Having a tendency to release a proton or to accept an electron pair from a donor. In aqueous solutions the pH is a measure of the acidity, i.e. an acidic solution is one in which the concentration of H_3O^+ exceeds that in pure water at the same temperature; i.e. the pH is lower than 7. A pH of 7 indicates a neutral solution.

acidic hydrogen A hydrogen atom in a molecule that enters into a dissociation equilibrium when the molecule is dissolved in a solvent. For example, in ethanoic acid (CH_3COOH) the acidic hydrogen is the one on the carboxyl group, –COOH:

$$CH_3COOH + H_2O \rightleftharpoons CH_3COO^- + H_3O^+.$$

acidity constant *See* dissociation constant.

acid value A measure of the free acid present in fats, oils, resins, plasticizers, and solvents, defined as the number of milligrams of potassium hydroxide required to neutralize the free acids in one gram of the substance.

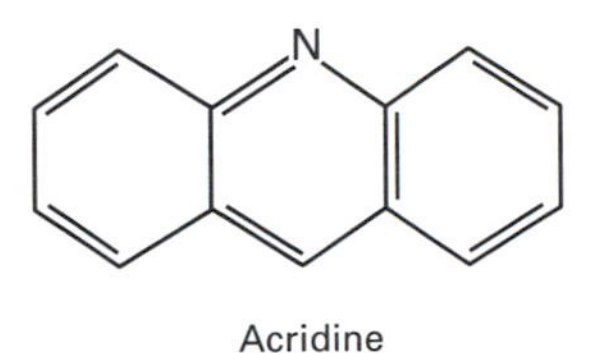

Acridine

acridine ($C_{12}H_9N$) A colorless crystalline heterocyclic compound with three fused rings. Derivatives of acridine are used as dyes and biological stains.

Acrilan (*Trademark*) A synthetic fiber that consists of a copolymer of 1-cyanoethene (acrylonitrile; vinyl cyanide) and ethenyl ethanoate (vinyl acetate). *See* acrylic resin.

acrolein *See* propenal.

acrylic acid *See* propenoic acid.

acrylic resin A synthetic resin made by polymerizing an amide, nitrile, or ester derivative of 2-propenoic acid (acrylic acid). Acrylic resins (known as 'acrylics') are used in a variety of ways. A common example is poly(methylmethacrylate), which is produced by polymerizing methyl methacrylate, $CH_2{:}CH(CH_3)COOCH_3$. This is the clear material sold as Plexiglas. Another example is the compound methyl 2-cyanoacrylate, $CH_2{:}CH(CN)COOCH_3$. This polymerizes very readily in air and is the active constituent of 'superglue'. In both these cases there is a double C=C bond conjugated with the carbonyl C=O bond and the polymerization has a free-radical mechanism. The free election is on the carbon atom next to the carbonyl group, which stabilizes the radical. Another example of an acrylic polymer is formed by free-radical polymerization of acrylonitrile (CH_2:CHCN) to give poly-(acrylonitrile). This is used in synthetic fibers (such as Acrilan). In this case the unpaired electron is on the carbon next to the –CN group. Acrylic resins are also used in paints.

acrylonitrile *See* propenonitrile.

actinic radiation Radiation that can cause a chemical reaction; for example, ultraviolet radiation is actinic.

actinomycin Any of a number of antibiotics produced by certain bacteria. The main one, *actinomycin D* (or *dactinomycin*), can bind between neighbouring base pairs in DNA, preventing RNA synthesis. It is used in the treatment of some cancers.

action spectrum A graph showing the effect of different wavelengths of radiation, usually light, on a given process. It is often similar to the ABSORPTION SPECTRUM of the substance that absorbs the radiation and can therefore be helpful in identifying that substance. For example, the action spectrum of photosynthesis is similar to the absorption spectrum of chlorophyll.

activated charcoal *See* charcoal.

activated complex The partially bonded system of atoms in the TRANSITION STATE of a chemical reaction.

activation energy Symbol: E_a The minimum energy a system must acquire before a chemical reaction can occur, regardless of whether the reaction is exothermic or endothermic. Activation energy is often represented as an energy barrier that has to be overcome if a reaction is to take place. *See also* Arrhenius equation; transition state.

activator *See* accelerator.

active mass *See* mass action.

active site 1. A site on the surface of a solid catalyst at which catalytic activity occurs or at which the catalyst is particularly effective.

2. The region of an ENZYME molecule that combines with and acts on the substrate. It consists of catalytic amino acids arranged in a configuration specific to a particular substrate or type of substrate. The ones that are in direct combination are the *contact amino acids*. Other amino acids may be further away but still play a role in the action of the enzyme. These are *auxilliary amino acids*. Binding of a regulatory compound to a separate site, known as the ALLOSTERIC SITE, on the enzyme molecule may change this configuration and hence the efficiency of the enzyme activity.

activity 1. Symbol: *a* Certain thermodynamic properties of a solvated substance are dependent on its concentration (e.g. its tendency to react with other substances). Real substances show departures from ideal behavior and a corrective concentration term – the activity – has to be introduced into equations describing real solvated systems.

2. Symbol: *A* The average number of atoms disintegrating per unit time in a radioactive substance.

activity coefficient Symbol: *f* A measure of the degree of deviation from ideality of a dissolved substance, defined as:

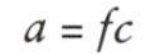

$$a = fc$$

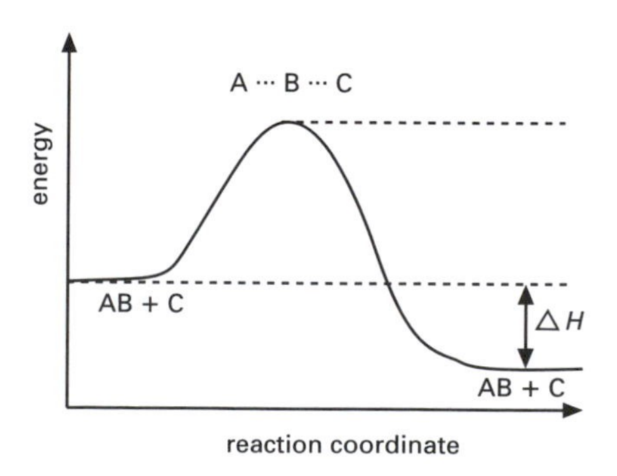

Activation energy

where *a* is the activity and *c* the concentration. For an ideal solute $f = 1$; for real systems *f* can be less or greater than unity.

acyclic Describing a compound that is not cyclic (i.e. a compound that does not contain a ring in its molecules).

acyl anhydride *See* acid anhydride.

acylating agent *See* acylation.

acylation Any reaction that introduces an acyl group (RCO–) into a compound. *Acylating agents* are compounds such as acyl halides (RCOX) and acid anhydrides (RCOOCOR), which react with such nucleophiles as H_2O, ROH, NH_3, and RNH_2. In these reactions a hydrogen atom of a hydroxyl or amine group is replaced by the RCO– group. In *acetylation* the acetyl group (CH_3CO–) is used. In *benzoylation* the benzoyl group (C_6H_5CO–) is used. Acylation is used to prepare crystalline derivatives of organic compounds to identify them (e.g. by melting point) and also to protect –OH groups in synthetic reactions.

acyl group

acyl halide (acid halide) A type of organic compound of the general formula RCOX, where X is a halogen (acyl chloride, acyl bromide, etc.).

Acyl halides can be prepared by the reaction of a carboxylic acid with a halogenating agent. Commonly, phosphorus halides are used (e.g. PCl_5) or a sulfur dihalide oxide (e.g. $SOCl_2$):

$$RCOOH + PCl_5 \rightarrow RCOCl + POCl_3 + HCl$$

$$RCOOH + SOCl_2 \rightarrow RCOCl + SO_2 + HCl$$

The acyl halides have irritating vapors and fume in moist air. They are very reactive to the hydrogen atom of compounds containing hydroxyl (–OH) or amine (–NH_2) groups. For example, the acyl halide ethanoyl chloride (acetyl chloride; CH_3COCl) reacts with water to give a carboxylic acid (ethanoic acid):

$$CH_3COCl + H_2O \rightarrow CH_3COOH + HCl$$

With an alcohol (e.g. ethanol) it gives an ester (ethyl ethanoate):

$$CH_3COCl + C_2H_5OH \rightarrow CH_3COOC_2H_5 + HCl$$

With ammonia it gives an amide (ethanamide; acetamide):

$$CH_3COCl + NH_3 \rightarrow CH_3CONH_2 + HCl$$

With an amine (e.g. methylamine) it gives an N-substituted amine (N-methyl ethanamide)

$$CH_3COCl + CH_3NH_2 \rightarrow CH_3CONH(CH_3)$$

See also acylation.

addition polymerization *See* polymerization.

addition reaction A reaction in which additional atoms or groups of atoms are introduced into an unsaturated compound, such as an alkene, alkyne, aldehyde, or ketone. A simple example is the addition of bromine across the double bond in ethene:

$$H_2C{:}CH_2 + Br_2 \rightarrow BrH_2CCH_2Br$$

Addition reactions can occur by addition of electrophiles or nucleophiles. *See* electrophilic addition; nucleophilic addition.

adduct *See* coordinate bond.

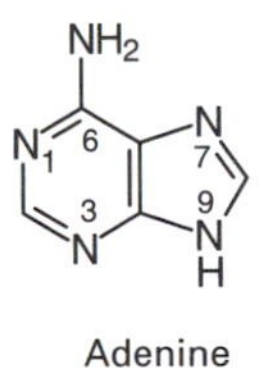

Adenine

adenine A nitrogenous base found in DNA and RNA. It is also a constituent of certain coenzymes, and when combined with the sugar ribose it forms the nucleoside adenosine found in AMP, ADP, and ATP. Adenine has a purine ring structure. *See also* DNA.

adenosine (adenine nucleoside) A NUCLEOSIDE formed from adenine linked to D-ribose with a β-glycosidic bond. It is widely found in all types of cell, either as the free nucleoside or in combination in nucleic

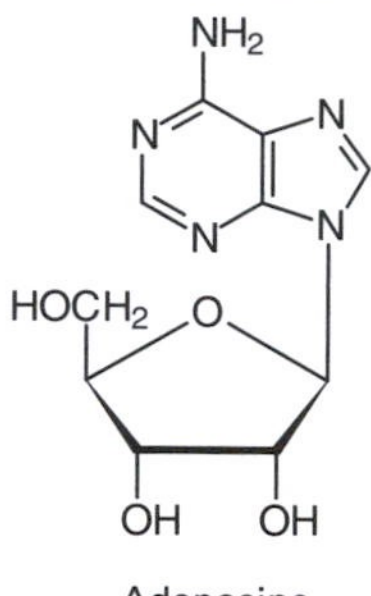

Adenosine

acids. Phosphate esters of adenosine, such as ATP, are important carriers of energy in biochemical reactions.

adenosine diphosphate *See* ADP.

adenosine monophosphate *See* AMP.

adenosine triphosphate *See* ATP.

adiabatic change A change for which no energy enters or leaves the system. In an adiabatic expansion of a gas, mechanical work is done by the gas as its volume increases and the gas temperature falls. For an ideal gas undergoing a reversible adiabatic change it can be shown that

$$pV^{\gamma} = K_1$$
$$T^{\gamma}p^{1-\gamma} = K_2$$
$$\text{and } TV^{\gamma-1} = K_3$$

where K_1, K_2, and K_3 are constants and γ is the ratio of the principal specific heat capacities. *Compare* isothermal change.

adipic acid *See* hexanedioic acid.

adjacent Designating atoms or bonds that are next to each other in a molecule.

ADP (adenosine diphosphate) A nucleotide consisting of adenine and ribose with two phosphate groups attached. *See also* ATP.

adrenalin *See* epinephrine.

adsorbate A substance that is adsorbed on a surface. *See* adsorption.

adsorbent Having a tendency to adsorb. As a noun the adsorbent is the substance on which adsorption takes place. *See* adsorption.

adsorption A process in which a layer of atoms or molecules of one substance forms on the surface of a solid or liquid. All solid surfaces take up layers of gas from the surrounding atmosphere. The adsorbed layer may be held by chemical bonds (*chemisorption*) or by weaker van der Waals forces (*physisorption*).
Compare absorption.

aerobic Describing a biochemical process that takes place only in the presence of free oxygen. *Compare* anaerobic.

aerobic respiration (oxidative metabolism) Respiration in which free oxygen is used to oxidize organic substrates to carbon dioxide and water, with a high yield of energy. Carbohydrates, fatty acids, and excess amino acids are broken down yielding acetyl CoA and the reduced coenzymes NADH and $FADH_2$. The acetyl coenzyme A enters a cyclic series of reactions, the KREBS CYCLE, with the production of carbon dioxide and further molecules of NADH and $FADH_2$. NADH and $FADH_2$ are passed to the ELECTRON-TRANSPORT CHAIN (involving cytochromes and flavoproteins), where they combine with atoms of free oxygen to form water. Energy released at each stage of the chain is used to form ATP during a coupling process. The substrate is completely oxidized and there is a high energy yield. There is a net production of 38 ATPs per molecule of glucose during aerobic respiration, a yield of about 19 times that of anaerobic respiration. Aerobic respiration is therefore the preferred mechanism of the majority of organisms. *See also* oxidative phosphorylation; respiration.

aerosol *See* sol.

affinity The extent to which one substance reacts with another in a chemical change.

afterdamp *See* firedamp.

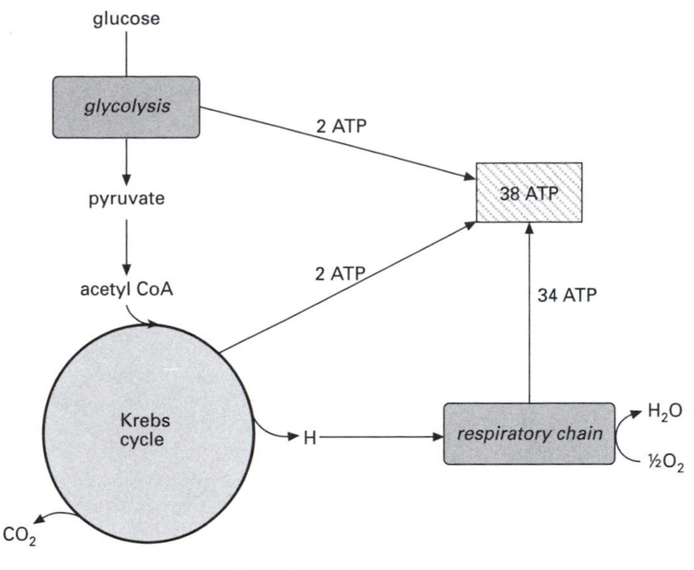

Aerobic respiration

agent orange A herbicide consisting of a mixture of two weedkillers (2,4-D and 2,4,5-T). It was designed for use in chemical warfare to defoliate trees in areas where an enemy may be hiding or to destroy enemy crops. Agent orange, so-called from the orange-colored canisters in which it was supplied, was first used by US forces during the Vietnam war. It contains traces of the highly toxic chemical DIOXIN, which causes cancers and birth defects.

air gas *See* producer gas.

alanine *See* amino acid.

albumen The white of an egg, which consists mainly of the protein ALBUMIN.

albumin A soluble protein that occurs in many animal fluids, such as blood serum and egg white.

alcohol A type of organic compound of the general formula ROH, where R is a hydrocarbon group. Examples of simple alcohols are methanol (CH_3OH) and ethanol (C_2H_5OH). Alcohols have the –OH group attached to a carbon atom that is part of an alkyl group. If the carbon atom is part of

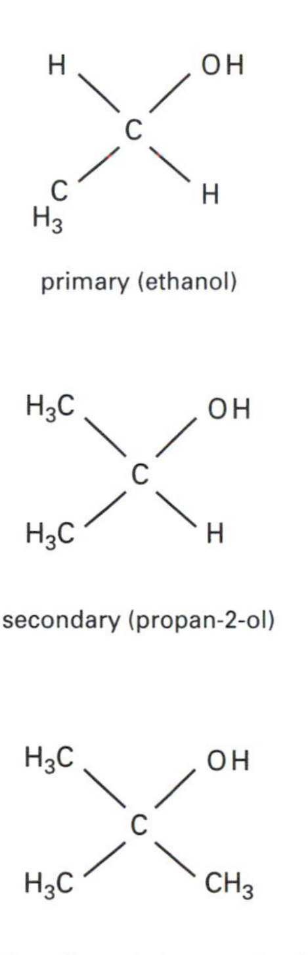

Alcohol

an aromatic ring, as in PHENOL, C_6H_5OH, the compound does not have the characteristic properties of alcohols. Phenylmethanol ($C_6H_5CH_2OH$) does have the characteristic properties of alcohols (in this case the carbon atom to which the –OH is attached is not part of the aromatic ring).

Alcohols can have more than one –OH group; those containing two, three, or more such groups are described as *dihydric*, *trihydric*, and *polyhydric* respectively (as opposed to alcohols containing one –OH group, which are *monohydric*). For example, ethane-1,2-diol (ethylene glycol; ($HOCH_2CH_2OH$) is a dihydric alcohol and propane-1,2,3-triol (glycerol; $HOCH_2CH(OH)CH_2OH$) is a trihydric alcohol. Dihydric alcohols are known as *diols*; trihydric alcohols as *triols*, etc. In general, alcohols are named by using the suffix -ol with the name of the parent hydrocarbon.

Alcohols are further classified according to the environment of the –C–OH grouping. If the carbon atom is attached to two hydrogen atoms, the compound is a *primary alcohol*. If the carbon atom is attached to one hydrogen atom and two other groups, it is a *secondary alcohol*. If the carbon atom is attached to three other groups, it is a *tertiary alcohol*. Alcohols can be prepared by:

1. Hydrolysis of haloalkanes using aqueous potassium hydroxide:
$$RI + OH^- \rightarrow ROH + I^-$$
2. Reduction of aldehydes by nascent hydrogen (e.g. from sodium amalgam in water):
$$RCHO + 2[H] \rightarrow RCH_2OH$$

The main reactions of alcohols are:

1. Oxidation by potassium dichromate(VI) in sulfuric acid. Primary alcohols give aldehydes, which are further oxidized to carboxylic acids:
$$RCH_2OH \rightarrow RCHO \rightarrow RCOOH$$
Secondary alcohols are oxidized to ketones.
$$R^1R^2CHOH \rightarrow R^1R^2CO$$
2. Formation of esters with acids. The reaction, which is reversible, is catalyzed by H^+ ions:
$$ROH + R'COOH \rightleftharpoons R'COOR + H_2O$$
3. Dehydration over hot pumice (400°C) to alkenes:
$$RCH_2CH_2OH - H_2O \rightarrow RCH{:}CH_2$$
4. Reaction with sulfuric acid. Two types of reaction are possible. With excess acid at 160°C dehdyration occurs to give an alkene:
$$RCH_2CH_2OH + H_2SO_4 \rightarrow$$
$$H_2 \quad {}_2CH_2.HSO_4$$
$$RCH_2CH_2.HSO_4 \rightarrow$$
$$RCH{:}CH_2 + H_2SO_4$$
With excess alcohol at 140°C an ether is formed:
$$2ROH \rightarrow ROR + H_2O$$

See also acetal; acylation; Grignard reagent.

aldaric acid *See* sugar acid.

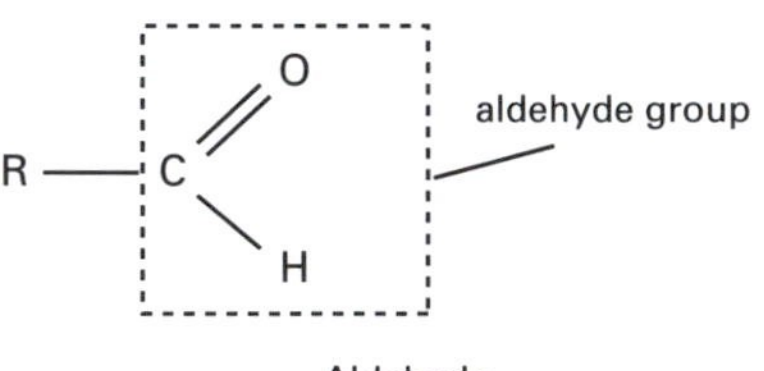

Aldehyde

aldehyde A type of organic compound with the general formula RCHO, where the –CHO group (the aldehyde group) consists of a carbonyl group attached to a hydrogen atom. Simple examples of aldehydes are methanal (formaldehyde; HCHO) and ethanal (acetaldehyde; CH_3-CHO).

Aldehydes are formed by oxidizing a primary alcohol; in the laboratory potassium dichromate(VI) is used in sulfuric acid. They can be further oxidized to carboxylic acids. Reduction (using a catalyst or nascent hydrogen from sodium amalgam in water) produces the parent alcohol. For example, oxidation of ethanol (C_2H_5OH) gives ethanal (acetaldehyde; CH_3CHO):
$$C_2H_5OH + [O] \rightarrow CH_3CHO + H_2O$$
Further oxidation gives ethanoic acid (acetic acid; CH_3COOH):
$$CH_3CHO + [O] \rightarrow CH_3COOH$$
The systematic method of naming aldehydes is to use the suffix -al with the

name of the parent hydrocarbon. For example: methane (CH_4) is the parent hydrocarbon of the alcohol methanol (CH_3OH), the aldehyde methanal (HCHO), and the carboxylic acid methanoic acid (HCOOH); ethane (C_2H_6) is the parent hydrocarbon of the alcohol ethanol (C_2H_5-OH), the aldehyde ethanal (CH_3CHO), and the carboxylic acid ethanoic acid (CH_3COOH); etc. An older method of naming aldehydes is based on the name of the related acid. For example, methanoic acid (HCOOH) has the traditional name 'formic acid' and the related aldehyde (HCHO) is traditionally called 'formaldehyde'. Similarly, ethanoic acid (CH_3OOH) is commonly known as 'acetic acid' and the aldehyde CH_3CHO is known as 'acetaldehyde'.

Reactions of aldehydes are:

1. Aldehydes are reducing agents, being oxidized to carboxylic acids in the process. These reactions are used as tests for aldehydes using such reagents as FEHLING'S SOLUTION and TOLLEN'S REAGENT (silver-mirror test).
2. They form addition compounds with hydrogen cyanide to give *cyanohydrins*. For example, propanal gives 2-hydroxybutanonitrile:
$$C_2H_5CHO + HCN \rightarrow C_2H_5CH(OH)CN$$
3. They form *bisulfite addition* compounds with the hydrogensulfite(IV) ion (bisulfite; HSO_3^-):
$$RCHO + HSO_3^- \rightarrow RCH(OH)(HSO_3)$$
4. They undergo condensation reactions with such compounds as hydrazine, hydroxylamine, and their derivatives.
5. With alcohols they form hemiacetals and ACETALS.
6. Simple aldehydes polymerize readily. Polymethanal or methanal trimer can be formed from METHANAL depending on the conditions. ETHANAL gives ethanal trimer or ethanal tetramer.

See also Cannizzaro reaction; condensation reaction; ketone.

Alder, Kurt (1902–1958) German organic chemist who is noted for the process known as the DIELS–ALDER REACTION. Particular cases of the reaction had been known since the 1900s but Alder and Otto Diels recognized that this mechanism is very common. They first reported their results in 1928. Alder and Diels shared the 1950 Nobel Prize for chemistry for this work.

alditol *See* sugar alcohol.

aldohexose An aldose SUGAR with six carbon atoms.

aldol A compound that contains both an aldehyde group (–CHO) and an alcohol group (–OH). *See* aldol reaction.

aldol reaction A reaction in which two molecules of aldehyde combine to give an *aldol* – i.e. a compound containing both aldehyde and alcohol functional groups. The reaction is base-catalyzed; the reaction of ethanal (acetaldehyde) refluxed with sodium hydroxide gives:
$$2CH_3CHO \rightarrow CH_3CH(OH)CH_2CHO$$
The mechanism is similar to that of the CLAISEN CONDENSATION: the first step is removal of a proton to give a carbanion, which subsequently attacks the carbon of the carbonyl group on the other molecule:
$$CH_3CHO + OH^- \rightarrow {}^-CH_2CHO + H_2O$$
$$CH_3CHO + {}^-CH_2CHO \rightarrow CH_3CH(OH)CH_2CHO.$$

aldonic acid *See* sugar acid.

aldopentose An aldose SUGAR with five carbon atoms.

aldose A SUGAR containing an aldehyde group (CHO) or a potential aldehyde group.

algin *See* alginic acid.

alginic acid (algin; $(C_6H_8O_6)_n$) A yellow-white organic solid that is found in brown algae. It is a complex polysaccharide and produces, in even very dilute solutions, a viscous liquid. Alginic acid has various uses, especially in the food industry as a stabilizer and texture agent.

alicyclic compound An aliphatic cyclic compound, such as cyclohexane or cyclopropane.

aliphatic compound An organic compound with properties similar to those of the alkanes, alkenes, and alkynes and their derivatives. Most aliphatic compounds have an open chain structure but some, such as cyclohexane and sucrose, have rings (these are described as *alicyclic*). The term is used in distinction to AROMATIC COMPOUNDS, which are similar to benzene.

alizarin (1,2-dihydroxyanthraquinone) An important orange-red organic compound used in the dyestuffs industry to produce red lakes. It occurs naturally in the root of the plant madder and may also be synthesized from anthraquinone.

alkali A water-soluble strong base. Strictly the term refers to the hydroxides of the alkali metals (group 1) only, but in common usage it refers to any soluble base. Thus borax solution may be described as mildly alkaline.

alkaloid One of a group of natural organic compounds found in plants. They contain oxygen and nitrogen atoms; most are poisonous. However, they include a number of important drugs with characteristic physiological effects, e.g. morphine, codeine, caffeine, cocaine, and nicotine.

alkane A type of hydrocarbon with general formula C_nH_{2n+2}. Alkanes are saturated compounds, containing no double or triple bonds. Systematic names end in -ane: methane (CH_4) and ethane (C_2H_6) are typical examples. The alkanes are fairly unreactive (their former name, the *paraffins*, means 'small affinity'). In ultraviolet radiation they react with halogens to give a mixture of substitution products. This involves a free-radical chain reaction and is important as a first step in producing other compounds from alkanes. There are a number of ways of preparing specific alkanes:

1. From a sodium salt of a carboxylic acid treated with sodium hydroxide:
   $$RCOO^-Na^+ + NaOH \rightarrow RH + Na_2CO_3$$
2. By reduction of a haloalkane with nascent hydrogen from the action of ethanol on a zinc–copper couple:
   $$RX + 2[H] \rightarrow RH + HX$$
3. By the WURTZ REACTION – i.e. sodium in dry ether on a haloalkane:
   $$2RX + 2Na \rightarrow 2NaX + RR$$
4. By the KOLBÉ ELECTROLYTIC METHOD:
   $$RCOO^- \rightarrow RR$$
5. By refluxing a haloalkane with magnesium in dry ether to form a GRIGNARD REAGENT:
   $$RI + Mg \rightarrow RMgI$$
   With acid this gives the alkane:
   $$RMgI + H \rightarrow RH$$

The main source of lower molecular weight alkanes is natural gas (for methane) and crude oil.

$$CH_3 - CH_2 - CH_2 - CH_2 - CH = CH_2$$

hex-1-ene

$$CH_3 - CH_2 - CH_2 - CH = CH - CH_3$$

hex-2-ene

$$CH_3 - CH_2 - CH = CH - CH_2 - CH_3$$

hex-3-ene

Alkene

alkene A type of aliphatic hydrocarbon containing one or more double bonds in the molecule. Alkenes with one double bond have the general formula C_nH_{2n}. The alkenes are unsaturated compounds. They can be obtained from crude oil by cracking alkanes. Systematic names end in -ene: examples are ethene (C_2H_4) and propene (C_3H_6), both of which are used in plastics production and as starting materials for the manufacture of many other organic chemicals. The former general name for an alkene was *olefin*.

The methods of synthesizing alkenes are:

1. The elimination of HBr from a haloalkane using an alcoholic solution of potassium hydroxide:

$$RCH_2CH_2Br + KOH \rightarrow KBr + H_2O + RCH{:}CH_2$$

2. The dehydration of an alcohol by passing the vapor over hot pumice (400°C):

$$RCH_2CH_2OH \rightarrow RCH{:}CH_2 + H_2O$$

The reactions of simple alkenes include:

1. Hydrogenation using a catalyst (usually nickel at about 150°C):

$$RCH{:}CH_2 + H_2 \rightarrow RCH_2CH_3$$

2. Addition reactions with halogen acids to give haloalkanes:

$$RCH{:}CH_2 + HX \rightarrow RCH_2CH_2X$$

The addition follows MARKOVNIKOFF'S RULE.

3. Addition reactions with halogens, e.g.

$$RCH{:}CH_2 + Br_2 \rightarrow RCHBrCH_2Br$$

4. Hydration using concentrated sulfuric acid, followed by dilution and warming:

$$RCH{:}CH_2 + H_2O \rightarrow RCH(OH)CH_3$$

5. Oxidation by cold potassium permanganate solutions to give diols:

$$RCH{:}CH_2 + H_2O + [O] \rightarrow RCH(OH)CH_2OH$$

6. Oxidation to form cyclic epoxides (oxiranes). Ethene can be oxidized in air using a silver catalyst to the cyclic compound epoxyethane (C_2H_4O). More generally peroxy carboxylic acids are used as the oxidizing agent.
7. Polymerization to polyethene (by the ZIEGLER PROCESS or PHILLIPS PROCESS).

See also oxo process; ozonolysis.

In general, addition to simple alkenes is ELECTROPHILIC ADDITION. Attack is by an electrophile on the pi orbital of the alkene. In the case of attack by a halogen acid (e.g. HBr), the initial reaction is by the (positive) hydrogen giving a positively charged intermediate ion (carbocation) and a Br^- ion. The Br^- ion then attacks the intermediate carbocation. In the case of a halogen (e.g. Br_2) the bromine acts as an electrophile to form an initial cyclic positively charged *bromonium ion* and a negative Br^- ion. The Br^- ion further attacks the bromonium ion to give the substituted product.

alkoxide An organic compound containing an ion of the type RO^-, where R is an alkyl group. Alkoxides can be made by the reaction of metallic sodium on an alcohol. For example, ethanol reacts with sodium to give sodium ethoxide:

$$2C_2H_5OH + 2Na \rightarrow 2C_2H_5O^-Na^+) + H_2$$

Alkoxides are ionic compounds containing an *alkoxide ion* (RO^-). They are named according to the parent alcohol. Thus, methanol (CH_3OH) gives methoxides CH_3O^-, ethanol (C_2H_5OH) gives ethoxides $C_2H_5O^-$, etc.

alkoxyalkane (diethyl ether) *See* ether.

alkylbenzene A type of organic hydrocarbon containing one or more alkyl groups substituted onto a benzene ring. Methylbenzene (toluene; $C_6H_5CH_3$) is the simplest example. Alkylbenzenes can be made by a FRIEDEL–CRAFTS REACTION or by the WURTZ REACTION. Industrially, large quantities of methylbenzene are made from crude oil.

Substitution of alkylbenzenes can occur at the benzene ring; the alkyl group directs the substituent into the 2- or 4-position. Substitution of hydrogen atoms on the alkyl group can also occur.

alkyl group A group obtained by removing a hydrogen atom from an alkane or other aliphatic hydrocarbon. For example, the methyl group (CH_3–) is derived from methane (CH_4).

alkyl halide *See* haloalkane.

alkyl sulfide A THIOETHER with the general formula RSR′, where R and R′ are alkyl groups.

alkyne A type of hydrocarbon containing one or more triple carbon–carbon bonds in its molecule. Alkynes with one triple bond have the general formula C_nH_{2n-2}. The alkynes are unsaturated compounds. The simplest member of the series is ethyne (acetylene; C_2H_2), which can be prepared by the action of water on calcium dicarbide.

$$CaC_2 + 2H_2O \rightarrow Ca(OH)_2 + C_2H_2$$

The alkynes were formerly called the *acetylenes*.

In general, alkynes can be made by the cracking of alkanes or by the action of a hot alcoholic solution of potassium hydroxide on a dibromoalkane, for example:

$$BrCH_2CH_2Br + KOH \rightarrow$$
$$KBr + CH_2{:}CHBr + H_2O$$
$$CH_2{:}CHBr + KOH \rightarrow$$
$$CHCH + KBr + H_2O$$

The main reactions of the alkynes are:

1. Hydrogenation with a catalyst (usually nickel at about 150°C):
$$C_2H_2 + H_2 \rightarrow C_2H_4$$
$$C_2H_4 + H_2 \rightarrow C_2H_6$$
2. Addition reactions with halogen acids:
$$C_2H_2 + HI \rightarrow H_2C{:}CHI$$
$$H_2C{:}CHI + HI \rightarrow CH_3CHI_2$$
3. Addition of halogens; for example, with bromine in tetrachloromethane:
$$C_2H_2 + Br_2 \rightarrow BrHC{:}CHBr$$
$$BrHC{:}CHBr + Br_2 \rightarrow Br_2HCCHBr_2$$
4. With dilute sulfuric acid at 60–80°C and mercury(II) catalyst, ethyne forms ethanal (acetaldehyde):
$$C_2H_2 + H_2O \rightarrow H_2C{:}C(OH)H$$
This enol form converts to the aldehyde:
$$CH_3COH$$
5. Ethyne polymerizes if passed through a hot tube to produce some benzene:
$$3C_2H_2 \rightarrow C_6H_6$$
6. Ethyne forms unstable dicarbides (acetylides) with ammoniacal solutions of copper(I) and silver(I) chlorides.

Addition to simple alkynes is ELECTROPHILIC ADDITION, as with ALKENES.

allosteric site A part of an enzyme separate from the active site to which a specific effector or modulator can be attached. This attachment is reversible and alters the activity of the enzyme. Allosteric enzymes possess an allosteric site in addition to their ACTIVE SITE. This site is as specific in its relationship to modulators as active sites are to substrates. *See* active site. Some iron-enzymatic proteins e.g. hemoglobin also undergo allosteric effects.

allyl group *See* propenyl group.

alpha amino acid *See* amino acid.

alpha helix A highly stable structure in which peptide chains are coiled to form a spiral. Each turn of the spiral contains approximately 3.6 amino-acid residues. The R group of these amino-acids extends outward from the helix and the helix is held together by hydrogen bonding between successive coils. If the alpha helix is stretched the hydrogen bonds are broken but reform on relaxation. The alpha helix is found in muscle protein and keratin. It is one of the two basic secondary structures of PROTEINS.

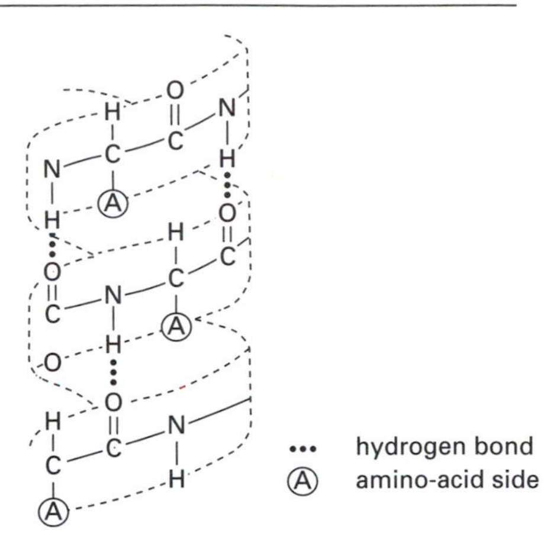

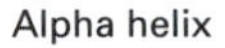

Alpha helix

alpha-naphthol test (Molisch's test) A standard test for carbohydrates in solution. Molisch's reagent, alpha-naphthol in alcohol, is mixed with the test solution. Concentrated sulfuric acid is added and a violet ring at the junction of the two liquids indicates the presence of carbohydrates.

alternating copolymer *See* polymerization.

aluminum trimethyl *See* trimethylaluminum.

amalgam An alloy of mercury with one or more other metals. Amalgams may be liquid or solid. An amalgam of sodium (Na/Hg) with water is used as a source of nascent hydrogen.

amatol A high explosive that consists of

a mixture of ammonium nitrate and TNT (trinitrotoluene).

amide **1.** A type of organic compound of general formulae $RCONH_2$ (primary), $(RCO)_2NH$ (secondary), and $(RCO)_3N$ (tertiary). Amides are crystalline solids and are basic in nature, some being soluble in water. They can be formed by reaction of ammonia with acid anhydrides:

$$(RCO)_2O + 2NH_3 \rightarrow RCONH_2 \quad {}^{-}NH_4^{+}$$

They can also be made by reacting ammonia with an acyl chloride:

$$RCOCl + 2NH_3 \rightarrow RCONH_2 + NH_4Cl$$

Reactions of amides include:

1. Reaction with hot acids to give carboxylic acids:
$$RCONH_2 + HCl + H_2O \rightarrow RCOOH + NH_4Cl$$
2. Reaction with nitrous acid to give carboxylic acids and nitrogen:
$$RCONH_2 + HNO_2 \rightarrow RCOOH + N_2 + H_2O$$
3. Dehydration by phosphorus(V) oxide to give a nitrile:
$$RCONH_2 - H_2O \rightarrow RCN$$

See also Hofmann degradation.

2. An inorganic salt containing the NH_2^- ion. Ionic amides are formed by the reaction of ammonia with certain reactive metals (such as sodium and potassium). Sodamide, $NaNH_2$, is a common example.

amination The introduction of an amino group ($-NH_2$) into an organic compound. An example is the conversion of an aldehyde or ketone into an amide by reaction with hydrogen and ammonia in the presence of a catalyst:

$$RCHO + NH_3 + H_2 \rightarrow RCH_2NH_2 + H_2O$$

amine A compound containing a nitrogen atom bound to hydrogen atoms or hydrocarbon groups. Amines have the general formula R_3N, where R can be hydrogen or an alkyl or aryl group. They can be prepared by reduction of amides or nitro compounds.

Amines are classified according to the number of organic groups bonded to the nitrogen atom: one, *primary*; two, *secondary*; three, *tertiary*. Since amines are basic they can form the quaternary ion, R_3NH^+. All three types, plus a quaternium salt, can be produced by the HOFMANN DEGRADATION (which occurs in a sealed vessel at 100°C):

$$RX + NH_3 \rightarrow RNH_3^+ X^-$$
$$RNH_3^+ X^- + NH_3 \rightleftharpoons RNH_2 + NH_4X$$
$$RNH_2 + RX \rightarrow R_2NH_2^+ X^-$$
$$R_2NH_2^+ X^- + NH_3 \rightleftharpoons R_2NH + NH_4X$$
$$R_2NH + RX \rightarrow R_3NH^+ X^-$$
$$R_3NH^+ X^- + NH_3 \rightleftharpoons R_3N + NH_4X$$
$$R_3N + RX \rightarrow R_4N^+X^-$$

Reactions of amines include:

1. Reaction with acids to form salts:
$$R_3N + HX \rightarrow R_3NH^+X^-$$
2. Reaction with acyl halides to give *N*-substituted amides (primary and secondary amines only):
$$RNH_2 + R'COCl \rightarrow R'CONHR + HX$$

See also amine salt.

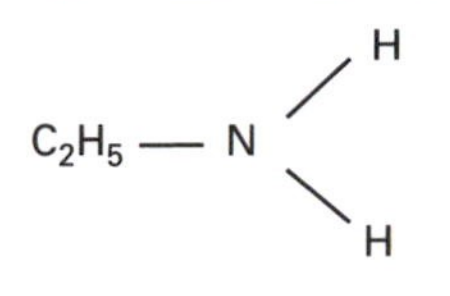

primary (ethylamine)

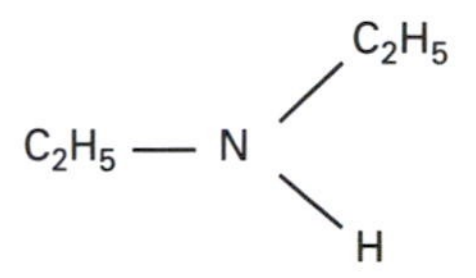

secondary (diethylamine)

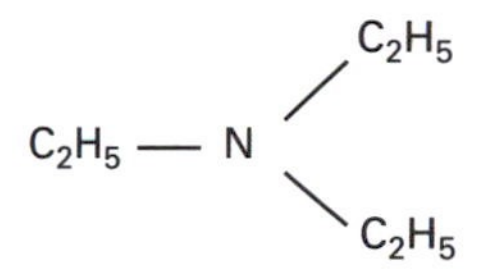

tertiary (triethylamine)

Amine

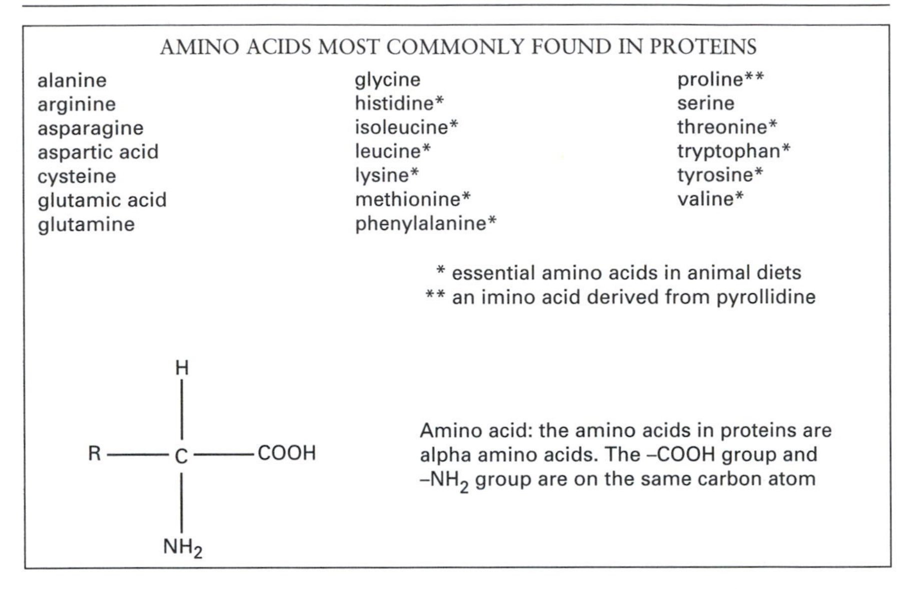

AMINO ACIDS MOST COMMONLY FOUND IN PROTEINS

alanine	glycine	proline**
arginine	histidine*	serine
asparagine	isoleucine*	threonine*
aspartic acid	leucine*	tryptophan*
cysteine	lysine*	tyrosine*
glutamic acid	methionine*	valine*
glutamine	phenylalanine*	

* essential amino acids in animal diets
** an imino acid derived from pyrollidine

Amino acid: the amino acids in proteins are alpha amino acids. The –COOH group and $-NH_2$ group are on the same carbon atom

amine salt A salt similar to an ammonium salt, but with organic groups attached to the nitrogen atom. For example, triethylamine ($(C_2H_5)_3N$) will react with hydrogen chloride to give triethylammonium chloride:

$$(C_2H_5)_3N + HCl \rightarrow (C_2H_3)_3NH^+Cl^-$$

Salts of this type may have four groups on the nitrogen atom. For example, with chloroethane, tetraethylammonium chloride can be formed:

$$(C_2H_5)_3N + C_2H_5Cl \rightarrow (C_2H_5)_4N^+Cl^-$$

Sometimes amine salts are named using the suffix '-ium'. For instance, aniline ($C_6H_5NH_2$) forms anilinium chloride $C_6H_5NH_3^+Cl^-$. Often insoluble alkaloids are used in medicine in the form of their amine salt (sometimes referred to as the 'hydrochloride').

amino acid A derivative of a carboxylic acid in which a hydrogen atom in an aliphatic acid has been replaced by an amino group. Thus, from ethanoic acid, the amino acid 2-aminoethanoic acid (glycine) is formed. The amino acids of special interest are those that occur as constituents of naturally occurring PEPTIDES and PROTEINS. These all have the $-NH_2$ and –COOH groups attached to the same carbon atom; i.e. they are *alpha amino acids*. All are white, crystalline, soluble in water (but not in alcohol), and, with the sole exception of the simplest member, all are optically active.

In the body the various proteins are assembled from the necessary amino acids and it is important therefore that all the amino acids should be present in sufficient quantities. In adult humans, twelve of the twenty amino acids can be synthesized by the body itself. Since these are not required in the diet they are known as *nonessential amino acids*. The remaining eight cannot be synthesized by the body and have to be supplied in the diet. They are known as *essential* amino acids.

aminobenzene *See* aniline.

aminoethane *See* ethylamine.

amino group The group $-NH_2$.

amino sugar A sugar in which a hydroxyl group (OH) has been replaced by an amino group (NH_2). Glucosamine (from glucose) occurs in many polysaccharides of vertebrates and is a major component of chitin. Galactosamine or chondrosamine (from galactose) is a major component of cartilage and glycolipids. Amino sugars are important components of bacterial cell walls.

aminotoluine *See* toluidine.

ammonia (NH_3) A colorless gas with a characteristic pungent odor. On cooling and compression it forms a colorless liquid, which becomes a white solid on further cooling. Ammonia is very soluble in water (a saturated solution at 0°C contains 36.9% of ammonia); the aqueous solution is alkaline and contains a proportion of free ammonia. Ammonia is also soluble in ethanol. It reacts with acids to form ammonium salts; for example, it reacts with hydrogen chloride to form ammonium chloride:

$$NH_3(g) + HCl(g) \rightarrow NH_4Cl(g)$$

See also amine salt.

ammoniacal Describing a solution in aqueous ammonia.

amount of substance Symbol: n A measure of the number of entities present in a substance. *See* mole.

AMP (adenosine monophosphate) A nucleotide consisting of adenine, ribose, and phosphate. *See* ATP.

amphiprotic Able to act as both an ACID and a base. For example, the amino acids are amphiprotic because they contain both acidic (–COOH) and basic (–NH_2) groups. *See also* amphoteric; solvent.

ampholyte ion *See* zwitterion.

amphoteric A material that can display both acidic and basic properties. The term is most commonly applied to the oxides and hydroxides of metals that can form both cations and complex anions. For example, zinc oxide dissolves in acids to form zinc salts and also dissolves in alkalis to form zincates, $[Zn(OH)_4]^{2-}$. Compounds such as the amino acids can also be described as amphoteric, although it is more usual to use the term AMPHIPROTIC.

amu *See* atomic mass unit.

amyl group *See* pentyl group.

amyl nitrite ($C_5H_{11}ONO$) A pale brown volatile liquid organic compound; a nitrous acid ester of 3-methylbutanol (isoamyl alcohol). It is used in medicine as an inhalant to dilate the blood vessels (and thereby prevent pain) in patients with angina pectoris.

amylopectin The water-insoluble fraction of STARCH.

amylose A polymer of GLUCOSE; a polysaccharide sugar that is found in STARCH.

anabolic steroid Any STEROID hormone or synthetic steroid that promotes growth and formation of new tissue. Anabolic steroids are used in the treatment of wasting diseases. They are also sometimes used in agriculture to boost livestock production. People also use them to build up muscles, although this is now generally outlawed in sporting activities.

anabolism All the metabolic reactions that synthesize complex molecules from more simple molecules. *See also* metabolism.

anaerobic Describing a biochemical process that takes place in the absence of free oxygen. *Compare* aerobic.

anaerobic respiration Respiration in which oxygen is not involved. It is found in yeasts, bacteria, and occasionally in muscle tissue. In this type of respiration the organic substrate is not completely oxidized and the energy yield is low. In the absence of oxygen in animal muscle tissue, glucose is degraded to pyruvate by GLYCOLYSIS, with the production of a small amount of energy and also lactic acid, which may be oxidized later when oxygen becomes available (*see* oxygen debt). FERMENTATION is an example of anaerobic respiration, in which certain yeasts produce ethanol and carbon dioxide as end products. Only two molecules of ATP are produced by this process. *Compare* aerobic respiration.

analysis The process of determining the constituents or components of a sample. There are two broad major classes of

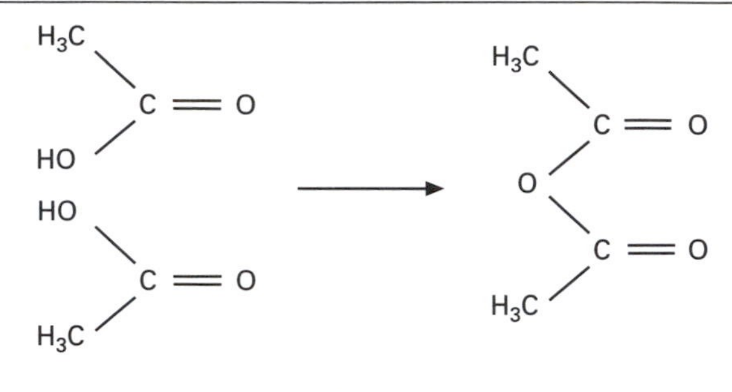

Anhydride

analysis, *qualitative analysis* – essentially answering the question 'what is it?' – and *quantitative analysis* – answering the question 'how much of such and such a component is present?' There is a large number of analytical methods that can be applied, depending on the nature of the sample and the purpose of the analysis. These include gravimetric, volumetric, and systematic qualitative analysis (classical wet methods); and instrumental methods, such as chromatographic, spectroscopic, nuclear, fluorescence, and polarographic techniques.

ångstrom Symbol Å A unit of length defined as 10^{-10} meter. The ångstrom was used for expressing wavelengths of light or ultraviolet radiation or for the sizes of molecules; the nanometer is now preferred.

anhydride A compound formed by removing water from an acid or, less commonly, a base. Many nonmetal oxides are anhydrides of acids: for example CO_2 is the anhydride of H_2CO_3 and SO_3 is the anhydride of H_2SO_4. Organic anhydrides are formed by removing H_2O from two carboxylic acid groups, giving compounds with the functional group –CO.O.CO–. These form a class of organic compounds called ACID ANHYDRIDES.

anhydrous Describing a substance that lacks moisture, or a salt with no water of crystallization.

aniline (aminobenzene; phenylamine; $C_6H_5NH_2$) A colorless oily substance made by reducing nitrobenzene ($C_6H_5NO_2$). Aniline is used for making dyes, pharmaceuticals, and other organic compounds.

animal starch *See* glycogen.

anion A negatively charged ion, formed by addition of electrons to atoms or molecules. In electrolysis anions are attracted to the positive electrode (the anode). *Compare* cation.

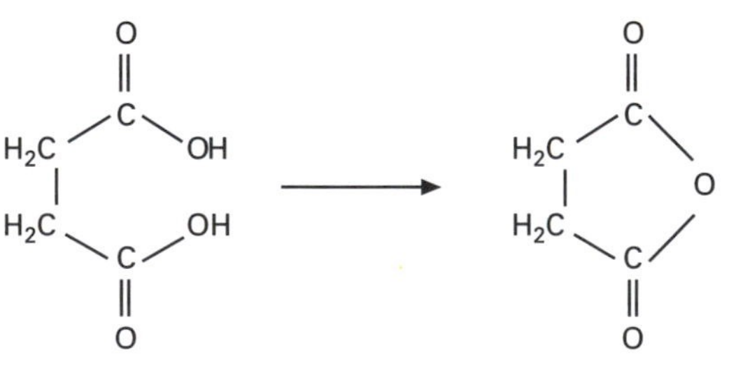

maleic acid

maleic anhydride

Anhydride: a cyclic anhydride

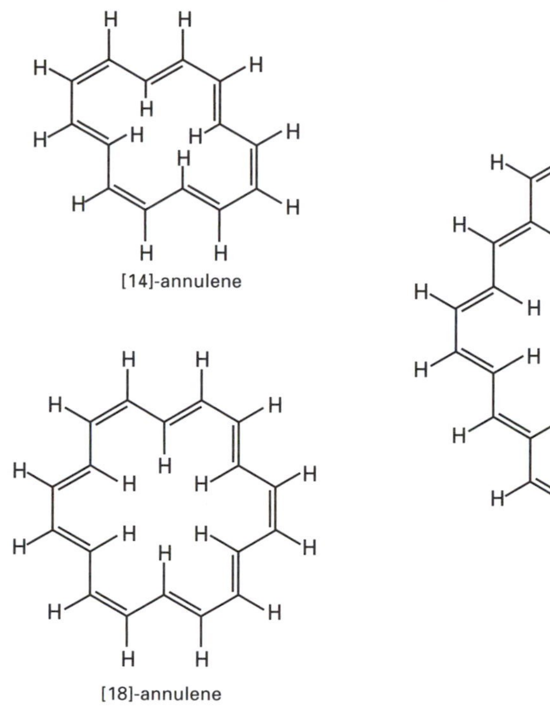

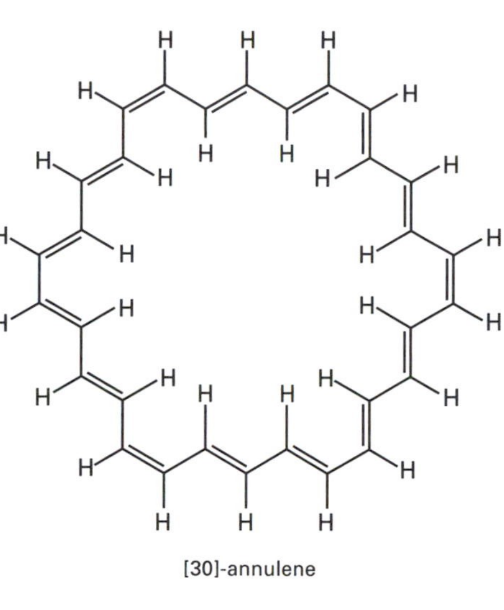

Annulene

anionic detergent *See* detergent.

anionic resin An ION-EXCHANGE material that can exchange anions, such as Cl^- and OH^-, for anions in the surrounding medium. Such resins are used for a wide range of analytical and purification purposes.

They are often produced by addition of a quaternary ammonium group ($N(CH_3)_4^+$) or a phenolic group ($-OH^-$) to a stable polyphenylethene resin. A typical exchange reaction is:

$$\text{resin–}N(CH_3)_4^+Cl^- + KOH \rightleftharpoons \text{resin–}N(CH_3)_4^+OH^- + KCl$$

Anionic resins can be used to separate mixtures of halide ions. Such mixtures can be attached to the resin and recovered separately by elution.

annulene A ring compound containing alternating double and single C–C bonds. The compound C_8H_8, having an eight-membered ring of carbon atoms, is the next annulene larger than benzene. It is not an AROMATIC COMPOUND because it is not planar and does not obey the Hückel rule. C_8H_8 is called *cyclo-octatetraene*. Higher annulenes are designated by the number of carbon atoms in the ring. [10]-annulene obeys the Hückel rule but is not aromatic because it is not planar as a result of interactions of the hydrogen atoms inside the ring. There is evidence that [18]-annulene, which is a stable red solid, has aromatic properties.

anode In electrolysis, the electrode that is at a positive potential with respect to the cathode. In any electrical system, such as a discharge tube or electronic device, the anode is the terminal at which electrons flow out of the system.

anomer Either of two isomeric forms of a cyclic sugar that differ in the disposition of the –OH group on the carbon next to the O atom of the ring (the *anomeric* *carbon*). Anomers are diastereoisomers. They are designated α– or β– according to whether the –OH is below or above the ring respectively. *See illustration at* sugar.

anomeric carbon *See* anomer.

anthocyanin One of a group of water-soluble pigments found dissolved in higher plant cell vacuoles. Anthocyanins are red, purple, and blue and are widely distributed, particularly in flowers and fruits, where they are important in attracting insects, birds, etc. They also occur in buds and sometimes contribute to the autumn colors of leaves. They are natural pH indicators, often changing from red to blue as pH increases, i.e. acidity decreases. Color may also be modified by traces of iron and other metal salts and organic substances, for example cyanin is red in roses but blue in the cornflower. *See* flavonoid.

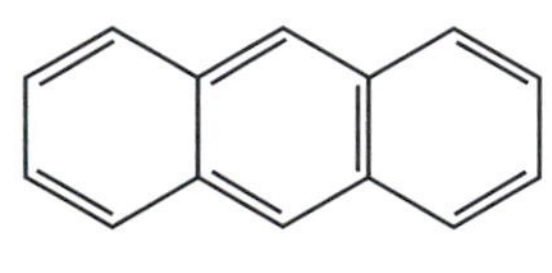

Anthracene

anthracene ($C_{14}H_{10}$) A white crystalline solid used extensively in the manufacture of dyes. Anthracene is found in the heavy- and green-oil fractions of crude oil and is obtained by fractional crystallization. Its structure is benzene-like, having three six-membered rings fused together. The reactions are characteristic of AROMATIC COMPOUNDS.

anthracite The highest grade of coal, with a carbon content of between 92% and 98%. It burns with a hot blue flame, gives off little smoke and leaves hardly any ash.

anthraquinone ($C_6H_4(CO)_2C_6H_4$) A colorless crystalline quinone used in producing dyestuffs such as alizarin.

antibonding orbital *See* orbital.

anticlinal conformation *See* conformation.

antiknock agent A substance added to gasoline to inhibit preignition or 'knocking'. A common example is lead tetraethyl, although use of this is discouraged in many countries for environmental reasons.

antioxidant A substance that inhibits oxidation. Antioxidants are added to such products as foods, paints, plastics, and rubber to delay their oxidation by atmospheric oxygen. Some work by forming chelates with metal ions, thus neutralizing the catalytic effect of the ions in the oxidation process. Other types remove intermediate oxygen free radicals. Naturally occurring antioxidants can limit tissue or cell damage in the body. These include vitamin E and β-carotene.

antiperiplanar conformation *See* conformation.

apoenzyme The protein part of a conjugate enzyme. It is an enzyme whose cofactor has been removed (e.g. via dialysis) rendering it catalytically inactive. When combined with its PROSTHETIC GROUP or coenzyme it forms a complete enzyme (HOLOENZYME).

aprotic *See* solvent.

aqueous Describing a solution in water.

arene An organic compound containing a benzene ring; i.e. an aromatic hydrocarbon or a derivative of an aromatic hydrocarbon.

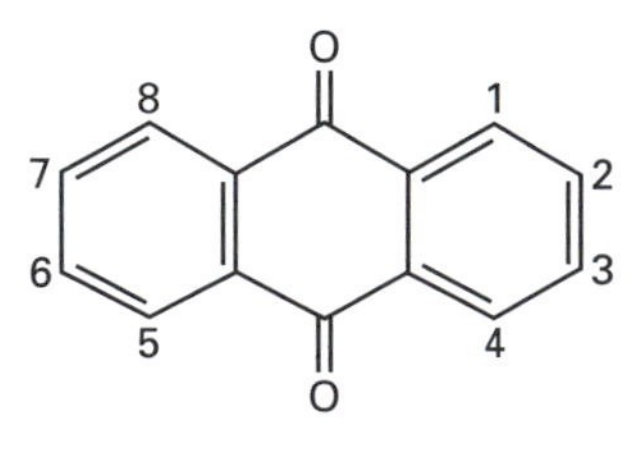

Anthraquinone

arginine *See* amino acid.

aromatic compound An organic compound with characteristic chemical reactions, usually containing BENZENE rings in its structure. Aromatic compounds, such as benzene, have a planar ring of atoms linked by alternate single and double bonds. The characteristic of aromatic compounds is that their chemical properties are not those expected for an unsaturated compound; they tend to undergo nucleophilic substitution of hydrogen (or other groups) on the ring, and addition reactions only occur under special circumstances.

The explanation of this behavior is that the electrons in the double bonds are delocalized over the ring, so that the six bonds are actually all identical and intermediate between single bonds and double bonds. The pi electrons are thus spread in a molecular orbital above and below the ring. The evidence for this delocalization in benzene is that the bond lengths between carbon atoms in benzene are all equal and intermediate in size between single and double bond lengths. Also, if two hydrogen atoms attached to adjacent carbon atoms are substituted by other groups, the compound has only one structure. If the bonds were different two isomers would exist. Benzene has a stabilization energy of 150 kJ mol^{-1} over the Kekulé structure. It is possible to characterize aromatic behavior by detecting a ring current in NMR. Certain heterocyclic molecules, such as PYRIDINE, also have aromatic properties.

The delocalization of the electrons in the pi orbitals of benzene accounts for the properties of benzene and its derivatives, which differ from the properties of alkenes and other aliphatic compounds. The phenomenon is called *aromaticity*. A definition of aromaticity is that it occurs in compounds that obey the *Hückel rule*: i.e. that there should be a planar ring with a total of $(4n + 2)$ pi electrons (where n is any integer). Using this rule as a criterion certain nonbenzene rings show aromaticity. Such compounds are called *nonbenzenoid aromatics*. Examples are the cyclopentadienyl ion $C_5H_5^-$ and the tropyllium ion $C_7H_7^+$. Other compounds that have a ring of atoms with alternate double and single bonds, but do not obey the rule (e.g. cyclooctatetraene, which has a nonplanar ring of alternating double and single bonds) are called *pseudoaromatics*.

Compare aliphatic compound. *See also* annulene.

aromaticity *See* aromatic compound.

Arrhenius equation An equation relating the rate constant of a chemical reaction and the temperature at which the reaction takes place:

$$k = A\exp(-E_a/RT)$$

where A is a constant, k the rate constant, T the thermodynamic temperature in kelvins, R the gas constant, and E_a the activation energy of the reaction.

Reactions proceed at different rates at different temperatures, i.e. the magnitude of the rate constant is temperature dependent. The Arrhenius equation is often written in a logarithmic form, i.e.

$$\log_e k = \log_e A - E/2.3RT$$

This equation enables the activation energy for a reaction to be determined. It is named for the Swedish chemist Svante August Arrhenius (1859–1927).

Arrhenius theory *See* acid.

aryl group An organic group derived by removing a hydrogen atom from an aromatic hydrocarbon or derivative. The phenyl group, C_6H_5–, is the simplest example.

ascorbic acid *See* vitamin C.

asparagine *See* amino acid.

aspartic acid *See* amino acid.

aspirin (acetylsalicylic acid; $C_9H_8O_4$) A colorless crystalline compound made by treating salicylic acid with ethanoyl hydride. It is used as an analgesic and antipyretic drug, and small doses are prescribed for adult patients at risk of heart attack or stroke. It should not be given to children.

association The combination of molecules of a substance with those of another to form more complex species. An example is a mixture of water and ethanol (which are termed *associated liquids*), the molecules of which combine via hydrogen bonding.

asymmetric atom *See* chirality; isomerism; optical activity.

atactic polymer *See* polymerization.

atmosphere A unit of pressure defined as 101 325 pascals (atmospheric pressure). The atmosphere is used in chemistry only for rough values of pressure; in particular, for stating the pressures used in high-pressure industrial processes.

atom The smallest part of an element that can exist as a stable entity. Atoms consist of a small dense positively charged nucleus, made up of neutrons and protons, with electrons in a cloud around this nucleus. The chemical reactions of an element are determined by the number of electrons (which is equal to the number of protons in the nucleus). All atoms of a given element have the same number of protons (the proton number). A given element may have two or more isotopes, which differ in the number of neutrons in the nucleus.

The electrons surrounding the nucleus are grouped into *shells* – i.e. main orbits around the nucleus. Within these main orbits there may be subshells. These correspond to atomic orbitals. An electron in an atom is specified by four quantum numbers:

1. The *principal quantum number* (n), which specifies the main energy levels. n can have values 1, 2, etc. The corresponding shells are denoted by letters K, L, M, etc., the K shell (n = 1) being the nearest to the nucleus. The maximum number of electrons in a given shell is $2n^2$.
2. The *orbital quantum number* (l), which specifies the angular momentum. For a given value of n, l can have possible values of n–1, n–2, ... 2, 1, 0. For instance, the M shell (n = 3) has three subshells with different values of l (0, 1, and 2). Sub-shells with angular momentum 0, 1, 2, and 3 are designated by letters s, p, d, and f.
3. The *magnetic quantum number* (m). This can have values $-l$, $-(l-1)$ ... 0 ... $+(l-1)$, $+l$. It determines the orientation of the electron orbital in a magnetic field.
4. The *spin quantum number* (m_s), which specifies the intrinsic angular momentum of the electron. It can have values $+\frac{1}{2}$ and $-\frac{1}{2}$.

Each electron in the atom has four quantum numbers and, according to the Pauli exclusion principle, no two electrons can have the same set of quantum numbers. This explains the electronic structure of atoms.

atomicity The number of atoms per molecule of an element. Helium, for example, has an atomicity of one, nitrogen two, and ozone three.

atomic mass unit (amu) Symbol: u A unit of mass used for atoms and molecules, equal to 1/12 of the mass of an atom of carbon-12. It is equal to $1.660\,33 \times 10^{-27}$ kg.

atomic number *See* proton number.

atomic orbital *See* orbital.

atomic weight *See* relative atomic mass (r.a.m.).

ATP (adenosine triphosphate) The universal energy carrier of living cells. Energy from respiration or, in photosynthesis, from sunlight is used to make ATP from ADP. It is then reconverted to ADP in various parts of the cell by enzymes known as *ATPases*, the energy released being used to drive three main cellular processes: mechanical work (muscle contraction and cellular movement); the active transport of molecules and ions; and the biosynthesis of other molecules. It can also be converted to light, electricity, and heat.

ATP is a nucleotide consisting of adenine and ribose with three phosphate

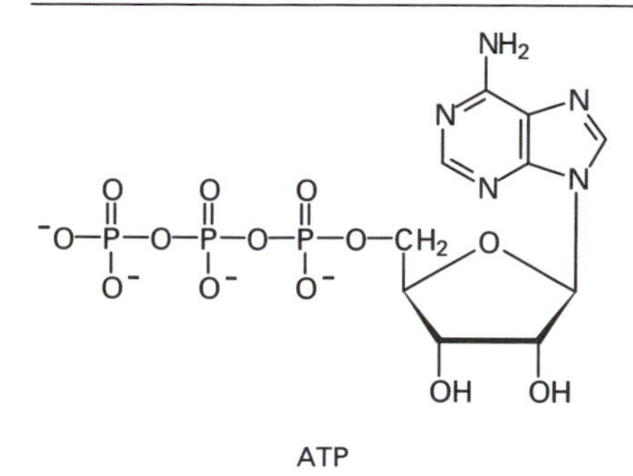

ATP

groups attached. Hydrolysis of the terminal phosphate bond releases energy (30.6 kJ mol^{-1}) and is coupled to an energy-requiring process. Further hydrolysis of ADP to AMP sometimes occurs, releasing more energy. The pool of ATP is small, but the faster it is used, the faster it is replenished. ATP is not transported around the body, but is synthesized where it is needed.

atto- Symbol: a A prefix denoting 10^{-18}. For example, 1 attometer (am) = 10^{-18} meter (m).

autocatalysis *See* catalyst.

autoclave An apparatus consisting of an airtight container whose contents are heated by high-pressure steam; the contents may also be agitated. Autoclaves are used for reactions between gases under pressure in industrial processing and for sterilizing objects.

auxin Any of a group of plant hormones, the most common naturally occurring one being indole acetic acid, IAA. Auxins are made continually in growing shoot and root tips. Synthetic auxins, cheaper and more stable than IAA, are employed in agriculture, horticulture, and research. These include indoles and naphthyls: e.g. NAA (naphthalene acetic acid) used mainly as a rooting and fruit setting hormone; phenoxyacetic acids, e.g. 2,4-D (2,4-dichlorophenoxyacetic acid) used as weed-killers and modifiers of fruit development; and more toxic and persistent benzoic auxins, e.g. 2,4,5-trichlorobenzoic acid, also formerly used as herbicides but now widely restricted.

Avogadro constant (Avogrado number) Symbol: N_A The number of particles in one mole of a substance. Its value is 6.022 52 × 10^{23} mol^{-1}.

Avogadro number *See* Avogadro constant.

Avogadro's law The principle that equal volumes of all gases at the same temperature and pressure contain equal numbers of molecules. It is often called *Avogadro's hypothesis*. It is strictly true only for ideal gases.

axial conformation *See* cyclohexane.

azeotrope (azeotropic mixture) A mixture of liquids for which the vapor phase has the same composition as the liquid phase. It therefore boils without change in composition and, consequently, without progressive change in boiling point.

The composition and boiling points of azeotropes vary with pressure, indicating that they are not chemical compounds. Azeotropes may be broken by distillation in the presence of a third liquid, by chemical reactions, adsorption, or fractional crystallization. *See* constant-boiling mixture.

azeotropic distillation A method used to separate mixtures of liquids that cannot be separated by simple distillation. Such a mixture is called an *azeotrope*. A solvent is added to form a new azeotrope with one of the components, and this is then removed and subsequently separated in a second column. An example of the use of azeotropic distillation is the dehydration of 96% ethanol to absolute ethanol. Azeotropic distillation is not widely used because of the difficulty of finding inexpensive nontoxic noncorrosive solvents that can easily be removed from the new azeotrope.

azeotropic mixture *See* azeotrope.

azide **1.** An organic compound of general formula RN_3.
2. An inorganic compound containing the ion N_3^-.

azine An organic heterocyclic compound that has a hexagonal ring containing carbon and nitrogen atoms. Pyridine (C_5H_5N) is the simplest example.

azo compound A type of organic compound of the general formula RN:NR′, where R and R′ are aromatic groups. Azo compounds can be formed by coupling a DIAZONIUM COMPOUND with an aromatic phenol or amine. Most are colored because of the presence of the *azo group* –N:N–.

azo dye An important type of dye used in acid dyes for wool and cotton. The dyes are azo compounds; usually sodium salts of sulfonic acids.

azo group *See* azo compound.

azulene ($C_{10}H_8$) A blue crystalline compound having a seven-membered ring fused to a five-membered ring. It converts to naphthalene on heating.

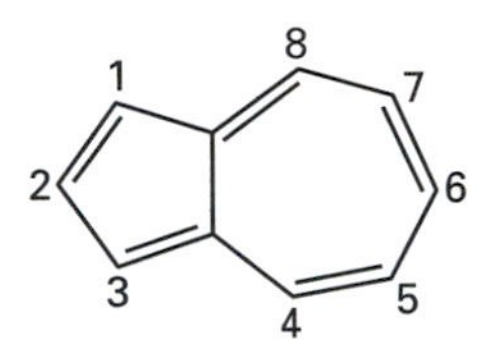

Azulene

backbiting A process that can occur in certain free-radical POLYMERIZATION reactions, in which a radical with an unpaired electron on the end of the chain converts into one in which the unpaired election is not at the end of the chain. For example, the radical

$RCH_2CH_2CH_2CH_2CH_2\ CH_2\bullet$

may convert into

$RCH_2CH\bullet CH_2CH_2\ CH_2CH_3$

Effectively, this involves a transfer of a hydrogen atom within the molecule. Typically, the free electron moves from the end of the chain to atom five, counting from the end. This is because the process involves a transition state with a six-membered ring. The new free radical is more stable than the original one. Further polymerization occurs at the new unpaired electron leading to the production of polymers with butyl ($CH_3CH_2CH_2CH_2$–) side chains.

Baeyer, Johann Friedrich Wilhelm Adolph von (1835–1917) German organic chemist. Baeyer worked mainly in organic synthesis and is noted for his study of the dye indigo. He started his work on indigo in 1865 and continued for 20 years; he determined the structure of indigo in 1883. The structure he postulated was correct (except for the stereochemistry of the double bond, which was subsequently shown by x-ray crystallography to be *trans*). Baeyer discovered a number of substances including barbituric acid. His later investigations on ring compounds and polyacetylenes led him to consider the stability of carbon–carbon bonds in cyclic compounds. This resulted in the *Baeyer strain theory*. Baeyer was awarded the 1905 Nobel Prize for chemistry for his work on indigo and aromatic compounds.

Baeyer–Villiger reaction A type of reaction in which a ketone reacts with a peroxy acid, with resulting production of an ester. For example,

R–CO–R → R–CO–O–R.

The reaction involves 'insertion' of an oxygen atom next to the carbonyl (CO) group. Typical peroxy acids used are trifluoroperethanoic acid (CF_3.CO.O.OH) and *meta*-chloroperbenzoic acid (*m*-CPBA; ClC_6H_4.CO.O.OH). The reaction was discovered in 1899 by the German chemists A. Baeyer and V. Villiger, and is commonly used in organic synthesis. In certain cases hydrogen peroxide (H_2O_2) can be used as the oxidizing agent. This is sometimes known as the *Dakin reaction*. The Baeyer–Villiger reaction is a type of rearrange-

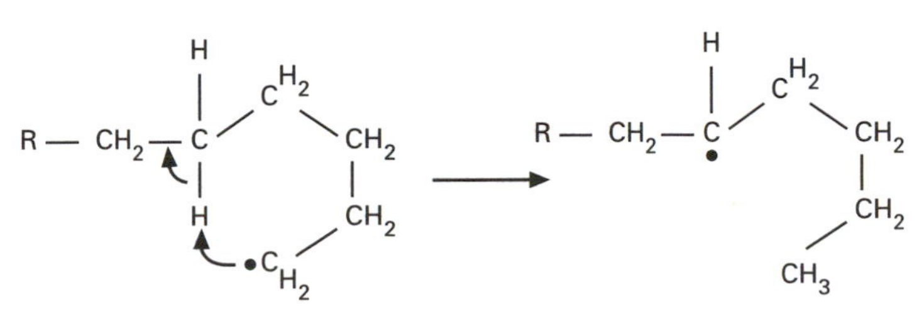

Backbiting

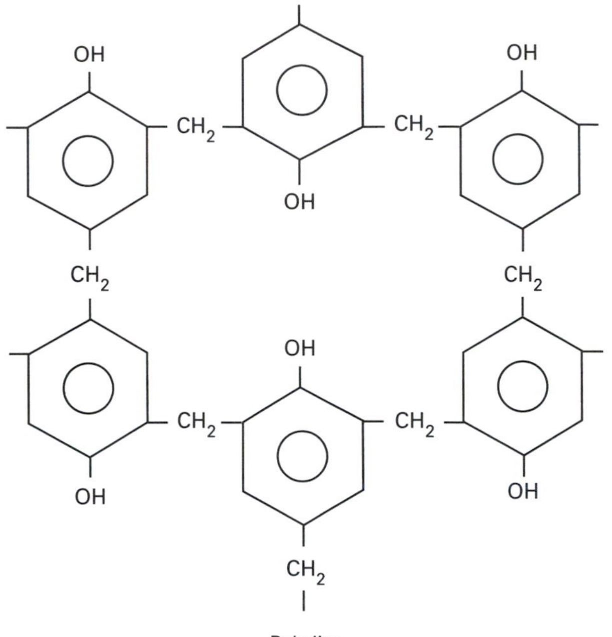

Bakelite

ment. For a peroxy acid X.CO.O.OH, there is an intermediate cation formed $R_2C^+(OH)(O.CO.X)$. The mechanism involves migration of a group R onto the oxygen of the peroxy acid group.

Bakelite (*Trademark*) A common thermosetting synthetic polymer formed by the condensation of phenol (C_6H_5OH) and methanal (formaldehyde, HCOH). It is an example of a *phenolic resin* (or *phenol–formaldehyde resin*), and was one of the first useful synthetic polymers. The reaction between phenol and methanal occurs under acid conditions and involves electrophilic substitution on the benzene ring to give a three-dimensional polymeric structure. Bakelite is named for the Belgian-born US chemist Leo Hendrik Baekeland (1863–1944), who discovered it in 1909.

ball mill A device commonly used in the chemical industry for grinding solid material. Ball mills usually have slowly rotating steel-lined drums containing steel balls. The material is crushed by the tumbling action of the balls in the drum. *Compare* hammer mill.

banana bond (bent bond) In strained-ring compounds the bond angles that would be produced by hybridization of orbitals are not equal to the angles obtained by joining the atomic centers. In such cases it is sometimes assumed that the bonding orbital is bent or banana-like in shape. For example, in cyclopropane the three carbon atoms are arranged in an equilateral triangle, and the bond angle is 60°. The sp^3 hybridization gives an angle of about 104° between the orbitals. Consequently, the orbitals overlap at an angle, giving a banana bond. The term 'banana bond' is also used in a quite separate sense for a multicenter bond of the type present in electron-deficient compounds such as diborane (B_2H_6).

band spectrum A SPECTRUM that appears as a number of bands of emitted or absorbed radiation. Band spectra are characteristic of molecules. Often each band can be resolved into a number of closely spaced lines. The different bands correspond to changes of electron orbit in the molecules and the closely spaced lines in each band, seen under higher resolution, are the result of different vibrational states of the molecule.

barrel A measurement of volume often used in the oil and chemical industries. One barrel is equal to 159 liters (about 29 US gallons).

Barton, Sir Derek Harold Richard (1918–98) British organic chemist noted for his work on the stereochemistry of organic molecules, particularly natural products. In a major paper published in 1950 he suggested that the rates of reactions in isomers are strongly influenced by the spatial orientations of their functional groups. This paper initiated the branch of organic chemistry known as conformational analysis. Barton studied many natural products, including phenols. In 1959 he developed a simple synthesis for the hormone aldosterone. He shared the 1969 Nobel Prize for chemistry with Norwegian chemist Odd Hassell.

base *See* acid.

base analog An unnatural purine or pyrimidine that can be incorporated into DNA, causing altered base pairing. Some base analogs are used therapeutically as anticancer drugs.

base-catalyzed reaction A reaction catalyzed by bases. Typical base-catalyzed reactions are the CLAISEN CONDENSATION and the ALDOL REACTION, in which the first step is abstraction of a proton to give a carbanion.

base pairing The linking together of the two helical strands of DNA by bonds between complementary bases, adenine pairing with thymine and guanine pairing with cytosine. The specific nature of base pairing enables accurate replication of the chromosomes and thus maintains the constant composition of the genetic material. In pairing between DNA and RNA the uracil of RNA pairs with adenine.

basic Acting as a base; having a tendency to release hydroxide ions (OH^-) in aqueous solution. A basic solution has an excess of OH^- ions over H^+ ions; i.e. a pH greater than 7.

batch process A manufacturing process in which the reactants are fed into the process in fixed quantities (batches), rather than in a continuous flow. At any particular instant all the material, from its preparation to the final product, has reached a definite stage in the process. Such processes present problems of automation and instrumentation and tend to be wasteful of energy. For this reason, batch processing is used on an industrial scale only when small quantities of valuable or strategic materials are required, e.g. specialist chemicals or pharmaceuticals. *Compare* continuous process.

Beckmann rearrangement A type of reaction in which the OXIME of a ketone is converted into an amide using a sulfuric acid catalyst. First discovered by the German chemist Ernst Beckmann (1853–1923), it is used in the manufacture of polyamides (*see* nylon).

Beckmann thermometer A type of mercury thermometer designed to measure small differences in temperature rather than scale degrees. Beckmann thermometers have a larger bulb than common thermometers and a stem with a small internal diameter, so that a range of 5°C covers about 30 centimeters in the stem. The mercury bulb is connected to the stem in such a way that the bulk of the mercury can be separated from the stem once a particular 5° range has been attained. The thermometer can thus be set for any particular range. The Beckmann thermometer has commonly been used for measuring such quantities as depression of freezing point and elevation of boiling point.

bent bond *See* banana bond.

benzaldehyde *See* benzenecarbaldehyde.

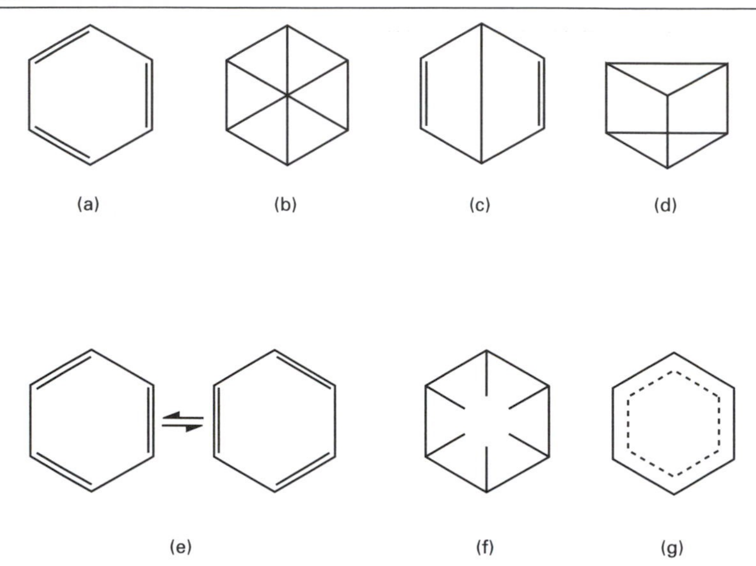

Benzene: early structures suggested for benzene: (a) Kekulé (1865); (b) Claus (1867); (c) Dewar (1867); (d) Ladenburg (1869); (e) Kekulé (1865); (f) Armstrong–Baeyr (1887); (g) Thiele (1899)

benzene (C_6H_6) A colorless liquid hydrocarbon with a characteristic odor. Benzene is a highly toxic compound and continued inhalation of the vapor is harmful. It was originally isolated from coal tar and for many years this was the principal source of the compound. Contemporary manufacture is from hexane; petroleum vapor is passed over platinum at 500°C and at a pressure of 10 atmospheres:

$$C_6H_{14} \rightarrow C_6H_6 + 4H_2$$

Benzene is the simplest aromatic hydrocarbon. *See* aromatic compound. The structure of benzene was the subject of considerable speculation in the 19th century. The basic problem – known as the *benzene problem* – was that of reconciling the formula of benzene, C_6H_6, with its chemical reactions. The empirical formula is the same as that of acetylene, C_2H_2, and

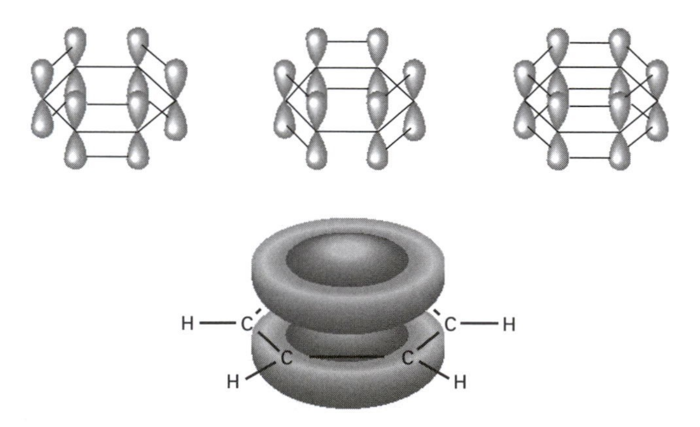

Benzene: in benzene, 6 p orbitals can combine in different ways to give delocalized molecular orbitals. The one of lowest energy has two donut-shaped areas above and below the ring of carbon atoms.

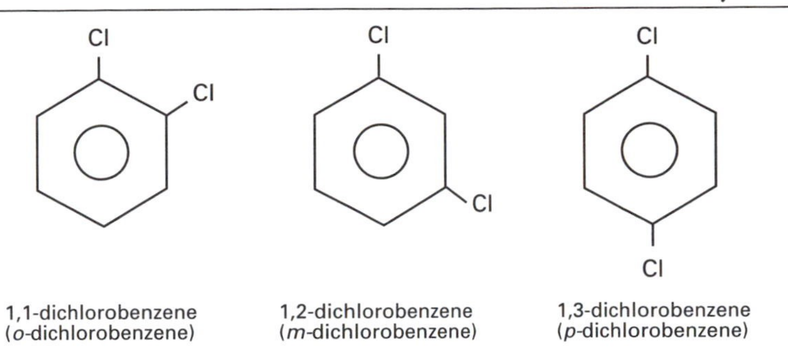

Benzene: disubstituted derivates of benzene

it might be expected that benzene would undergo similar reactions. However, benzene does not show the usual behavior of a compound containing double or triple bonds.

For example, acetylene adds bromine to yield CHBr:CHBr and eventually $CHBr_2$-$CHBr_2$. Benzene, with an iron bromide catalyst, suffers displacement of one of its hydrogen atoms to yield C_6H_5Br. This type of activity in which substitution reactions occur indicates that benzene might be saturated.

Benzene, however, does not always act as a saturated compound. In sunlight bromine is added to give $C_6H_6Cl_6$ and hydrogen can also be added with a nickel catalyst to yield cyclohexane, C_6H_{12}. A number of different formulae were put forward to try to explain the properties. In 1865 the German chemist August Kekulé (1829–96) suggested a structure with alternate double and single bonds in a hexagonal ring. To account for the fact that benzene has only three disubstitution products, he further proposed that the positions of the bonds oscillate so that two molecules are in equilibrium. This structure – the *Kekulé formula* – is the one often used in formulae of compounds containing benzene rings.

The modern idea of aromaticity is based not on equilibrium between Kekulé structures but on RESONANCE between them. The bonds in benzene have characters between double and single bonds: the carbon atoms are held together by six single bonds and the remaining six electrons, from the double bonds, are delocalized over the ring. This is the reason benzene has all its C-C bonds of the same length and undergoes ELECTROPHILIC SUBSTITUTION reactions.

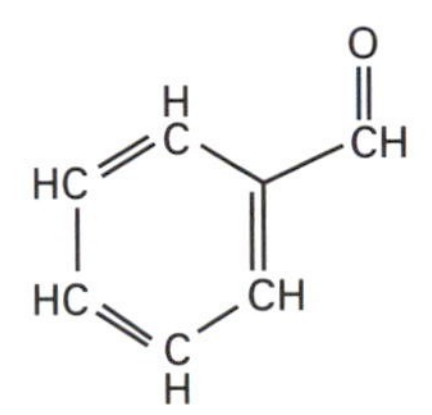

Benzenecarbaldehyde (benzaldehyde)

benzenecarbaldehyde (benzaldehyde; C_6H_5CHO) A yellow oily ALDEHYDE with a distinct almondlike odor (the compound occurs in almond kernels). Benzenecarbaldehyde may be synthesized in the laboratory by the usual methods of aldehyde synthesis. It is used as a food flavoring and in the manufacture of dyes and antibiotics, and can be readily manufactured by the chlorination of methylbenzene (toluene) on the methyl group and the subsequent hydrolysis of dichloromethylbenzene:

$$C_6H_5CH_3 + Cl_2 \rightarrow C_6H_5CHCl_2$$
$$C_6H_5CHCl_2 + 2H_2O \rightarrow C_6H_5CH(OH)_2 + 2HCl$$
$$C_6H_5CH(OH)_2 \rightarrow C_6H_5CHO + H_2O$$

benzenecarbonyl chloride (benzoyl chloride; C_6H_5COCl) A liquid acyl chloride used as a benzoylating agent. *See* acylation.

benzenecarbonyl group (benzoyl group) The group C_6H_5CO-.

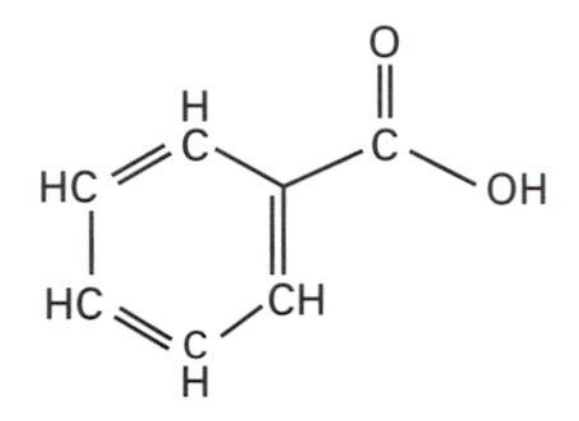

Benzenecarboxylic acid (benzoic acid)

benzenecarboxylic acid (benzoic acid; C_6H_5COOH) A white crystalline carboxylic acid found naturally in some plants. It is used as a food preservative. The carboxyl group (–COOH) directs further substitution onto the benzene ring in the 3 position.

benzene-1,2-dicarboxylic acid (phthalic acid; $C_6H_4(COOH)_2$) A white crystalline aromatic acid. On heating it loses water to form *phthalic anhydride*, which is used to make dyestuffs and polymers.

benzene-1,4-dicarboxylic acid (terephthalic acid; $C_6H_4(COOH)_2$) A colorless crystalline organic acid used to produce Dacron and other polyesters.

benzene-1,3-diol (resorcinol; $C_6H_4(OH)_2$) A white crystalline phenol used in the manufacture of dyestuffs and celluloid.

benzene-1,4-diol (hydroquinone; quinol; $C_6H_4(OH)_2$) A white crystalline phenol used in making dyestuffs. *See also* quinone.

benzene ring The cyclic hexagonal arrangement of six carbon atoms that are characteristic of benzene and its derivatives. *See* aromatic compound; benzene.

benzenesulfonic acid ($C_6H_5SO_2OH$) A white crystalline sulfonic acid made by sulfonation of benzene. Any further substitution onto the benzene ring is directed into the 3 position.

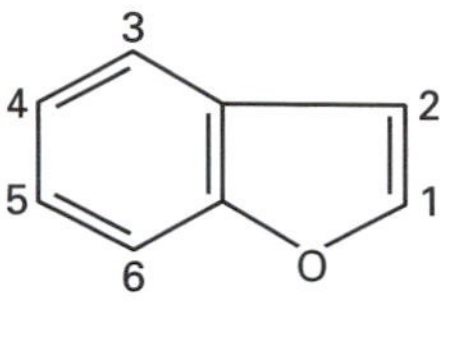

Benzfuran (coumarone)

benzfuran (coumarone; C_8H_6O) A crystalline compound having a benzene ring fused to a furan ring.

benzilic acid rearrangement A reaction in which *benzil* (1,2-diphenylethan-1,2-dione) is treated with hydroxide and then with acid to give *benzilic acid* (2-hydroxy-2,2-diphenylethanoic acid):

$$C_6H_5.CO.CO.C_6H_5 \rightarrow (C_6H_5)_2C(OH).COOH$$

The reaction, which involves migration of a phenyl group (C_6H_5-) from one carbon atom to another, was the first rearrangement reaction to be described (by German chemist Justus von Liebig in 1828).

benzoic acid *See* benzenecarboxylic acid.

benzole A mixture of mainly aromatic hydrocarbons obtained from coal.

benzopyrene *See* benzpyrene.

benzoquinone *See* quinone.

benzoylation The introduction of a benzoyl group (benzenecarbonyl group) into a compound. *See* acylation.

benzoyl chloride *See* benzenecarbonyl chloride.

benzoyl group *See* benzenecarbonyl group.

benzpyrene (benzopyrene; $C_{20}H_{12}$) A cyclic aromatic hydrocarbon with a structure consisting of five fused benzene rings. It occurs in coal tar and is produced by incomplete combustion of some organic compounds. Benzpyrene, which is present in tobacco smoke, has marked carcinogenic properties.

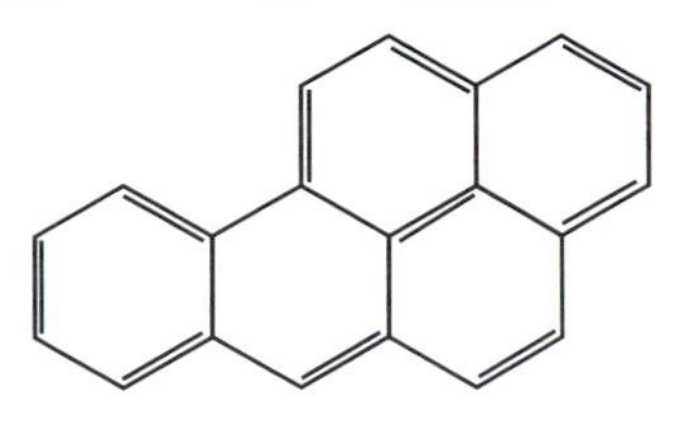

Benzpyrene

benzpyrrole *See* indole.

benzyl alcohol *See* phenylmethanol.

benzyl group The group $C_6H_5CH_2$–.

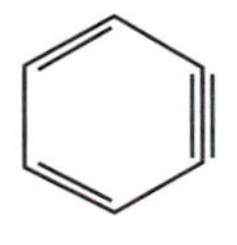

Benzyne

benzyne (C_6H_4) A short-lived intermediate present in some reactions. The ring of six carbon atoms contains two double bonds and one triple bond (the systematic name is *1,2-didehydrobenzene*).

Bergius process A process formerly used for making hydrocarbon fuels from coal. A mixture of powdered coal, heavy oil, and a catalyst was heated with hydrogen at high pressure.

beta-pleated sheet A type of PROTEIN structure in which polypeptide chains run close to each other and are held together by hydrogen bonds at right angles to the main chain. The structure is folded in regular 'pleats'. Fibres having this type of structure are usually composed of amino acids with short side chains. The chains may run in the same direction (parallel) or opposite directions (antiparallel). It is one of the two basic secondary structures of proteins.

bi- Prefix meaning 'two'. For example, biphenyls are compounds that have two phenyl groups joined together, as in C_6H_5–C_6H_5. The prefix is also commonly used in naming inorganic compounds to indicate the presence of hydrogen; for instance, sodium bisulfate ($NaHSO_4$) is sodium hydrogensulfate, etc.

bicarbonate *See* hydrogencarbonate.

bimolecular Describing a reaction or a step in a reaction that involves two molecules, ions, etc. For example the decomposition of hydrogen iodide,

$$2HI \rightarrow H_2 + I_2$$

takes place between two molecules and is therefore a bimolecular reaction. All bimolecular reactions are second order, but some second-order reactions are not bimolecular. *See also* order.

binary compound A chemical com-

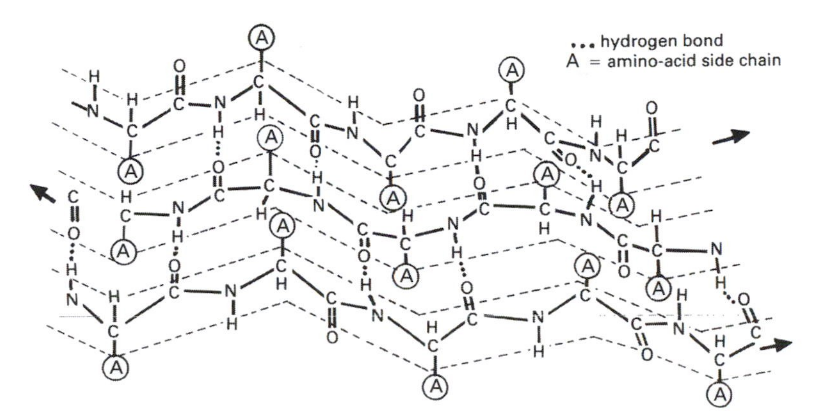

Beta-pleated sheet

pound formed from only two elements. Water (H_2O) and sodium chloride ($NaCl$) are examples.

bioassay An experimental technique for measuring quantitatively the strength of a biologically active chemical by its effect on a living organism. For example, the vitamin activity of certain substances can be measured using bacterial cultures. The increase in bacterial numbers is compared against that achieved with known standards for vitamins.

biochemical oxygen demand (BOD) The amount of oxygen taken from natural water by microorganisms that decompose organic waste matter in the water. It is therefore a measure of the quantity of organic pollutants present. The biochemical oxygen demand is determined by measuring the amount of oxygen in a sample of water, storing the sample, and then making the measurement again five days later.

biochemistry The study of chemical compounds and reactions occurring in living organisms.

biodegradable *See* pollution.

biosynthesis The series of reactions by which organisms obtain the various compounds needed for life.

biotechnology The application of technology to biological processes for industrial, agricultural, and medical purposes. For example, bacteria such as *Penicillium* and *Streptomycin* are used to produce antibiotics and fermenting yeasts produce alcohol in beer and wine manufacture. Recent developments in genetic engineering have enabled the large-scale production of hormones, blood serum proteins, and other medically important products. Genetic modification of farm crops, and even livestock, offers the prospect of improved protection against pests, or products with novel characteristics, such as new flavors or extended storage properties. *See also* enzyme technology.

biotin A water-soluble vitamin generally found, together with vitamins in the B group, in the VITAMIN B COMPLEX. It is widely distributed in natural foods, egg yolk, kidney, liver, and yeast being good sources. Biotin is required as a coenzyme for carboxylation reactions in cellular metabolism.

biphenyl ($C_6H_5C_6H_5$) An organic compound having a structure in which two phenyl groups are joined by a C–C bond. *See also* polychlorinated biphenyl.

bipyridyl *See* dipyridyl.

bisulfite addition compound *See* aldehyde.

bitumen *See* tar.

biuret ($H_2NCONHCONH_2$) A colorless crystalline organic compound made by heating urea (carbamide). It is used in a chemical test for proteins. *See also* protein; urea.

bivalent (divalent) Having a valence of two.

block copolymer *See* polymerization.

boat conformation *See* cyclohexane.

BOD *See* biochemical oxygen demand.

boiling The process by which a liquid is converted into a gas or vapor by heating at its *boiling point*. At this temperature the vapor pressure of the liquid is equal to the external pressure, and bubbles of vapor can form within the liquid. The boiling point is always the same for a particular liquid at a given pressure (for reference purposes usually taken as standard pressure). *See also* elevation of boiling point.

boiling point-composition diagram A diagram for a two-component liquid system representing both the variation of the boiling point and the composition of the vapor phase as the liquid-phase composition is varied.

Boltzmann constant Symbol: *k* The constant 1.380 54 J K^{-1}, equal to the gas constant (*R*) divided by the Avogadro constant (N_A).

bomb calorimeter A device for measuring the energy released during the combustion of substances (e.g. foods and fuels). It consists of a strong sealed insulated container in which a known amount of the substance is ignited in an atmosphere of pure oxygen. The substance undergoes complete combustion at constant volume and the resultant rise in temperature can be used to calculate the energy released by the reaction. Such energy values (*calorific values*) are often quoted in joules per kilogram (J kg^{-1}), formerly in calories.

bond *See* chemical bond.

bond energy The energy involved in forming a chemical bond. For methane, for instance, the energy of the C–H bond is one quarter of the energy involved for the process

$$C + 4H \rightarrow CH_4$$

It is thus one quarter of the heat of atomization.

The *bond dissociation energy* is a different quantity to the bond energy. It is the energy required to break a particular bond in a compound, e.g.:

$$CH_4 \rightarrow CH_3 + H$$

More formally, the *bond enthalpy* can be used.

bonding orbital *See* orbital.

bond length The length of a chemical bond; the distance between the centers of the nuclei of two atoms joined by a chemical bond. Bond lengths may be measured by electron or x-ray diffraction.

Bosch process The reaction

$$CO + H_2O \rightarrow CO_2 + H_2$$

using WATER GAS passed over a hot catalyst. It was used by Carl Bosch (1874–1940) to produce hydrogen for the Haber process for making ammonia.

Boyle's law At a constant temperature, the pressure of a fixed mass of a gas is inversely proportional to its volume: i.e.

$$pV = K$$

where *K* is a constant. The value of *K* depends on the temperature and on the nature of the gas. The law holds strictly only for ideal gases. Real gases follow Boyle's law at low pressures and high temperatures. *See* gas laws.

Brady's reagent *See* 2,4-dinitrophenylhydrazine.

Bragg equation An equation used to deduce the crystal structure of a material using data obtained from x-rays directed at its surface. The conditions under which a crystal will reflect a beam of x-rays with maximum intensity is:

$$n\lambda = 2d\sin\theta$$

where θ is the angle of incidence and reflection (the *Bragg angle*) that the x-rays make with the crystal planes, *n* is a small integer, λ is the wavelength of the x-rays, and *d* is the distance between the crystal planes.

branched chain *See* chain.

bromine A deep red, moderately reactive element (symbol Br) belonging to the halogens; i.e. group 17 (formerly VIIA) of the periodic table. Bromine is a liquid at room temperature (mercury is the only other element with this property). It occurs in small amounts in seawater, salt lakes, and salt deposits but is much less abundant than chlorine. A number of organobromine compounds are important commercially. At one time the main use of bromine was as 1,2-dibromoethane. This was added to gasoline to combine with the lead produced by decomposition of the antiknock agent lead tetraethyl. This use has declined with the reduction in use of leaded gasoline for environmental reasons. Quantities of bromine are used in polybrominated diphenyl ethers (PBDEs), which are effective flame retardents in plastics. A number of bromine HALONS are also important.

bromoethane (ethyl bromide; C_2H_5Br) A colorless volatile compound, used as a refrigerant. It can be made from ethene and hydrogen bromide.

bromoform *See* tribromomethane.

bromomethane (methyl bromide; CH_3Br) A colorless volatile compound used as a solvent. It can be made from methane and bromine.

Brønsted acid *See* acid.

Brønsted base *See* acid.

buckminsterfullerene An allotrope of carbon containing clusters of 60 carbon atoms bound in a highly symmetric polyhedral structure. The C_{60} polyhedron has a combination of pentagonal and hexagonal faces similar to the panels on a soccer ball. The molecule was named for the American architect Richard Buckminster Fuller (1895–1983) because its structure resembles a geodesic dome (invented by Fuller). The C_{60} polyhedra are informally called *bucky balls*. The original method of making the allotrope was to fire a high-power laser at a graphite target. This also produces less stable carbon clusters, such as C_{70}. It can be produced more conveniently using an electric arc between graphite electrodes in an inert gas. The allotrope is soluble in benzene, from which it can be crystallized to give yellow crystals. This solid form is known as *fullerite*.

The discovery of buckminsterfullerene led to a considerable amount of research into its properties and compounds. Particular interest has been shown in trapping metal ions inside the carbon cage to form enclosure compounds. Buckminsterfullerene itself is often simply called *fullerene*. The term also applies to derivatives of buckminsterfullerene and to similar cluster (e.g. C_{70}). Carbon structures similar to that in C_{60} can also form small tubes, known as *bucky tubes*.

bucky ball *See* buckminsterfullerene.

bucky tube *See* buckminsterfullerene.

buffer A solution in which the pH remains reasonably constant when acids or alkalis are added to it; i.e. it acts as a buffer against (small) changes in pH. Buffer solutions generally contain a weak acid and one of its salts derived from a strong base; e.g. a solution of ethanoic acid and sodium ethanoate. If an acid is added, the H^+ reacts with the ethanoate ion (from dissociated sodium ethanoate) to form undissociated ethanoic acid; if a base is added the OH^- reacts with the ethanoic acid to form water and the ethanoate ion. The effectiveness of the buffering action is determined by the concentrations of the acid–anion pair:

$$K = [H^+][CH_3COO^-]/[CH_3COOH]$$

where K is the dissociation constant.

Phosphate, oxalate, tartrate, borate, and carbonate systems can also be used for buffer solutions.

bumping Violent boiling of a liquid caused when bubbles form at a pressure above atmospheric pressure.

Bunsen burner A gas burner consisting of a vertical metal tube with an adjustable air-inlet hole at the bottom. Gas is allowed into the bottom of the tube and the gas–air mixture is burnt at the top. With too little air the flame is yellow and sooty. Correctly adjusted, the burner gives a flame with a pale blue inner cone of incompletely burnt gas, and an almost invisible outer flame where the gas is fully oxidized and reaches a temperature of about 1500°C.

burette A piece of apparatus used for the addition of variable volumes of liquid in a controlled and measurable way. The burette is a long cylindrical graduated tube of uniform bore fitted with a stopcock and a small-bore exit jet, enabling a drop of liquid at a time to be added to a reaction vessel. Similar devices are used to introduce measured volumes of gas at regulated pressure in the investigation of gas reactions.

buta-1,3-diene (butadiene; CH_2:CH-CH:CH_2) A colorless gas made by catalytic dehydrogenation of butane. It is used in the manufacture of synthetic rubber. Buta-1,3-diene is conjugated and to some extent the pi electrons are delocalized over the whole of the molecule. The molecule can exist in cis and trans forms. *See also* Diels–Alder reaction.

butadiene *See* buta-1,3-diene.

butanal (butyraldehyde; C_3H_7CHO) A colorless liquid aldehyde.

butane (C_4H_{10}) A gaseous alkane obtained either from the gaseous fraction of crude oil or by the 'cracking' of heavier fractions. It is the fourth member of the homologous series of alkanes. Butane is easily liquefied under pressure and its main use is as a portable supply of fuel (bottle gas). It is also used in the industrial production of buta-1,3-diene. The isomeric hydrocarbon $CH_3CH(CH_3)CH_3$ (2-methylpropane) is known as *isobutane*.

butanedioic acid (succinic acid) A crystalline carboxylic acid, $HOOC(CH_2)_2$-COOH, that occurs in amber and certain plants. It forms during the fermentation of sugar (sucrose).

butanoic acid (butyric acid; C_3H_7-COOH) A colorless liquid carboxylic acid. Esters of butanoic acid are present in butter.

butanol (butyl alcohol; C_4H_9OH) Either of two alcohols that are derived from butane: the primary alcohol butan-1-ol ($CH_3(CH_2)_2CH_2OH$) and the secondary alcohol butan-2-ol ($CH_3CH(OH)CH_2$-CH_3). Both are colorless volatile liquids used as solvents.

butanone (methyl ethyl ketone; CH_3-COC_2H_5) A colorless volatile liquid ketone. It is manufactured by the catalytic oxidation of butane and used as a solvent.

butenedioic acid Either of two isomers. *Transbutenedioic acid* (*fumaric acid*) is a crystalline compound found in certain plants. *Cisbutenedioic acid* (*maleic acid*) is used in the manufacture of synthetic resins. It can be converted into the trans isomer by heating at 120°C. The cis form, when strongly heated, loses water to give a cyclic acid anhydride (*maleic anhydride*).

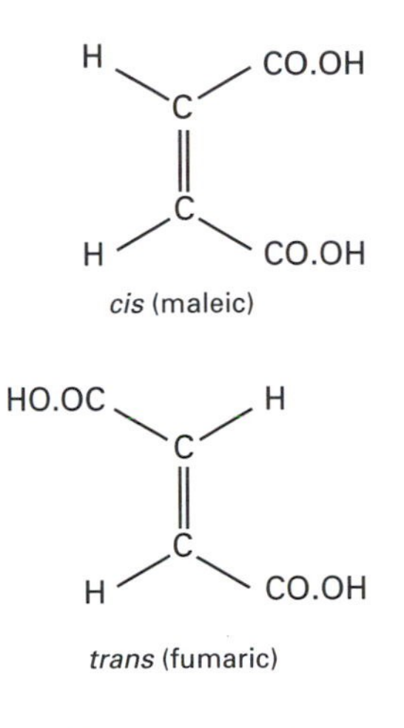

Butenedioic acid

Butlerov, Aleksandr Mikhailovich (1828–86) Russian organic chemist who was one of the main pioneers of the concept of structure for compounds. He first put forward his ideas in 1861. Butlerov predicted that tertiary alcohols exist and synthesized tertiary butanol. In 1876 he proposed the basics of the concept of tautomerism. He also studied formaldehyde (methanal) and its polymerization to various sugars.

butyl alcohol *See* butanol.

butyl group The straight-chain alkyl group $CH_3(CH_2)_2CH_2$–.

butyl rubber A type of synthetic rubber made by copolymerizing isobutylene (2-methylpropene, CH_3:$C(CH_3)_2$) with small amounts of isoprene (methylbuta-1,3-diene, CH_3:$C(CH_3)CH$:CH_2).

Before the introduction of tubeless tires butyl rubber was used for inner tubes because it is impervious to air. Subsequently halogenated butyl rubbers were developed (*halobutyls*), which could be cured at higher temperature and vulcanized with other rubbers. Both *chlorobutyls* and *bromobutyls* are manufactured. These types of rubber are used in tubeless tires bonded to the inner surface of the tire. Other uses are in sealants, hoses, and pond liners.

butyraldehyde *See* butanal.

butyric acid *See* butanoic acid.

by-product A substance obtained during the manufacture of a main chemical product. For example, propanone was formerly manufactured from propan-1-ol, but is now obtained as a by-product in the manufacture of phenol by the CUMENE PROCESS.

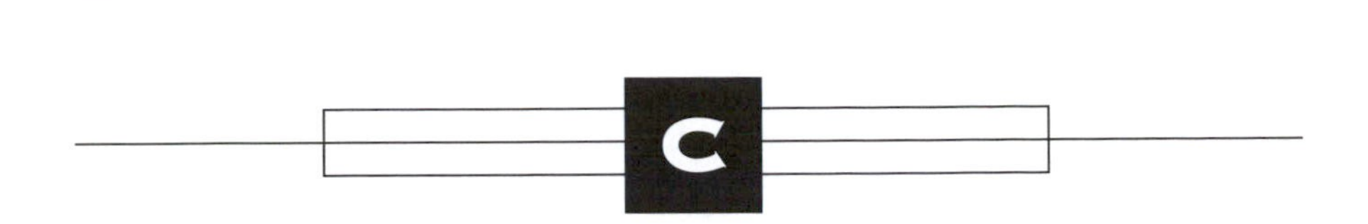

cadaverine An amine, $H_2N[CH_2]_5NH_2$, produced from lysine in decaying meat and fish.

caffeine An alkaloid found in certain plants, especially tea and coffee. It is a stimulant of the central nervous system and a diuretic. The systematic name is *1,3,7-trimethylxanthine*.

cage compounds *See* clathrate.

Cahn–Ingold–Prelog system *See* CIP system.

calciferol *See* vitamin D.

calcium acetylide *See* calcium carbide.

calcium carbide (calcium acetylide; calcium dicarbide; Ca C_2) A white solid that can be produced by heating coke with calcium oxide at high temperature (2000°C). It contains the dicarbide ion $^-C{\equiv}C^-$ and reacts with water to give ethyne (acetylene; C_2H_2):

$$CaC_2 + 2H_2O \rightarrow Ca(OH)_2 + C_2H_2$$

Formerly it was an important source of ethyne.

calcium dicarbide *See* calcium carbide.

calixarene *See* host–guest chemistry.

calorie Symbol: cal A unit of energy approximately equal to 4.2 joules. It was formerly defined as the energy needed to raise the temperature of one gram of water by one degree Celsius. Because the specific thermal capacity of water changes with temperature, this definition is not precise. The mean or thermochemical calorie (cal_{TH}) is defined as 4.184 joules. The international table calorie (cal_{IT}) is defined as 4.1868 joules. Formerly the mean calorie was defined as one hundredth of the heat needed to raise one gram of water from 0°C to 100°C, and the 15°C calorie as the heat needed to raise it from 14.5°C to 15.5°C.

calorific value The energy content of a substance, defined as the energy released in burning unit mass. Calorific values are measured using a BOMB CALORIMETER. They are used to express the efficiency of fuels (in megajoules per kilogram). Calorific values are also applied to foods (in kilojoules per gram or in calories). Here they measure the energy produced when the food is oxidized in metabolism.

calorimeter A device or apparatus for measuring thermal properties such as specific heat capacity, calorific value, etc. *See* bomb calorimeter.

Calvin, Melvin (1911–97) American chemist. Calvin worked out the biological mechanisms that occur in photosynthesis. He used techniques such as chromatography and radioisotopes to study the reactions of photosynthesis that do not require light. He found that there is a series of reactions, now known as the *Calvin cycle*. Calvin summarized his findings in a work entitled *The Path of Carbon in Photosynthesis* (1957). Calvin won the 1961 Nobel Prize for chemistry for his work on photosynthesis. He continued to work on various problems associated with photosynthesis.

Calvin cycle (reductive pentose-phosphate cycle) *See* photosynthesis.

cAMP *See* cyclic AMP.

camphor ($C_{10}H_{16}O$) A naturally-occurring white organic compound with a characteristic penetrating odor. It is a cyclic compound and a ketone, formerly obtained from the wood of the camphor tree but now made synthetically. Camphor is used as a platicizer for celluloid and as an insecticide against clothes moths.

cane sugar *See* sucrose.

Cannizzaro reaction The reaction of aldehydes to give alcohols and acid anions in the presence of strong bases. The aldehydes taking part in the Cannizzaro reaction do not have hydrogen atoms on the carbon attached to the aldehyde group. For instance, in the presence of hot aqueous sodium hydroxide:

$$NaOH + 2C_6H_5CHO \rightarrow C_6H_5CH_2OH + C_6H_5COO^-Na^+$$

This reaction is a disproportionation, involving both oxidation (to acid) and reduction (to alcohol). Another example is the reaction of methanal to give methanol and methanoate ions.

$$NaOH + 2HCHO \rightarrow CH_3OH + HCOO^-Na^+$$

The reaction was first described by the Italian chemist Stanislao Cannizzaro (1826– 1910) in 1853.

canonical form *See* resonance.

caproic acid *See* hexanoic acid.

caprolactam ($C_6H_{11}NO$) A white crystalline substance used in the manufacture of NYLON.

carbamide *See* urea.

carbanion An intermediate in an organic reaction in which one carbon atom carries a negative charge. Carbanions may be formed by abstracting a hydrogen ion from a C–H bond using a base, e.g. from ethanal to form $^-CH_2CHO$ (*see* aldol reaction). They can also be formed from organometallic compounds in which the carbon atom is bonded to an electropositive metal.

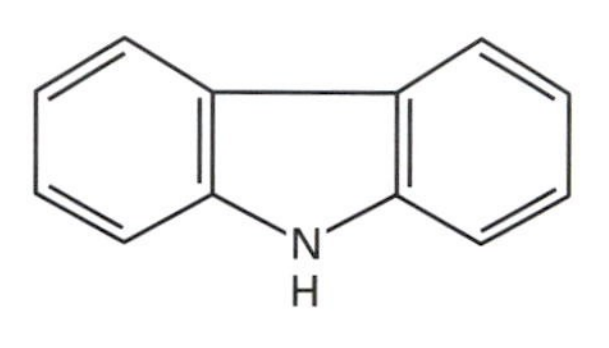

Carbazole

carbazole ($C_{12}H_9N$) A white crystalline compound used in the manufacture of dyestuffs.

carbene A transient species of the form RR′C:, with two valence electrons that do not form bonds. The simplest example is *methylene*, H_2C:. Some complex carbenes can be isolated but most are short-lived intermediates in reactions. In a carbene the carbon atom has two valence electrons that are not involved in bonding, and carbene species are highly electrophilic. Typically they can attack carbon–carbon double bonds to produce cyclopropane derivatives. Also they can attack single bonds in insertion reactions. For example, they can insert into O–H bonds:

$$R\text{–}O\text{–}H + R_2C: \rightarrow R\text{–}O\text{–}C(R_2)\text{–}H$$

They can also insert into C–H bonds:

$$R\text{–}H + R_2C: \rightarrow R\text{–}C(R_2)\text{–}H$$

Reactions like these make carbenes important 'reagents' in organic synthesis and various methods have been developed for generating them in the reaction medium.

All carbenes can exist in two possible states. The carbon atom has one s orbital

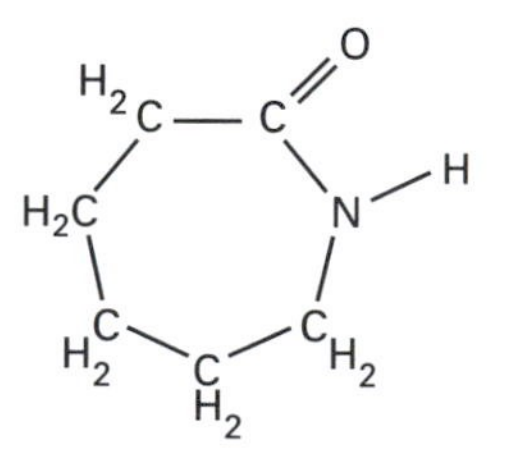

Caprolactam

and three p orbitals available for bonding. In carbenes sp^2 hybridization occurs and the carbon atom has three trigonal sp^2 hybrid orbitals in a plane, with one p orbital at right angles to the plane. Two of the hybrid sp^2 orbitals each contain one electron, and form sigma bonds with other carbon atoms. The two remaining valence electrons may be distributed in one of two possible ways. One, the *triplet state*, has one electron in a sp^2 orbital and the other in the p orbital. The other, the *singlet state*, has both electrons in the sp^2 orbital, with an empty p orbital. A triplet-state carbene can be detected by electron spin resonance (ESR) because there are two unpaired electrons. Carbenes in the singlet state do not have unpaired electrons and are not detectable by ESR. Moreover, in a carbene R_2C:, the R–C–R bond angle will be larger if the carbene is in the triplet state than if it is in the singlet state. In the singlet state the nonbonding electrons form a LONE PAIR, which occupies more space than a single electron. In general, the triplet state (with two unpaired electrons) is more stable than the singlet state. However, there is not a large energy difference between the two states and, depending on the groups attached to the carbon, the singlet state may be the more stable for certain carbenes. Also, the state that is actually formed as an intermediate may depend on the mechanism of production. The less stable form of the carbene may be produced and its chemical behavior depends on how quickly it reacts or converts to the more stable state.

The electronic state of the carbene – whether it is singlet or triplet – may have important consequences for how it reacts. For example, a singlet carbene adding to a double bond to produce a cyclopropane ring does so in one stage, so any stereochemistry of the double bond is preserved in the product. If the carbene has a triplet state it acts as a diradical, and the reaction proceeds in two stages. Any stereochemistry about the double bond will not be preserved in the final product

carbenium ion *See* carbocation.

carbide A compound of carbon with a more electropositive element. The carbides of the elements are classified into:

1. *Ionic carbides*, which contain the carbide ion C^{4-}. An example is aluminum carbide, Al_4C_3. Compounds of this type react with water to give methane (they were formerly also called *methanides*). The *dicarbides* are ionic carbon compounds that contain the dicarbide ion $^-C:C^-$. The best-known example is calcium dicarbide, CaC_2, also known as CALCIUM CARBIDE, or simply *carbide*. Compounds of this type give ethyne with water. They were formerly called *acetylides* or *ethynides*. Ionic carbides are formed with very electropositive metals. They are crystalline.
2. *Covalent carbides*, which have giant-molecular structures, as in silicon carbide (SiC) and boron carbide (B_4C_3). These are hard high-melting solids. Other covalent compounds of carbon (CO_2, CS_2, CH_4, etc.) have covalent molecules.
3. *Interstitial carbides*, which are interstitial compounds of carbon with transition metals. Titanium carbide (TiC) is an example. These compounds are all hard high-melting solids, with metallic properties. Some carbides (e.g. nickel carbide Ni_3C) have properties intermediate between those of interstitial and ionic carbides.

carbocation An ion with a positive charge in which the charge is mostly localized on a carbon atom. There are two types. *Carbonium ions* have five bonds to the carbon atom and a complete outer shell of 8 electrons. The simplest example would be the carbonium ion CH_5^+, which could be regarded as formed by adding H^+ to methane, CH_4, in the same way that the ammonium ion, NH_4^+, is formed from ammonia, NH_3. There is, however, a difference between ammonia and methane in that ammonia has a lone pair of electrons, which it can donate in forming the NH_4^+ ion. The carbonium ion CH_5^+ (and similar ions) is a transient species, produced in the gas phase by electron bombardment of organic compounds and detected in a mass spectrum. Its shape is that of a carbon atom with three hydrogens in a plane and one hydrogen above and one below (a trigonal bipyramid).

Carbenium ions have three bonds to the central carbon and are planar, with the bonds directed toward the corners of a triangle (sp^2 hybridization). They have six electrons in the outer shell of carbon and a vacant p orbital. Carbenium ions are important intermediates in a number of organic reactions, notably the S_N1 mechanism of NUCLEOPHILIC SUBSTITUTION. It is possible to produce stable carbenium ions in salts of the type $(C_6H_5)_3C^+Cl^-$, which are orange-red solids. In these the triphenylmethyl cation is stabilized by delocalization over the three phenyl groups. It is also possible to produce carbenium ions using SUPERACIDS.

carbocyclic compound A compound, such as benzene or cyclohexane, that contains a ring of carbon atoms in its structure.

carbohydrate Any of a class of compounds occurring widely in nature and having the general formula $C_x(H_2O)_y$. (Note that although the name suggests a hydrate of carbon these compounds are in no way hydrates and have no similarities to classes of hydrates.) Carbohydrates are generally divided into two main classes: SUGARS and POLYSACCHARIDES.

Carbohydrates are both stores of energy and structural elements in living systems; plants having typically 15% carbohydrate and animals about 1% carbohydrate. The body is able to build up polysaccharides from simple units (anabolism) or break the larger units down to more simple units for releasing energy (catabolism).

carbolic acid A former name for phenol (hydroxybenzene; C_6H_5OH).

carbon The first element of group 14 (formerly IVA) of the periodic table. Carbon is a universal constituent of living matter and the principal deposits of carbon compounds are derived from living sources; i.e., carbonates (chalk and limestone) and fossil fuels (coal, oil, and gas). It also occurs in the mineral dolomite. The element forms only 0.032% by mass of the Earth's crust. Minute quantities of elemental carbon also occur as the allotropes graphite and diamond. A third allotrope, BUCKMINSTERFULLERENE (C_{60}), also exists.

Naturally occurring carbon has the isotopic composition ^{12}C (98.89%), ^{13}C (1.11%) and ^{14}C (minute traces in the upper atmosphere produced by slow neutron capture by ^{14}N atoms). ^{14}C is used for radiocarbon dating because of its long half-life of 5730 years.

carbonate A salt of carbonic acid (containing the ion CO_3^{2-}).

carbonation **1.** The solution of carbon dioxide in a liquid under pressure, as in carbonated soft drinks.
2. The addition of carbon dioxide to compounds, e.g. the insertion of carbon dioxide into Grignard reagents.

carbon black A finely divided form of carbon produced by the incomplete combustion of such hydrocarbon fuels as natural gas or petroleum oil. It is used as a black pigment in inks and as a filler for rubber in tire manufacture.

carbon cycle The circulation of carbon compounds in the environment, one of the major natural cycles of an element. Carbon dioxide in the air is used by green plants in photosynthesis (in which it is combined with water to form sugars and starches). Plants are eaten by animals which exhale carbon dioxide, or when plants and animals die their remains decompose with the production of carbon dioxide. Some plants are burned or converted to fossil fuels which are burned, again with the formation of carbon dioxide.

carbon dating (radiocarbon dating) A method of dating – measuring the age of (usually archaeological) materials that contain matter of living origin. It is based on the fact that ^{14}C, a beta emitter of half-life approximately 5730 years, is being formed continuously in the atmosphere as a result of cosmic-ray action. The ^{14}C becomes incorporated into living organisms. After death of the organism the amount of radioactive carbon decreases exponentially by radioactive decay. The ratio of ^{12}C to ^{14}C is thus a measure of the time elapsed since the death of the organic material. The method is most valuable for specimens of up to 20 000 years old, though it has been modified to measure ages up to 70 000 years. For ages of up to about 8000 years the carbon time scale has been calibrated by dendrochronology; i.e. by measuring the ^{12}C:^{14}C ratio in tree rings of known age.

carbon dioxide (CO_2) A colorless odorless nonflammable gas formed when carbon burns in excess oxygen. It is also produced by respiration. Carbon dioxide is present in the atmosphere (0.03% by volume) and is converted in plants to carbohydrates by photosynthesis. In the laboratory it is made by the action of dilute acid on metal carbonates. Industrially, it is obtained as a by-product in certain processes, such as fermentation or the manufacture of lime. The main uses are as a refrigerant (solid carbon dioxide, called *dry ice*) and in fire extinguishers and carbonated drinks. Increased levels of carbon dioxide in the atmosphere from the combustion of fossil fuels are thought to contribute to the greenhouse effect.

Carbon dioxide is the anhydride of the weak acid carbonic acid, which is formed in water:

$$CO_2 + H_2O \rightleftharpoons H_2CO_3$$

carbon disulfide (CS_2) A colorless poisonous flammable liquid made from methane (natural gas) and sulfur. The pure compound is virtually odorless, but CS_2 usually has a revolting smell because of the presence of other sulfur compounds. It is used as a solvent and in the production of xanthates in making viscose rayon.

carbon fibers Fibers of graphite, which are used, for instance, to strengthen polymers. They are made by heating stretched textile fibers and have an orientated crystal structure.

carbonic acid (H_2CO_3) A dibasic acid formed in small amounts in solution when carbon dioxide dissolves in water:

$$CO_2 + H_2O \rightleftharpoons H_2CO_2$$

It forms two series of salts: hydrogencarbonates (HCO_3^-) and carbonates (CO_3^{2-}). The pure acid cannot be isolated.

carbonium ion *See* carbocation.

carbonize (carburize) To convert an or-

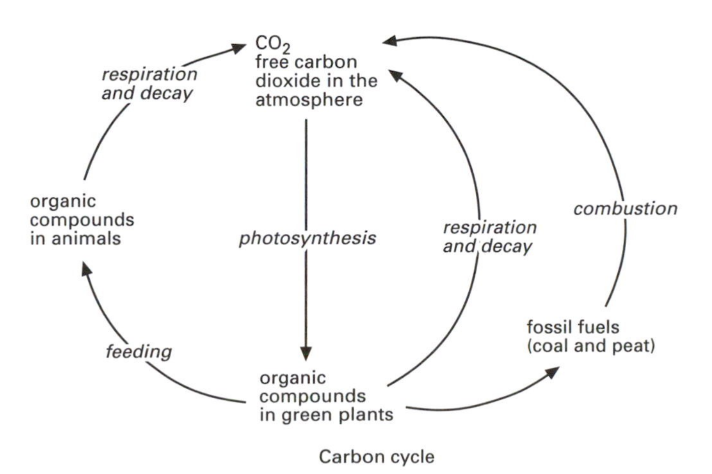

Carbon cycle

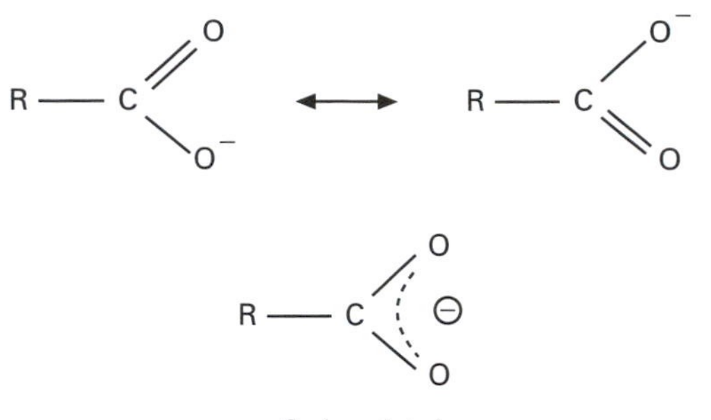

Carboxylate ion

ganic compound into carbon by incomplete oxidation at high temperature.

carbon monoxide (CO) A colorless flammable toxic gas formed by the incomplete combustion of carbon. In the laboratory it can be made by dehydrating methanoic acid with concentrated sulfuric acid:

$$HCOOH - H_2O \rightarrow CO$$

Industrially, it is produced by the oxidation of carbon or of natural gas, or by the water-gas reaction. It is a powerful reducing agent and is used in metallurgy.

Carbon monoxide is neutral and only sparingly soluble in water. It is not the anhydride of methanoic acid, although under extreme conditions it can react with sodium hydroxide to form sodium methanoate. It forms metal carbonyls with transition metals, and its toxicity is due to its ability to form a complex with hemoglobin (in preference to oxygen).

carbon tetrachloride *See* tetrachloromethane.

carbonyl A complex in which carbon monoxide ligands are coordinated to a metal atom. A common example is tetracarbonyl nickel(0), $Ni(CO)_4$.

carbonyl chloride (phosgene; $COCl_2$) A colorless toxic gas with a choking smell. It is used as a chlorinating agent and to make polyurethane plastics and insecticides; it was formerly employed as a war gas.

carbonyl group The group –C=O. It occurs in aldehydes (RCO.H), ketones (RR′CO), carboxylic acids (RCO.OH), and in carbonyl complexes of transition metals. The group is polar, with negative charge on the oxygen.

carboxyhemoglobin A complex formed when carbon monoxide coordinates to the iron atom in hemoglobin molecules. The product is very stable and hemoglobin has a much greater affinity for carbon monoxide than for oxygen. The toxic effect of carbon monoxide is due to its ability to block hemoglobin as an oxygen carrier.

carboxylate ion The ion $-COO^-$, produced by ionization of a carboxyl group. In a carboxylate ion the negative charge is generally delocalized over the O–C–O grouping and the two C–O bonds have the same length, intermediate between that of a double C=O and a single C–O.

carboxyl group The organic group –CO.OH, present in carboxylic acids.

carboxylic acid A type of organic compound containing the CARBOXYL GROUP. Simple carboxylic acids have the general formula RCOOH. Many carboxylic acids occur naturally in plants and (in the form of esters) in fats and oils, hence the alternative name *fatty acids*. Carboxylic acids with one COOH group are *monobasic*, those with two, *dibasic*, and those with three, *tribasic*. The methods of preparation are:

1. Oxidation of a primary alcohol or an aldehyde:
   $$RCH_2OH + 2[O] \rightarrow RCOOH + H_2O$$

2. Hydrolysis of a nitrile using dilute hydrochloric acid:

$$RCN + HCl + 2H_2O \rightarrow RCOOH + NH_4Cl$$

The acidic properties of carboxylic acids are due to the carbonyl group, which attracts electrons from the C–O and O–H bonds. The CARBOXYLATE ION formed, $R–COO^-$, is also stabilized by delocalization of electrons over the O–C–O grouping.

Other reactions of carboxylic acids include the formation of ESTERS and the reaction with phosphorus(V) chloride to form ACYL HALIDES.

carburize *See* carbonize.

carbylamine reaction *See* isocyanide test.

carcinogen Any substance that causes living tissues to become cancerous. Chemical carcinogens include many organic compounds, e.g. hydrocarbons in tobacco smoke, as well as inorganic ones, e.g. asbestos. Carcinogenic physical agents include ultraviolet light, x-rays, and radioactive materials. Some viruses (e.g. hepatitis B) are also carcinogens. Many carcinogens are mutagenic, i.e. they cause changes in the DNA; dimethylnitrosamine, for example, methylates the bases in DNA. A potential carcinogen may therefore be identified by determining whether it causes mutations.

Carius method A method in quantitative analysis for determining the amounts of halogens, phosphorus, and sulfur in organic compounds. The compound is heated with concentrated nitric acid and silver nitrate in a sealed tube. The silver compounds produced are separated and weighed.

Carnot cycle The idealized reversible cycle of four operations occurring in a perfect heat engine. These are the successive adiabatic compression, isothermal expansion, adiabatic expansion, and isothermal compression of the working substance. The cycle returns to its initial pressure, volume, and temperature, and transfers energy to or from mechanical work. The efficiency of the Carnot cycle is the maximum attainable in a heat engine. *See* Carnot's principle.

Carnot's principle (Carnot theorem) The principle that efficiency of any heat engine cannot be greater than that of a reversible heat engine operating over the same temperature range. It follows directly from the second law of thermodynamics, and means that all reversible heat engines have the same efficiency, independent of the working substance. If heat is absorbed at temperature T_1 and given out at T_2, then the Carnot efficiency is $(T_1 - T_2)/T_1$.

Carnot theorem *See* Carnot's principle.

carotene A carotenoid pigment, examples being lycopene and α- and β-carotene. The latter compounds are important in animal diets as a precursor of vitamin A. *See* carotenoids; photosynthetic pigments.

carotenoid Any of a group of yellow, orange, or red pigments comprising the CAROTENES and XANTHOPHYLLS. They are found in all photosynthetic organisms, where they function mainly as accessory pigments in photosynthesis, and in some animal structures, e.g. feathers. They contribute, with anthocyanins, to the autumn colors of leaves since the green pigment chlorophyll, which normally masks the carotenoids, breaks down first. They are also found in some flowers and fruits, e.g. tomato. Carotenoids have three absorption peaks in the blue-violet region of the spectrum.

Carotenes are hydrocarbons. The most widespread is β-carotene. This is the orange pigment of carrots whose molecule is split into two identical portions to yield vitamin A during digestion in vertebrates. *Xanthophylls* resemble carotenes but contain oxygen. *See also* photosynthetic pigments.

Carothers, Wallace Hume (1896–1937) American organic chemist. Carothers is best known for having discovered nylon.

He also produced neoprene, a synthetic rubber, in 1931. He did so by treating vinylacetylene with hydrochloric acid. This produced the monomer chlorobutadiene which readily polymerizes to give the polymer neoprene. In the search to find artifical versions of silk and cellulose he used many types of condensation polymers. In 1935 he discovered the polyamide usually known as nylon by the condensation of adipic acid and hexamethylenediamine. Carothers suffered from depression and committed suicide in 1937. He therefore did not live to see the commercial production of nylon in 1940.

carrier gas The gas used to carry the sample in GAS CHROMATOGRAPHY.

casein A phosphorus-containing protein that occurs in milk and cheese. It is easily digested by young mammals and is their major source of protein and phosphorus. The protein has been used for making certain items, such as billiard balls and buttons, and has also been used to make glue. These uses have declined because of competition from synthetic polymers.

catabolism All the metabolic reactions that break down complex molecules to simpler compounds. The function of catabolic reactions is to provide energy. *See also* metabolism.

catalyst A substance that alters the rate of a chemical reaction without itself being changed chemically in the reaction. The catalyst can, however, undergo physical change; for example, large lumps of catalyst can, without loss in mass, be converted into a powder. Small amounts of catalyst are often sufficient to increase the rate of reaction considerably. A *positive catalyst* increases the rate of a reaction and a *negative catalyst* reduces it. *Homogeneous catalysts* are those that act in the same phase as the reactants (i.e. in gaseous and liquid systems). For example, nitrogen(II) oxide gas will catalyze the reaction between sulfur(IV) oxide and oxygen in the gaseous phase. *Heterogeneous catalysts* act in a different phase from the reactants. For example, finely divided nickel (a solid) will catalyze the hydrogenation of oil (liquid).

In increasing a reaction rate a catalyst provides a new pathway for which the rate-determining step has a lower activation energy than in the uncatalyzed reaction. A catalyst does not change the products in an equilibrium reaction and their concentration is identical to that in the uncatalyzed reaction; i.e. the position of the equilibrium remains unchanged. The catalyst simply increases the rate at which equilibrium is attained.

In *autocatalysis*, one of the products of the reaction itself acts as a catalyst. In this type of reaction the reaction rate increases with time to a maximum and finally slows down. For example, in the hydrolysis of ethyl ethanoate, the ethanoic acid produced catalyzes the reaction.

catalytic converter A device fitted to the exhaust system of gasoline-fuelled vehicles to remove pollutant gases from the exhaust. It consists of a honeycomb structure (to provide maximum area) coated with platinum, palladium, and rhodium catalysts. Such devices can convert carbon monoxide to carbon dioxide, oxides of nitrogen to nitrogen, and unburned fuel to carbon dioxide and water.

catalytic cracking The conversion, using a catalyst, of long-chain hydrocarbons from the refining of petroleum into more useful shorter-chain compounds such as those occurring in kerosene and gasoline.

catalytic reaction A chemical reaction that occurs at a measurable rate only in the presence of a catalyst.

catechol (1,2-dihydroxybenzene) A colourless crystalline PHENOL. It is used in photographic developing.

catecholamine Any of a group of important amines that contain a catechol ring in their molecules. They are neurotransmitters and hormones. Examples are epinephrine and norepinephrine.

catenation The formation of chains of atoms in molecules.

cathode In electrolysis, the electrode that is at a negative potential with respect to the anode. In any electrical system, such as a discharge tube or electronic device, the cathode is the terminal at which electrons enter the system.

cation A positively charged ion, formed by removal of electrons from atoms or molecules. In electrolysis, cations are attracted to the negatively charged electrode (the cathode). *Compare* anion.

cationic detergent *See* detergent.

cationic resin An ION-EXCHANGE material that can exchange cations, such as H^+ and Na^+, for ions in the surrounding medium. Such resins are used for a wide range of purification and analytical purposes.

They are often produced by adding a sulfonic acid group ($-SO_3^-H^+$) or a carboxylate group ($-COO^-H^+$) to a stable polyphenylethene resin. A typical exchange reaction is:

$$\text{resin}-SO_3^-H^+ + NaCl = \text{resin}-SO_3^-Na^+ + HCl$$

They have been used to great effect to separate mixtures of cations of similar size having the same charge. Such mixtures can be attached to cationic resins and progressive elution will recover them in order of decreasing ionic radius.

cell A system having two plates (electrodes) in a conducting liquid (electrolyte). An *electrolytic cell* is used for producing a chemical reaction by passing a current through the electrolyte (i.e. by electrolysis). A *voltaic* (or *galvanic*) *cell* produces an e.m.f. by chemical reactions at each electrode. Electrons are transferred to or from the electrodes, giving each a net charge.

cellulose A POLYSACCHARIDE $(C_6H_{10}O_5)_n$ of glucose, which is the main constituent of the cell walls of plants. It is obtained from wood pulp.

cellulose acetate (cellulose ethanoate) A polymeric substance made by acetylating cellulose using a mixture of ethanoic acid, ethanoic anhydride, and sulfuric acid. It is used in plastics, in acetate film, and in acetate rayon.

cellulose ethanoate *See* cellulose acetate.

cellulose trinitrate (guncotton; nitrocellulose) A highly flammable substance made by treating cellulose with a nitric–sulfuric acid mixture. Cellulose trinitrate is used in explosives and in lacquers. It is an ester of nitric acid (i.e. not a true nitro compound).

Celsius scale A temperature scale in which the temperature of melting pure ice is taken as 0° and the temperature of boiling water 100° (both at standard pressure). The *degree Celsius* (°C) is equal to the kelvin. This was known as the *centigrade scale* until 1948, when the present name became official. It is named for the Swedish astronomer Anders Celsius (1701–44). Celsius' original scale (1742) was inverted (i.e. had 0° as the steam temperature and 100° as the ice temperature). *See also* temperature scale.

centi- Symbol: c A prefix denoting 10^{-2}. For example, 1 centimeter (cm) = 10^{-2} meter (m).

centigrade scale *See* Celsius scale.

centrifugal pump A device commonly used for transporting fluids around a chemical plant. Centrifugal pumps usually have a set of blades rotating inside a fixed circular casing. As the blades rotate, the fluid is impelled out of the pump along a pipe. Centrifugal pumps do not produce high pressures but they have the advantage of being relatively cheap because they are simple in design, have no valves, and work at high speeds. In addition they are not damaged if a blockage develops. *Compare* displacement pump.

centrifuge An apparatus for rotating a container at high speeds, used to increase the rate of sedimentation of suspensions or the separation of two immiscible liquids. *See also* ultracentrifuge.

CFC Chlorofluorocarbon. *See* halocarbon.

c.g.s. system A system of units that uses the centimeter, the gram, and the second as the base mechanical units. Much early scientific work used this system, but it has now almost been abandoned in favor of SI UNITS.

chain When two or more atoms form bonds with each other in a molecule, a chain of atoms results. This chain may be a *straight chain*, in which each atom is added to the end of the chain, or it may be a *branched chain*, in which the main chain of atoms has one or more smaller *side chains* branching off it.

chain reaction A self-sustaining chemical reaction consisting of a series of steps, each of which is initiated by the one before it. An example is the reaction between hydrogen and chlorine:

$$Cl_2 \rightarrow 2Cl\bullet$$
$$H_2 + Cl\bullet \rightarrow HCl + H\bullet$$
$$H\bullet + Cl_2 \rightarrow HCl + Cl\bullet$$
$$2H\bullet \rightarrow H_2$$
$$2Cl\bullet \rightarrow Cl_2$$

The first stage, chain initiation, is the dissociation of chlorine molecules into atoms; this is followed by two chain propagation reactions. Two molecules of hydrogen chloride are produced and the ejected chlorine atom is ready to react with more hydrogen. The final steps, chain termination, stop the reaction. Chain reactions are important in certain types of free-radical POLYMERIZATION reactions.

chair conformation *See* cyclohexane.

chalcogens The elements of group 16 of the periodic table: oxygen, sulfur, selenium, tellurium, and polonium.

charcoal An amorphous form of carbon made by heating wood or other organic material in the absence of air. *Activated charcoal* is charcoal heated to drive off absorbed gas. It is used for absorbing gases and for removing impurities from liquids.

Chardonnet, Louis-Marie-Hilaire Bernigaud, Comte de (1839–1924) French organic chemist. Chardonnet is best known for inventing rayon. This was the first type of artificial silk to be produced. In 1884 he produced nitrocellulose (guncotton) by treating a pulp made from mulberry leaves with a mixture of nitric acid and sulphuric acid. The resulting cellulose compound was dissolved in a mixture of alcohol and ether, with this solution then being forced into cold water through capillary tubes. Nitrocellulose was very inflammable, so Chardonnet sought a non-inflammable fibre. This culminated in the development of rayon in 1889.

Charles' law For a given mass of gas at constant pressure, the volume increases by a constant fraction of the volume at 0°C for each Celsius degree rise in temperature. The constant fraction (α) has almost the same value for all gases – about 1/273 – and Charles' law can be written in the form

$$V = V_0(1 + \alpha_v\theta)$$

where V is the volume at temperature θ°C and V_0 the volume at 0°C. The constant α_v is the thermal expansivity of the gas. For an ideal gas its value is 1/273.15.

A similar relationship exists for the pressure of a gas heated at constant volume:

$$p = p_0(1 + \alpha_p\theta)$$

Here, α_p is the pressure coefficient. For an ideal gas

$$\alpha_p = \alpha_v$$

although they differ slightly for real gases. It follows from Charles' law that for a gas heated at constant pressure,

$$V/T = K$$

where T is the thermodynamic temperature and K is a constant. Similarly, at constant volume, p/T is a constant.

Charles' volume law is sometimes called *Gay-Lussac's law* after its independent discoverer.

chelate A metal coordination complex in which one ligand coordinates at two or more points to the same metal ion. The resulting complex contains rings of atoms that include the metal atom. An example of a *chelating agent* is 1,2-diaminoethane ($H_2NCH_2CH_2NH_2$), which can coordinate both its amine groups to the same atom. It is an example of a *bidentate ligand* (having two 'teeth'). EDTA, which can form up to six bonds, is another example of a chelating agent. The word chelate comes from the Greek word meaning 'claw'.

chemical bond A link between atoms that leads to an aggregate of sufficient stability to be regarded as an independent molecular species. Chemical bonds include covalent bonds, electrovalent (ionic) bonds, coordinate bonds, and metallic bonds. Hydrogen bonds and van der Waals forces are not usually regarded as true chemical bonds.

chemical combination, laws of A group of chemical laws developed during the late 18th and early 19th centuries, which arose from the recognition of the importance of quantitative (as opposed to qualitative) study of chemical reactions. The laws are:
1. the law of conservation of mass (matter);
2. the law of constant (definite) proportions;
3. the law of multiple proportions;
4. the law of equivalent (or reciprocal) proportions,

These laws played a significant part in Dalton's development of his atomic theory (1808).

chemical conversions *See* unit processes.

chemical dating A method of using chemical analysis to find the age of an archaeological specimen in which compositional changes have taken place over time. For example, the determination of the amount of fluorine in bone that has been buried gives an indication of its age because phosphate in the bone has gradually been replaced by fluoride ions from groundwater. Another dating technique depends on the fact that, in living organisms, amino acids are optically active. After death a slow racemization reaction occurs and a mixture of L- and D-isomers forms. The age of bones can be accurately determined by measuring the relative amounts of L- and D-amino acids present.

chemical engineering The branch of engineering concerned with the design and maintenance of a chemical plant and its ability to withstand extremes of temperature and pressure, corrosion, and wear. It enables laboratory processes producing grams of material to be converted into a large-scale plant producing tonnes of material. Chemical engineers plan large-scale chemical processes by linking together the appropriate unit processes and by studying such parameters as heat and mass transfer, separations, and distillations.

chemical equation A method of representing a chemical reaction using chemical formulae. The formulae of the reactants are given on the left-hand side of the equation, with the formulae of the products on the right. The two halves are separated by a directional arrow or arrows (or an equals sign). A number preceding a formula (called a *stoichiometric coefficient*) indicates the number of molecules of that substance involved. The equation must balance – that is, the number of atoms of any one element must be the same on both sides of the equation.

chemical equilbrium *See* equilibrium.

chemical formula *See* formula.

chemical reaction A process in which one or more elements or chemical compounds (the reactants) react to produce a different substance or substances (the products).

chemical shift *See* nuclear magnetic resonance.

chemiluminescence The emission of light during a chemical reaction.

chemisorption *See* adsorption.

chiral Having the property of CHIRALITY. For example, lactic acid is a chiral compound because it has two possible structures that cannot be superposed. *See* optical activity.

chirality The property of existing in left- and right-handed forms; i.e. forms that are not superposable. In chemistry the term is applied to the existence of optical isomers. *See* optical activity.

chirality element A part of a molecule that causes it to display chirality. The most common type of element is a *chirality center*, which is an atom attached to four different atoms or groups. This is also referred to as an *asymmetric atom*. Less commonly a molecule may have a *chirality axis*, as in the case of certain substituted allenes of the type $R_1R_2C=C=CR_3R_4$. In this form of compound the R_1 and R_2 groups do not lie in the same plane as the R_3 and R_4 groups because of the nature of the double bonds. The chirality axis lies along the C=C=C chain. It is also possible to have molecules that contain a *chirality plane*. *See* optical activity.

chitin A nitrogen-containing heteropolysaccharide found in some animals and the cell walls of most fungi. It is a polymer of *N*-acetylglucosamine. It consists of many glucose units, in each of which one of the hydroxyl groups has been replaced by an acetylamine group (CH_3CONH). The outer covering of arthropods, the cuticle, is impregnated in its outer layers with chitin, which makes the exoskeleton more rigid. It is associated with protein to give a uniquely tough yet flexible and light skeleton, which also has the advantage of being waterproof. The chitinous plates are thinner for bending and flexibility or thicker for stiffness as required. The plates cannot grow once laid down and are broken down at each molt. Chitin is also found in the hard parts of several other groups of animals.

chloral *See* trichloroethanal.

chloral hydrate *See* trichloroethanal.

chloramine (NH_2Cl) A colorless liquid made by reacting ammonia with sodium chlorate(I) (NaOCl). It is formed as an intermediate in the production of hydrazine. Chloramine is unstable and changes explosively into ammonium chloride and nitrogen trichloride.

chloride *See* halide.

chlorination **1.** Treatment with chlorine; for instance, the use of chlorine to disinfect water.
2. *See* halogenation.

chlorine A green reactive gaseous element belonging to the halogens; i.e. group 17 (formerly VIIA) of the periodic table. It occurs in sea-water, salt lakes, and underground deposits of halite, NaCl. It accounts for about 0.055% of the Earth's crust. Chlorine is strongly oxidizing and can be liberated from its salts only by strong oxidizing agents, such as manganese(IV) oxide, potassium permanganate(VII), or potassium dichromate; note that sulfuric acid is not sufficiently oxidizing to release chlorine from chlorides. Industrially, chlorine is prepared by the electrolysis of brine and in some processes chlorine is recovered by the high-temperature oxidation of waste hydrochloric acid. Chlorine is used in large quantities, both as the element, to produce chlorinated organic solvents, and for the production of polyvinyl chloride (PVC), the major thermoplastic in use today, and in the form of hypochlorites for bleaching.

The solubility of inorganic metal chlorides is such that they are not an environmental problem unless the metal ion itself is toxic but many organochlorine compounds are sufficiently stable for the accumulated residues of chlorine-containing pesticides to present a severe problem in some areas. This arises because they can accumulate in food chains and concentrate in the tissues of higher animals (*see* DDT).

Symbol: Cl; m.p. –100.38°C; b.p. –33.97°C; d. 3.214 kg m^{-3} (0°C); p.n. 17; r.a.m. 35.4527.

chloroacetic acid *See* chloroethanoic acid.

chlorobenzene (monochlorobenzene; C_6H_5Cl) A colorless liquid made by the catalytic reaction of chlorine with benzene or by the RASHIG PROCESS. It can be converted to phenol by reaction with sodium hydroxide under extreme conditions (300°C and 200 atmospheres pressure). It is also used in the manufacture of other organic compounds.

chloroethane (ethyl chloride; C_2H_5Cl) A gaseous compound made by the addition of hydrogen chloride to ethene. It is used as a refrigerant and a local anesthetic.

chloroethanoic acid (chloroacetic acid; $CH_2ClCOOH$) A colorless crystalline solid made by substituting one of the hydrogen atoms of the methyl group of ethanoic acid with chlorine, using red phosphorus. It is a stronger acid than ethanoic acid because of the electron-withdrawing effect of the chlorine atom. *Dichloroethanoic acid* (dichloroacetic acid, $CHCl_2COOH$) and *trichloroethanoic acid* (trichloroacetic acid, CCl_3COOH) are made in the same way. The acid strength increases with the number of chlorine atoms present.

chloroethene (vinyl chloride; $H_2C{:}CHCl$) A gaseous organic compound used in the manufacture of PVC (polyvinyl chloride). Chloroethene is manufactured by the reaction between ethyne and hydrogen chloride using a mercury(II) chloride catalyst:

$$C_2H_2 + HCl \rightarrow H_2C{:}CHCl$$

An alternative source, making use of the ready supply of ethene, is via dichloroethane:

$$H_2C{:}CH_2 + Cl_2 \rightarrow CH_2Cl.CH_2Cl \rightarrow H_2C{:}CHCl$$

chlorofluorocarbon *See* halocarbon.

chloroform *See* trichloromethane.

chloromethane (methyl chloride; CH_3Cl) A colorless flammable haloalkane gas made by chlorination of methane. It is used as a refrigerant and a local anesthetic.

chlorophyll A pigment present in plants that acts as a catalyst in the photosynthesis of carbohydrates from carbon dioxide and water. There are four types, known as chlorophylls *a*, *b*, *c*, and *d*. The chlorophylls are PORPHYRINS containing magnesium.

chloroprene ($CH_2{:}CClCH{:}CH_2$) A colorless conjugated diene used in the manufacture of synthetic chlorinated rubbers (such as neoprene). The systematic name is 2–chlorobuta-1,3–diene.

cholecalciferol *See* vitamin D.

choline An amino alcohol often classified as a member of the vitamin B complex. It can be synthesized in humans from lecithin by putrefaction in the bowel, but is required as an essential nutrient for some animals and microorganisms. It acts to disperse fat from the liver or prevent its excess accumulation. Its ester acetylcholine functions in the transmission of nerve impulses.

chromatography A technique used to separate or analyze complex mixtures. A number of related techniques exist; all depend on two phases: a *mobile phase*, which may be a liquid or a gas, and a *stationary phase*, which is either a solid or a liquid held by a solid. The sample to be separated or analyzed is carried by the mobile phase through the stationary phase. Different components of the mixture are absorbed or dissolved to different extents by the stationary phase, and consequently move along at different rates. In this way the components are separated. There are many different forms of chromatography depending on the phases used and the nature of the partition process between mobile and stationary phases. The main classification is into *column chromatography* and *planar chromatography*.

A simple example of column chromatography is in the separation of liquid mixtures. A vertical column is packed with an absorbent material, such as alumina (aluminum oxide) or silica gel. The sample is introduced into the top of the column and washed down it using a solvent. This process is known as *elution*; the solvent used is the *eluent* and the sample being separated is the *eluate*. If the components are colored, visible bands appear down the column as the sample separates out. The components are separated as they emerge from the bottom of the column. In this particular example of chromatography the partition process is adsorption on the particles of alumina or silica gel. Column chromatography can also be applied to mixtures of gases. *See* gas chromatography. In the other main type of chromatography, planar chromatography, the stationary phase is a flat sheet of absorbent material. *See* paper chromatography; thin-layer chromatography.

Components of the mixture are held back by the stationary phase either by adsorption (e.g. on the surface of alumina) or because they dissolve in it (e.g. in the moisture within chromatography paper).

chromophore A group of atoms in a molecule that is responsible for the color of the compound. Usually a chromophore is a group of atoms having delocalized electrons.

cinnamic acid *See* 3-phenylpropenoic acid.

CIP system (Cahn–Ingold–Prelog system) A method of producing a sequence rule used in the absolute description of stereoisomers in the *R–S* convention (*see* optical activity) or the E-Z CONVENTION. The rule is to consider the atoms that are bound directly to a chiral center (or to a double bond). The group in which this atom has the highest proton number has the highest priority. So, for example, in $HCClBr(NH_2)$, the order of priority is $Br > Cl > NH_2 > H$. If two atoms are the same, the substituents are considered, with the substituents of highest proton number taking precedence. So in $C(NH_2)(NO_2)(CH_3)(C_2H_6)$ the order is $NO_2 > NH_2 > C_2H_6 > CH_3$. The system is named after the British chemists Robert Cahn (1899–1981) and Sir Christopher Ingold (1893–1970) and the Bosnian–Swiss chemist Vladimir Prelog (1906–).

cis- Designating an isomer with groups that are adjacent. *See* isomerism.

cis-trans isomerism *See* isomerism.

citric acid A white crystalline dibasic carboxylic acid important in plant and animal cells. It is present in many fruits, especially citrus fruits. The systematic name is 2-hydroxypropane-1,2,3-tricarboxylic acid. The formula is:

$$HOOCCH_2C(OH)(COOH)CH_2COOH$$

Claisen condensation A reaction in which two molecules of ester combine to give a keto-ester – a compound containing a ketone group and an ester group. The reaction is base-catalyzed by sodium ethoxide; the reaction of ethyl ethanoate refluxed with sodium ethoxide gives:

$$2CH_3.CO.OC_2H_5 \rightarrow CH_3.CO.CH_2.CO.OC_2H_5 + C_2H_5OH$$

The mechanism is similar to that of the ALDOL REACTION, the first step being formation of a carbanion from the ester:

$$CH_3COC_2H_5 + {}^-OC_2H_5 \rightarrow {}^-CH_2COC_2H_5 + C_2H_5OH$$

This attacks the carbon atom of the carbonyl group on the other ester molecule, forming an intermediate anion that decomposes to the keto-ester and the ethanoate ion. It is named for the German organic chemist Ludwig Claisen (1851–1930) who described it in 1890.

clathrate (enclosure compound) A substance in which small (guest) molecules are trapped within the lattice of a crystalline (host) compound. Clathrates are formed when suitable host compounds are crystallized in the presence of molecules of the appropriate size. Although the term 'clathrate compound' is often used, they are not true compounds; no chemical bonds are formed, and the guest molecules interact by weak van der Waals forces. The clathrate is maintained by the cagelike lattice of the host. The host lattice must be broken down, for example, by heating or dissolution in order to release the guest. This should be compared with zeolites, in which the holes in the host lattice are large enough to permit entrance or emergence of the guest without breaking bonds in the host lattice. Quinol forms many clathrates, e.g. with SO_2; water (ice) forms a clathrate with xenon.

CoA *See* coenzyme A.

coagulation The association of particles (e.g. in colloids) into clusters. *See* flocculation.

coal A black mineral that consists mainly of carbon, used as a fuel and as a source of organic chemicals. It is the fossilized remains of plants that grew in the Carboniferous and Permian periods and were buried and subjected to high pressures underground. There are various types of coal, classified according to their increasing carbon content.

coal gas A fuel gas made by heating coal in a limited supply of air. It consists mainly of hydrogen and methane, with some carbon monoxide (which makes the gas highly poisonous). Coal tar and coke are formed as by-products. Coal gas was a major fuel in some countries in the 19th and early 20th century. *See* coal tar.

coal tar Tar produced by heating coal in the absence of oxygen. It is a mixture of many organic compounds (e.g. benzene, toluene, and naphthalene) and also contains free carbon.

cocaine An alkaloid obtained from the dried leaves of a South American shrub *Erythroxylon coca*. It is a stimulant and narcotic. Its use is restricted in many countries.

codeine A derivative of morphine, methylmorphine. It is less potent than morphine and is used as an analgesic. It is a controlled substance in the USA.

coenzyme Any of a group of molecules (which are small compared to the size of an enzyme) that enable enzymes to carry out their catalytic activity. Examples include nicotinamide adenine dinucleotide (NAD) and ubiquinone (coenzyme Q). Some coenzymes are capable of catalyzing reactions in the absence of an enzyme but the rate of reaction is never as high as when a catalyst is present. A coenzyme is not a true catalyst because it undergoes chemical change during the reaction.

coenzyme A (CoA) A complex nucleotide containing an active –SH group that is readily acetylated to $CoAS–COCH_3$ (acetyl CoA). Acetyl CoA is the source of the two-carbon units that feed into the KREBS CYCLE. It is produced from glycolysis and the breakdown of fatty acids and some amino acids. It is also a key intermediate in the biosynthesis of lipids and other anabolic reactions.

coenzyme Q *See* ubiquinone.

cofactor A nonprotein substance that helps an enzyme to carry out its activity. Cofactors may be cations or organic molecules, known as coenzymes. Unlike enzymes they are, in general, stable to heat. When a catalytically active enzyme forms a complex with a cofactor a *holoenzyme* is produced. An enzyme without its cofactor is termed an *apoenzyme*.

coherent units A system or subset of units (e.g. SI units) in which the derived units are obtained by multiplying or dividing together base units, with no numerical factor involved.

collagen The protein of fibrous connective tissues, present in bone, skin, and cartilage. It is the most abundant of all the proteins in the higher vertebrates. Collagen contains about 35% glycine, 11% alanine, 12% proline and small percentages of other amino acids. The amino acid sequence is remarkably regular with almost every third residue being glycine. Collagen is chemically inert (and insoluble) which suggests that its reactive side groups are immobilized by ionic bonding. Collagen fibrils are highly complex and have a variety of orientations depending on the biological function of the particular type of connective tissue. The secondary structure of collagen is that of a triple helix of peptide chains. Its tertiary structure is one of three alpha helices in a 'super helix', which is responsible for its high tensile strength and therefore its role in support tissues.

colligative properties A group of properties of solutions that depends on the number of particles present, rather than the nature of the particles. Colligative properties include:
1. The lowering of vapor pressure.
2. The elevation of boiling point.
3. The lowering of freezing point.
4. Osmotic pressure.

The explanation of these closely related phenomena depends on intermolecular forces and the kinetic behavior of the particles, which is qualitatively similar to those used in deriving the kinetic theory of gases.

collimator An arrangement for producing a parallel beam of radiation for use in a spectrometer or other instrument. A system of lenses and slits is utilized.

colloid A heterogeneous system in which the interfaces between phases, though not visibly apparent, are important factors in determining the system properties. The three important attributes of colloids are:
1. They contain particles, commonly made up of large numbers of molecules, forming the distinctive unit or *disperse phase*.
2. The particles are distributed in a continuous medium (the *continuous phase*).
3. There is a stabilizing agent, which has an affinity for both the particle and the medium; in many cases the stabilizer is a polar group.

Particles in the disperse phase typically have diameters in the range 10^{-6}–10^{-4} mm. Milk, rubber, and emulsion paints are typical examples of colloids. *See also* sol.

colorimetric analysis Quantitative analysis in which the concentration of a colored solute is measured by the intensity of the color. The test solution can be compared against standard solutions.

column chromatography *See* chromatography; gas chromatography.

combustion A reaction with oxygen with the production of heat and light. The combustion of solids and liquids occurs when they release flammable vapor, which reacts with oxygen in the gas phase. Combustion reactions usually involve a complex sequence of free-radical chain reactions. The light is produced by excited atoms, molecules, or ions. In highly luminous flames it comes from small incandescent particles of carbon.

Sometimes the term is also applied to slow reactions with oxygen, and also to reactions with other gases (for example, certain metals 'burn' in chlorine).

complex (coordination compound) A type of compound in which molecules or ions form coordinate bonds with a metal atom or ion. The coordinating species (called *ligands*) have lone pairs of electrons, which they can donate to the metal atom or ion. They are molecules such as ammonia or water, or negative ions such as Cl^- or CN^-. The resulting complex may be neutral or it may be a *complex ion*. For example:

$$Cu^{2+} + 4NH_3 \rightarrow [Cu(NH_3)_4]^{2+}$$
$$Fe^{3+} + 6CN^- \rightarrow [Fe(CN)_6]^{3-}$$
$$Fe^{2+} + 6CN^- \rightarrow [Fe(CN)_6]^{4-}$$

The formation of such coordination complexes is typical of transition metals. Often the complexes contain unpaired electrons and are paramagnetic and colored. *See also* chelate; sandwich compound.

component One of the separate chemical substances in a mixture in which no chemical reactions are taking place. For example, a mixture of ice and water has one

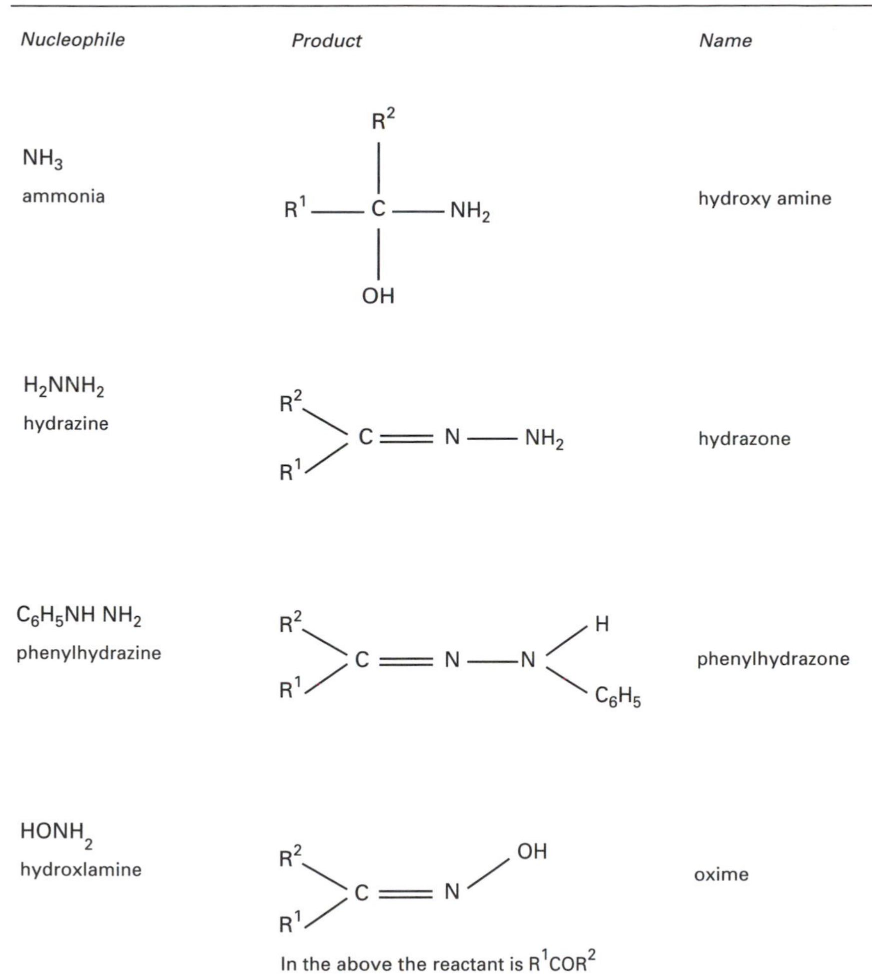

Condensation reaction: some reactions of aldehydes and ketones

component; a mixture of nitrogen and oxygen has two components. When chemical reactions occur between the substances in a mixture, the number of components is defined as the number of chemical substances present minus the number of equilibrium reactions taking place. Thus, the system: $N_2 + 3H_2 \rightleftharpoons 2NH_3$ is a two-component system.

compound A chemical combination of atoms of different elements to form a substance in which the ratio of combining atoms remains fixed and is specific to that substance. The constituent atoms cannot be separated by physical means; a chemical reaction is required for the compounds to be formed or to be changed. The existence of a compound does not necessarily imply that it is stable. Many compounds have lifetimes of less than a second. *See also* mixture.

concentrated Denoting a solution in which the amount of solute in the solvent is relatively high. The term is always relative; for example, whereas concentrated sulfuric acid may contain 96% H_2SO_4, concentrated potassium chlorate may contain as little as 10% $KClO_3$. *Compare* dilute.

concentration The amount of substance per unit volume or mass in a solution. *Molar concentration* is amount of substance (in moles) per cubic decimeter (liter). *Mass concentration* is mass of solute per unit volume. *Molal concentration* is amount of substance (in moles) per kilogram of solute.

concerted reaction A reaction that takes place in a single stage rather than as a series of simple steps. In a concerted reaction there is a transition state in which bonds are forming and breaking at the same time. An example is the S_N2 mechanism in NUCLEOPHILIC SUBSTITUTION. *See also* pericyclic reaction.

condensation The conversion of a gas or vapor into a liquid or solid by cooling.

condensation polymerization *See* polymerization.

condensation reaction A reaction in which addition of two molecules occurs followed by elimination of a smaller molecule, usually water. Condensation reactions (addition–elimination reactions) are characteristic of ALDEHYDES and KETONES reacting with a range of nucleophiles. There is typically nucleophilic addition at the C atom of the carbonyl group followed by elimination of water.

conducting polymer A type of organic polymer that conducts electricity like a metal. Conducting polymers are crystalline substances containing conjugated unsaturated carbon–carbon bonds. In principle, they provide lighter and cheaper alternatives to metallic conductors.

conductiometric titration A titration in which measurement of the electrical conductance is made continuously throughout the addition of the titrant and well beyond the equivalence point. This is in place of traditional end-point determination by indicators. The operation is carried out in a conductance cell, which is part of a resistance bridge circuit. The method depends on the fact that ions have different ionic mobilities, H^+ and OH^- having particularly high values. The method is especially useful for weak acid–strong base and strong acid–weak base titrations for which color-change titrations are unreliable.

configuration **1.** The arrangement of electrons about the nucleus of an atom. Configurations are represented by symbols, which contain:

1. An integer, which is the value of the principal quantum number (shell number).
2. A lower-case letter representing the value of the azimuthal quantum number (l), i.e.

   s means $l = 0$, p means $l = 1$,
   d means $l = 2$, f means $l = 3$.
3. A numerical superscript giving the number of electrons in that particular set; for example, $1s^2$, $2p^3$, $3d^5$.

The ground state electronic configuration (i.e. the most stable or lowest energy state) may then be represented as follows, for example, He, $1s^2$; N, $1s^22s^22p^5$. However, elements are commonly abbreviated by using an inert gas to represent the 'core', e.g. Zr has the configuration $[Kr]4d^25s^2$.

2. The arrangement of atoms or groups in a molecule.

conformation A particular shape of molecule that arises through the normal rotation of its atoms or groups about single bonds. Any of the possible conformations that may be produced is called a *conformer* (or *rotamer*), and there will be an infinite number of these possibilities, differing in the angle between certain atoms or groups on adjacent carbon atoms. Sometimes, the term 'conformer' is applied more strictly to possible conformations that have minimum energies – as in the case of the boat and chair conformations of CYCLOHEXANE. In considering conformations about a single bond, it is convenient to consider the *dihedral angle* between a bond from one carbon atom and a bond from the other carbon. This is the angle between the bonds as viewed along the C–C bond. It is also called the *torsion angle*. Conformations are also visualized using *Newman projection* diagrams. In these the molecule is viewed along a particular bond. The nearer atoms is represented by a point with bonds drawn to this point. The further atom is represented by a large circle, with bonds drawn to the edge of this circle. A conformation in which bonds on the back atom would be hidden by bonds in front is drawn slightly displaced. In the case of ethane, the maximum energy is the *eclipsed conformation*, in which the dihedral angle is O°. The minimum is the *staggered conformation*, with a dihedral angle of 60°.

If different atoms are attached to the carbon atoms, the conformational analysis is more complicated. For example, the compound $XH_2C–CH_2X$, where X is another group (e.g. Me), has conformers known as *syn-periplanar*, *synclinal* (or *gauche*), *anticlinal*, and *anti-periplanar* (*see illustration*).

Conformational analysis is also important in the structures of nonplanar rings. *See* cyclohexane.

conjugate acid *See* acid.

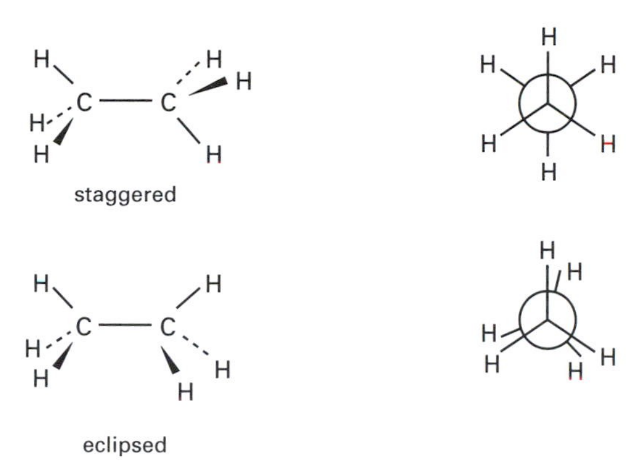

Conformation: the conformations of ethane resulting from rotation about the C–C bond. The diagrams on the right are Newman projections, with the molecule viewed along the C–C bond.

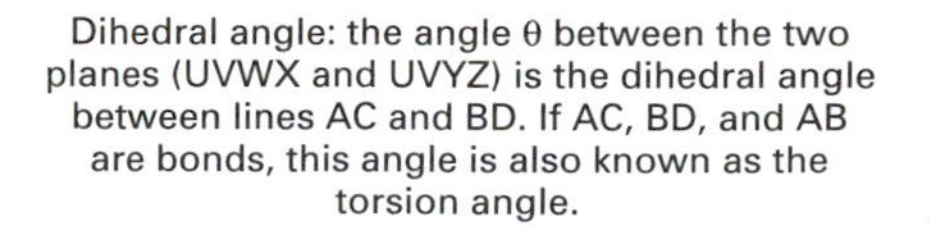

Dihedral angle: the angle θ between the two planes (UVWX and UVYZ) is the dihedral angle between lines AC and BD. If AC, BD, and AB are bonds, this angle is also known as the torsion angle.

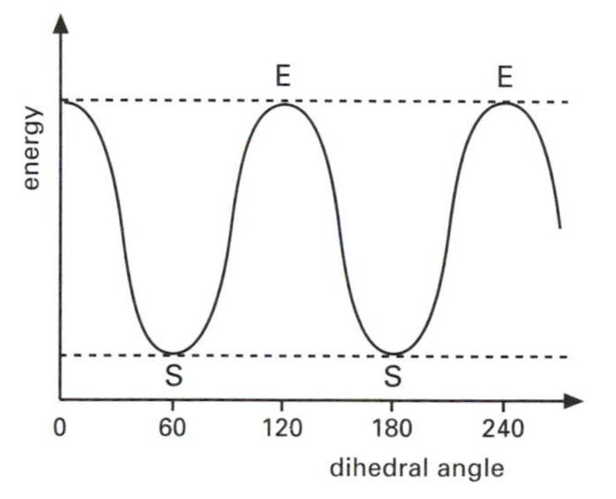

Conformation: the way in which energy changes with dihedral angle as a result of rotation about the double bond in ethane. E indicates an eclipsed conformation and S indicates a staggered conformation.

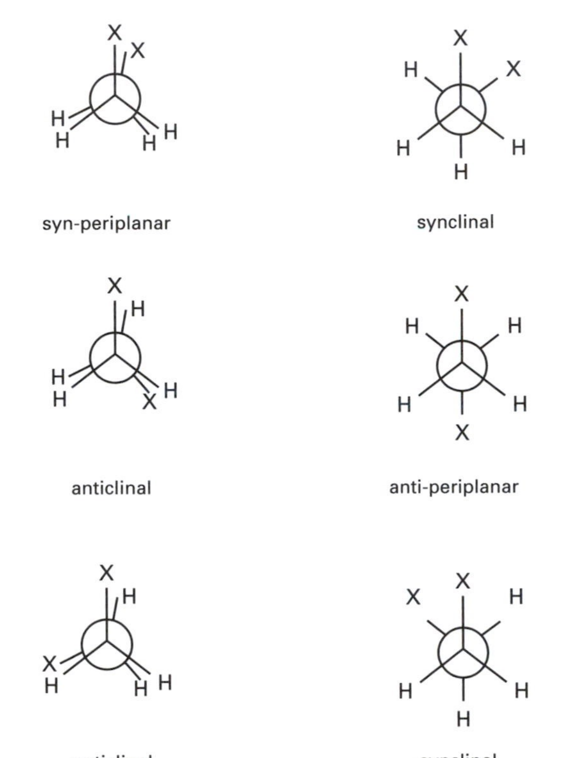

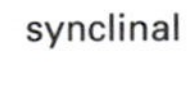

Conformation: the conformations of a disubstituted ethane CH_2XCH_2X for rotation about the C–C bond.

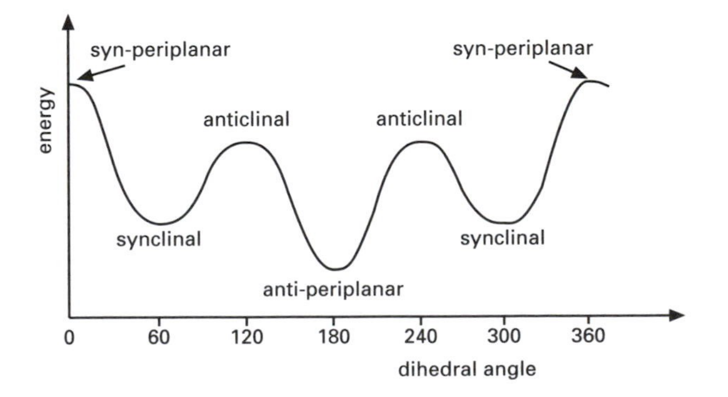

Conformation: the way in which energy changes with dihedral angle in CH_2XCH_2X for rotation about the C–C bond.

conjugate base *See* acid.

conjugated Describing compounds that have alternating double and single bonds in their structure. For example, but-1,3-ene ($H_2C{:}CHCH{:}CH_2$) is a typical conjugated compound. In such compounds there is delocalization of the electrons in the double bonds over part of the molecule.

conjugated protein A protein that on hydrolysis yields not only amino acids but also other organic and inorganic substances. They are simple proteins combined with nonprotein groups (prosthetic groups). *See also* glycoprotein; lipoprotein; phosphoprotein.

conservation of energy, law of Enunciated by Helmholtz in 1847, this law states that in all processes occurring in an isolated system the energy of the system remains constant. The law does of course permit energy to be converted from one form to another (including mass, since energy and mass are equivalent).

conservation of mass, law of Formulated by Lavoisier in 1774, this law states that matter cannot be created or destroyed. Thus in a chemical reaction the total mass of the products equals the total mass of the reactants (the term 'mass' must include any solids, liquids, and gases – including air – that participate).

constant-boiling mixture A general observation for most liquids is that the vapor phase above a liquid is richer in the more volatile component (a deviation from Raoult's law). Consequently most liquid mixtures show a regular increase in the boiling point as the liquid is progressively distilled. In distillation, a point is reached at which a constant boiling mixture or AZEOTROPE distills over. Further attempts to fractionate the distillate do not lead to a change in composition. An example of an azeotropic mixture of minimum boiling point is water (b.p. 100°C) and ethanol (b.p. 78.3°C), the azeotrope being 4.4% water and boiling at 78.1°C.

constant composition, law of *See* constant proportions; law of.

constant proportions, law of Formulated by Proust in 1779 after the analysis of a large number of compounds, the principle that the proportion of each element in a compound is fixed or constant. It follows that the composition of a pure chemical compound is independent of the method of preparation. It is also called the *law of definite proportions* and the *law of constant composition*.

continuous phase *See* colloid.

continuous process A manufacturing process in which the raw materials are constantly fed into the plant. These react as they flow through the equipment to give a continuing flow of product. At any point, only a small amount of material is at a particular stage in the process but material at all stages of the reaction is present. The fractional distillation of crude oil is an example of a continuous process. Such processes are relatively easy to automate and can therefore be used to manufacture a product cheaply. The disadvantages of continuous processing are that it usually caters for a large demand and the plant is expensive to install and cannot normally be used to make other things. *Compare* batch process.

continuous spectrum A spectrum composed of a continuous range of emitted or absorbed radiation. Continuous spectra are produced in the infrared and visible regions by hot solids. *See also* spectrum.

coordinate bond (dative bond) A covalent bond in which the bonding pair is visualized as arising from the donation of a lone pair from one species to another species, which behaves as an electron acceptor. The definition includes such examples as the 'donation' of the lone pair of the ammonia molecule to H^+ (an acceptor) to form NH_4^+ or to Cu^{2+} to form $[Cu(NH_3)_4]^{2+}$.

The donor groups are known as Lewis bases and the acceptors are either hydrogen ions or Lewis acids. Simple combinations, such as $H_3N \rightarrow BF_3$, are known as *adducts*. *See also* complex.

coordination compound *See* complex.

coordination number The number of coordinate bonds formed to a metal atom or ion in a complex.

copolymer *See* polymerization.

Cornforth, Sir John Warcup (1917–) Australian organic chemist. Cornforth is best known for his work on the problem of how the steroid cholesterol is synthesized in a cell. To investigate this problem he used the three isotopes of hydrogen – normal hydrogen (H-one), deuterium (H-two) and tritium (H-three) – and observed the different speeds of reactions found with these isotopes to infer how cholesterol was formed. He shared the 1975 Nobel Prize for chemistry with Vladimir PRELOG for this work. Cornforth has synthesized a number of other compounds including alkenes and oxazoles.

corn rule *See* optical activity.

coumarin (1,2-benzopyrone; $C_9H_6O_2$) A colorless crystalline compound with a pleasant odor, used in making perfumes. On hydrolysis with sodium hydroxide it forms *coumarinic acid*.

coumarinic acid *See* coumarin.

coumarone *See* benzfuran.

Couper, Archibald Scott (1831–92) 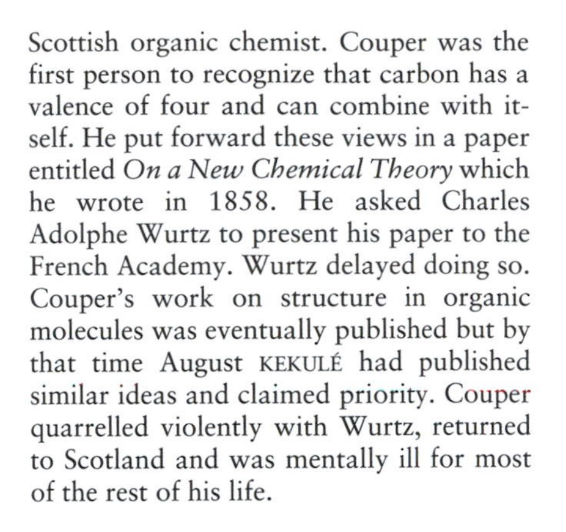
Scottish organic chemist. Couper was the first person to recognize that carbon has a valence of four and can combine with itself. He put forward these views in a paper entitled *On a New Chemical Theory* which he wrote in 1858. He asked Charles Adolphe Wurtz to present his paper to the French Academy. Wurtz delayed doing so. Couper's work on structure in organic molecules was eventually published but by that time August KEKULÉ had published similar ideas and claimed priority. Couper quarrelled violently with Wurtz, returned to Scotland and was mentally ill for most of the rest of his life.

coupling A chemical reaction in which two groups or molecules join together. An example is the formation of AZO COMPOUNDS.

covalent bond A bond formed by the sharing of an electron pair between two atoms. The covalent bond is conventionally represented as a line, thus H–Cl indicates that between the hydrogen atom and the chlorine atom there is an electron pair formed by electrons of opposite spin implying that the binding forces are strongly localized between the two atoms. Molecules are combinations of atoms bound together by covalent bonds; covalent bonding energies are of the order 10^3 kJ mol^{-1}.

Modern bonding theory treats the electron pairing in terms of the interaction of electron (atomic) ORBITALS and describes the covalent bond in terms of both 'bonding' and 'anti-bonding' molecular orbitals.

covalent crystal A crystal in which the atoms present are covalently bonded. They

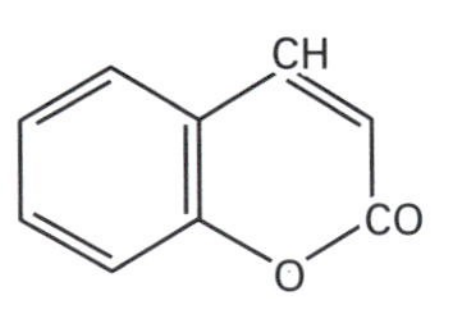

coumarin

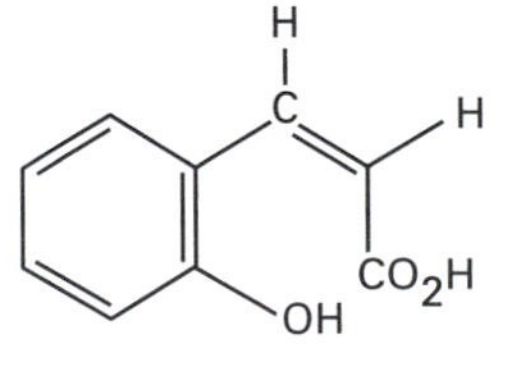

coumarinic acid

Coumarin

are sometimes referred to as giant lattices or macromolecules. The best-known completely covalent crystal is diamond.

covalent radius The radius an atom is assumed to have when involved in a covalent bond. For homonuclear diatomic molecules (e.g. Cl_2) this is simply half the measured internuclear distance. For heteroatomic molecules substitutional methods are used. For example, the internuclear distance of bromine fluoride (BrF) is about 180 pm, therefore using 71 pm for the covalent radius of fluorine (from F_2) we get 109 pm for bromine. The accepted value is 114 pm.

creosote A colorless oily liquid containing phenols and distilled from wood tar, used as a disinfectant. The name is also given to *creosote oil*, a dark brown liquid distilled from coal tar and used for preserving timber. It also consists of phenols, mixed with some methylphenols.

cresol *See* methylphenol.

Crick, Francis Harry Compton (1916–) British molecular biologist. Crick is best known for determining the structure of DNA with James WATSON in 1953. Based on a combination of model building, previous knowledge of the physical and chemical features of DNA and the x-ray diffraction photographs of Rosalind FRANKLIN they found the famous double helix structure. Together with Sydney Brenner, he worked on the problem of the genetic code. Crick put forward the Central Dogma of molecular genetics. This asserts that genetic information passes from DNA to RNA protein. It was subsequently shown that sometimes information can flow from RNA to DNA. In his later years Crick worked on how the mind works. He gave an account of this work in *The Astonishing Hypothesis* (1994). In 1988 his autobiography *What Mad Pursuit* was published. Crick, Watson, and Maurice WILKINS won the 1962 Nobel Prize for medicine for their work on DNA.

critical point The conditions of temperature and pressure under which a liquid being heated in a closed vessel becomes indistinguishable from the gas or vapor phase. At temperatures below the critical temperature (T_c) the substance can be liquefied by applying pressure; at temperatures above T_c this is not possible. For each substance there is one critical point; for example, for carbon dioxide it is at 31.1°C and 73.0 atmospheres.

critical pressure The lowest pressure needed to bring about liquefaction of a gas at its critical temperature.

critical temperature The temperature below which a gas can be liquefied by applying pressure and above which no amount of pressure is sufficient to bring about liquefaction. Some gases have critical temperatures above room temperature (e.g. carbon dioxide 31.1°C and chlorine 144°C) and have been known in the liquid state for many years. Liquefaction proved much more difficult for those gases (e.g. oxygen –118°C and nitrogen –146°C) that have very low critical temperatures.

critical volume The volume of one mole of a substance at its critical point.

cross linkage An atom or short chain joining two longer chains in a polymer.

crown ether A compound that has a large ring composed of $-CH_2-CH_2-O-$ units. For example, 18-crown-6 has the formula $C_{12}H_{24}O_6$ (six CH_2CH_2O units). The rings of these compounds are not planar – the name comes from the shape of the molecule. The oxygen atoms of these cyclic ethers can coordinate to central metal ions or to other positive ions (e.g. NH_4^+). The crown ethers have a number of uses in analysis, separation of mixtures, and as catalysts. *Cryptands* are similar compounds in which the ether chains are linked by nitrogen atoms to give a three-dimensional cage structure. They are similar in action to crown ethers but generally form more strongly bound complexes. *See also* host–guest chemistry.

crude oil *See* petroleum.

cryoscopic constant *See* depression of freezing point.

cryptand *See* crown ether.

crystal A solid substance that has a definite geometric shape. A crystal has fixed angles between its faces, which have distinct edges. The crystal will sparkle if the faces are able to reflect light. The constant angles are caused by the regular arrangements of particles (atoms, ions, or molecules) in the crystal. If broken, a large crystal will form smaller crystals.

In crystals, the atoms, ions, or molecules of the substance form a distinct regular array in the solid state. The faces and their angles bear a definite relationship to the arrangement of these particles.

crystal habit The shape of a crystal. The habit depends on the way in which the crystal has grown; i.e. the relative rates of development of different faces.

crystalline Denoting a substance that forms crystals. Crystalline substances have a regular internal arrangement of atoms, even though they may not exist as geometrically regular crystals. For instance, lead (and other metals) are crystalline. Such substances are composed of accumulations of tiny crystals.

crystallite A small crystal that has the potential to grow larger. It is often used in mineralogy to describe specimens that contain accumulations of many minute crystals of unknown chemical composition and crystal structure.

crystallization The process of forming crystals. When a substance cools from the gaseous or liquid state to the solid state, crystallization occurs. Crystals will also form from a solution saturated with a solute.

crystallography The study of the formation, structure, and properties of crystals. *See also* x-ray crystallography.

crystalloid A substance that is not a colloid and which will therefore not pass through a semipermeable membrane. *See* colloid; semipermeable membrane.

crystal structure The particular repeating arrangement of atoms, molecules, or ions in a crystal. 'Structure' refers to the internal arrangement of particles, not the external appearance.

crystal system A classification of crystals based on the shapes of their unit cell. If the unit cell is a parallelopiped with lengths a, b, and c and the angles between these edges are α (between b and c), β (between a and c), and γ (between a and b), then the classification is:
cubic: $a = b = c$; $\alpha = \beta = \gamma = 90°$
tetragonal: $a = b \neq c$; $\alpha = \beta = \gamma = 90°$
orthorhombic: $a \neq b \neq c$; $\alpha = \beta = \gamma = 90°$
hexagonal: $a = b \neq c$; $\alpha = \beta = 90°$; $\gamma = 120°$
trigonal: $a = b \neq c$; $\alpha = \beta = \gamma \neq 90°$
monoclinic: $a \neq b \neq c$; $\alpha = \gamma = 90° \neq \beta$
triclinic: $a \neq b \neq c$; $\alpha \neq \beta \neq \gamma$
The orthorhombic system is also called the *rhombic* system.

CS gas ((2-chlorobenzylidine-)-malanonitrile; $C_6H_4ClCH{:}C(CN)_2$) A white organic compound that is a nasal irritant used in powder form as a tear gas for riot control.

cumene process An industrial process for the manufacture of phenol from isopropylbenzene (cumene), which is itself made by passing benzene vapor and propene over a phosphoric acid catalyst (250°C and 30 atmospheres):

$$C_6H_6 + CH_2{:}CH(CH_3) \rightarrow C_6H_5CH(CH_3)_2$$

The isopropylbenzene is oxidized by air to a 'hydroperoxide':

$$C_6H_5C(CH_3)_2\text{–O–O–H}$$

This is hydrolyzed by dilute acid to phenol (C_6H_5OH) and propanone (CH_3COCH_3), which is a valuable by-product.

curie Symbol: Ci A unit of radioactivity, equivalent to the amount of a given radioactive substance that produces 3.7×10^{10} disintegrations per second, the number of disintegrations produced by one gram of radium.

cyanide *See* nitrile.

cyanocobalamin (vitamin B_{12}) One of the water-soluble B-group of vitamins. It has a complex organic ring structure at the center of which is a single cobalt atom. Foods of animal origin are the only important dietary source. A deficiency in humans leads to the development of pernicious anemia since the vitamin is required for the development of red blood cells. *See also* vitamin B complex.

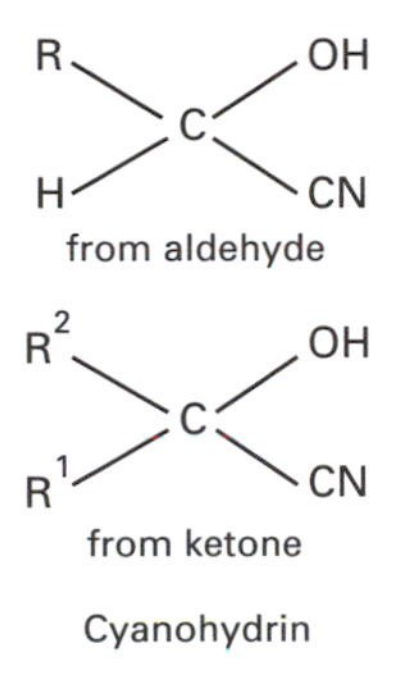

Cyanohydrin

cyanohydrin An addition compound formed between an aldehyde or ketone and hydrogen cyanide. The general formula is RCH(OH)(CN) (from an aldehyde) or RR′C(OH)(CN) (from a ketone). Cyanohydrins are easily hydrolyzed to hydroxycarboxylic acids. For instance, the compound 2-hydroxypropanonitrile (CH_3-CH(OH)(CN)) is hydrolyzed to 2-hydroxypropanoic acid (CH_3CH(OH)(COOH)).

cyclic AMP (cAMP; adenosine-3′,5′-monophosphate) A form of adenosine monophosphate (*see* AMP) formed from ATP in a reaction catalyzed by the enzyme adenylate cyclase. It has many functions, acting as an enzyme activator, genetic regulator, chemical attractant, secondary messenger, and as a mediator in the activity of many hormones, including epinephrine, norepinephrine, vasopressin, ACTH, and the prostaglandins.

cyclic compound A compound containing a ring of atoms. If the atoms forming the ring are all the same the compound is *homocyclic*; if different atoms are involved it is *heterocyclic*.

cyclization Any reaction in which a straight-chain compound is converted into a cyclic compound.

cycloaddition *See* pericyclic reaction.

cycloalkane A saturated cyclic hydrocarbon comprising a ring of carbon atoms, each carrying two hydrogen atoms, general formula C_nH_{2n}. Cyclopropane (C_3H_6), and cyclobutane (C_4H_8) both have strained rings and are highly reactive. Other cycloalkanes have similar properties to the alkanes, although they are generally less reactive than their corresponding alkane.

cyclohexadiene-1,4-dione *See* quinone.

cyclohexane (C_6H_{12}) A colorless liquid alkane that is commonly used as a solvent and in the production of hexanedioic acid (adipic acid) for the manufacture of nylon. Cyclohexane is manufactured by the reformation of longer chain hydrocarbons present in crude-oil fractions. It is also interesting from a structural point of view,

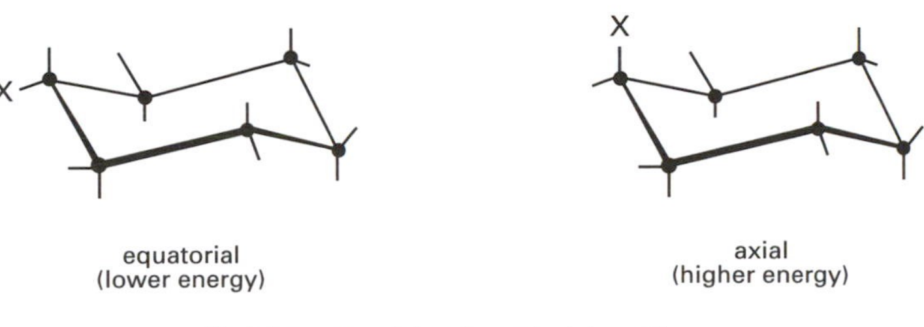

Cyclohexane: axial and equatorial positions

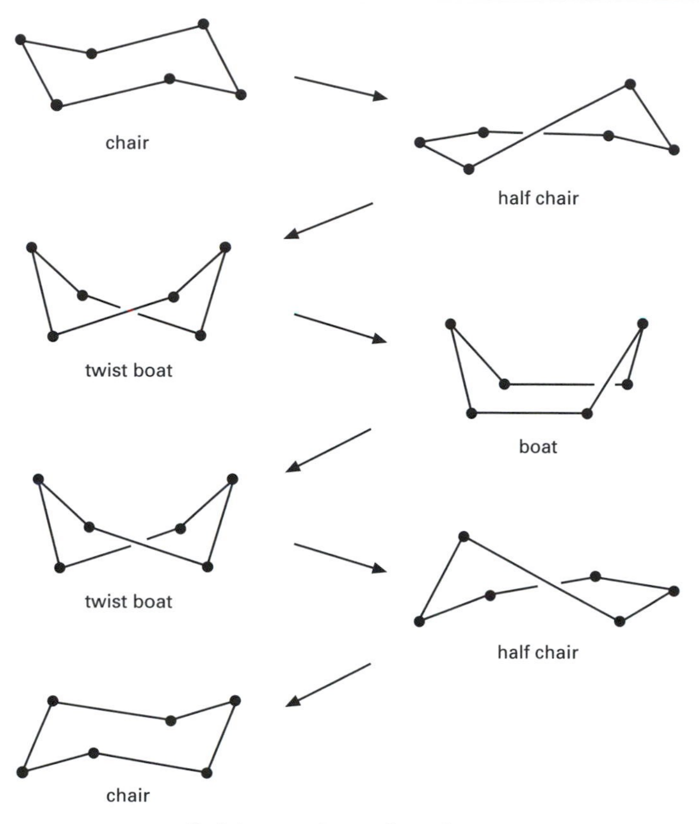

Cyclohexane: ring conformations

existing as a 'puckered' six-membered ring, having all bonds between carbon atoms at 109.9° (the tetrahedral angle). The molecule undergoes rapid interconversion between two *chair conformations*, which are energetically equivalent, passing through a *boat conformation* of higher energy. In passing from a chair to a boat, the cyclohexane ring passes through a *half-chair conformation*, which is the CONFORMATION of highest energy. This converts to a *twist-boat conformation*, which has a higher energy than the chair but lower than the true boat.

If cyclohexane has a substituent, there are also two different chair conformations, corresponding to whether the substituent is *axial* or *equatorial* (*see illustration*).

cyclonite A high explosive made from hexamine.

cyclo-octatetraene *See* annulene.

cyclopentadiene A cyclic hydrocarbon made by cracking petroleum. The molecules have a five-membered ring containing two carbon-carbon double bonds and one CH_2 group. It forms the negative *cyclopentadienyl ion* $C_5H_5^-$, present in SANDWICH COMPOUNDS and is a nonbenzenoid aromatic. *See* aromatic compound.

cyclopentadienyl ion *See* cyclopentadiene.

cysteine *See* amino acid.

cystine A compound formed by the joining of two cysteine amino acids through a –S–S– linkage (a *cystine link*). Bonds of this type are important in forming and maintaining the tertiary structure of proteins.

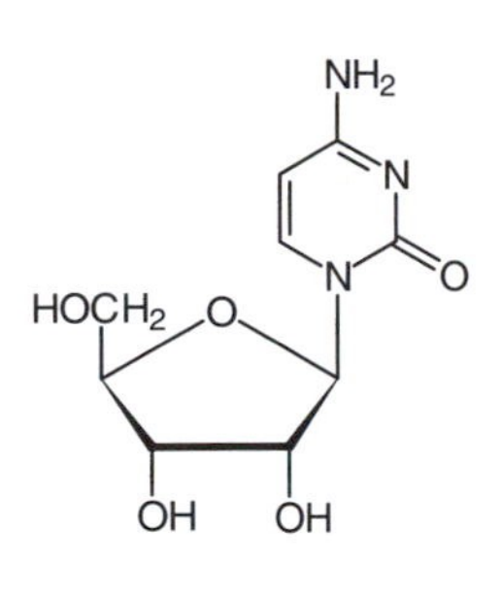

Cytidine

cytidine (cytosine nucleoside) A nucleoside formed when cytosine is linked to D-ribose via a β-glycosidic bond.

cytochrome Any of a group of conjugated proteins containing heme, that act as intermediates in the electron-transport chain. There are four main classes, designated *a*, *b*, *c*, and *d*.

cytokinin One of a class of plant hormones concerned with the stimulation of cell division, nucleic acid metabolism, and root-shoot interactions. Cytokinins are often purine derivatives: e.g. *kinetin* (6-furfuryl aminopurine), an artificial cytokinin commonly used in experiments; and zeatin, found in maize cobs.

cytosine A nitrogenous base found in DNA and RNA. Cytosine has the pyrimidine ring structure.

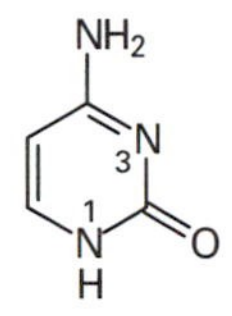

Cytosine

2,4-D (2,4-dichlorophenoxyacetic acid) A synthetic auxin used as a potent selective weedkiller. Monocotyledenous species with narrow erect leaves (e.g. cereals and grasses) are generally resistant to 2,4-D while dicotyledenous plants are often very susceptible. The compound is thus used for controlling weeds in cereal crops and lawns. *See* auxin.

Dalton's law (of partial pressures) The principle that the pressure of a mixture of gases is the sum of the partial pressures of each individual constituent. The *partial pressure* of a certain amount of a gas in a mixture is the pressure that it would exert if it alone were present in the container. Dalton's law is strictly true only for ideal gases. In real gases there are effects caused by intermolecular forces. It is named for the British chemist John Dalton (1766–1844), who proposed it in 1803.

dative bond *See* coordinate bond.

d-block elements The transition elements of the first, second, and third long periods of the periodic table, i.e. Sc to Zn, Y to Cd, and La to Hg. They are so called because in general they have inner d-levels with configurations of the type $(n-1)d^xns^2$ where $x = 1–10$.

DDT (dichlorodiphenyltrichloroethane; $(ClC_6H_4)_2CH(CCl_3)$) A colorless crystalline organic compound that was once widely used as an insecticide. It is very stable and tends to accumulate in the soil, and passes up the food chain to accumulate in the fatty tissues of carnivorous animals. Its systematic name is 1,1-bis(4-chlorophenyl)-2,2,2-trichloroethane.

deactivation A reduction in the reactivity of a substance or in the activity of a catalyst.

de Broglie wave A wave associated with a particle, such as an electron or proton. In 1924, Louis de Broglie suggested that, since electromagnetic waves can be described as particles (photons), particles of matter could also have wave properties. The wavelength (λ) has the same relationship to momentum (p) as in electromagnetic radiation:

$$\lambda = h/p$$

where h is the Planck constant. *See also* quantum theory.

debye Symbol: D A unit of electric dipole moment equal to $3.335\,64 \times 10^{-30}$ coulomb meter. It is used in expressing the dipole moments of molecules. The unit is named for the Dutch-born physical chemist Peter Debye (1884–1966).

Debye–Hückel theory A theory of the behavior (e.g. conductivity) of ions in dilute solutions of electrolytes. It assumes that electrolytes in dilute solution are completely dissociated into ions but takes into account interionic attraction and repulsion. Agreement between the theory and experiment occurs only with very dilute solutions (less than 10^{-3}M).

deca- Symbol: da A prefix denoting 10. For example, 1 decameter (dam) = 10 meters (m).

decahydronaphthalene *See* decalin.

decalin (decahydronaphthalene; $C_{10}H_{18}$) A liquid hydrocarbon made by the hydro-

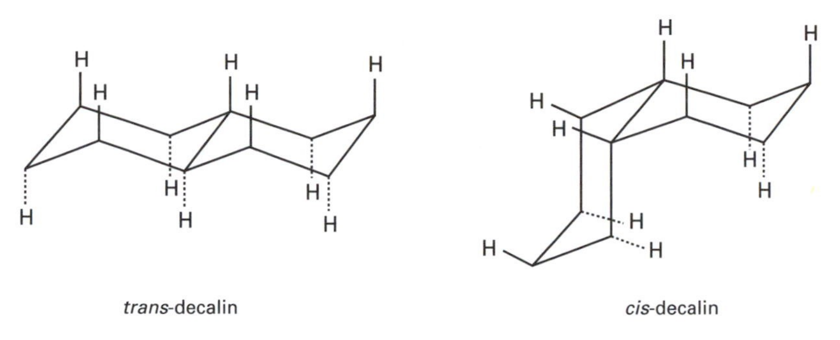

trans-decalin

cis-decalin

Decalin

genation of naphthalene at high temperature and pressure. There are two isomers.

decant To pour off the liquid above a sediment.

decay **1.** The spontaneous breakdown of a radioactive isotope. *See* half life; radioactivity.
2. The transition of excited atoms, ions, molecules, etc., to a state of lower energy.

deci- Symbol: d A prefix denoting 10^{-1}. For example, 1 decimeter (dm) = 10^{-1} meter (m).

decomposition The process in which a compound is broken down into compounds with simpler molecules.

decrepitation The process in which a crystalline solid emits a crackling noise on heating, usually because of loss of water of crystallization.

definite proportions, law of *See* constant proportions; law of.

degassing The removal of dissolved or absorbed gases from liquids or solids, either on heating or in a vacuum.

degenerate Describing different quantum states that have the same energy. For instance, the five d orbitals in a transition-metal atom all have the same energy but different values of the magnetic quantum number *m*. Differences in energy occur if a magnetic field is applied or if the arrangement of ligands around the atom is not symmetrical. The degeneracy is then said to be 'lifted'.

De Gennes, Pierre Gilles (1932–) French physicist. De Gennes is a versatile theoretical physicist who has made important contributions to the theory of liquid crystals and polymers. In particular, he has shown that although these forms of matter do not have order in the same sense as solid crystals they do have order that characterizes them. This enabled him to analyze them by using concepts such as order parameters and scaling taken from the theory of phase transitions. De Gennes gave an account of his work on liquid crystals in the book *The Physics of Liquid Crystals* (1974) and on polymers in *Scaling Concepts of Polymer Physics* (1979). In 1991 he won the Nobel Prize for physics for his contributions to liquid crystals and polymers.

degradation A type of chemical reaction involving the decomposition of a molecule into simpler molecules, usually in stages. The HOFMANN DEGRADATION of amides is an example.

degrees of freedom The independent ways in which particles can take up energy. In a monatomic gas, such as helium or argon, the atoms have three translational degrees of freedom (corresponding to motion in three mutually perpendicular directions). The mean energy per atom for each degree of freedom is $kT/2$, where k is the

Boltzmann constant and T the thermodynamic temperature; the mean energy per atom is thus $3kT/2$.

A diatomic gas has in addition two rotational degrees of freedom (about two axes perpendicular to the bond) and one vibrational degree (along the bond). The rotations also each contribute $kT/2$ to the average energy. The vibration contributes kT ($kT/2$ for kinetic energy and $kT/2$ for potential energy). Thus, the average energy per molecule for a diatomic molecule is $3kT/2$ (translation) + kT (rotation) + kT (vibration) = $7kT/2$.

Linear triatomic molecules also have two significant rotational degrees of freedom; nonlinear molecules have three. For nonlinear polyatomic molecules, the number of vibrational degrees of freedom is $3N - 6$, where N is the number of atoms in the molecule.

The molar energy of a gas is the average energy per molecule multiplied by the Avogadro constant. For a monatomic gas it is $3RT/2$, etc.

dehydration **1.** Removal of water from a substance.
2. Removal of the elements of water (i.e. hydrogen and oxygen in a 2:1 ratio) from a compound to form a new compound. An example is the dehydration of propanol to propene over hot pumice:

$$C_3H_7OH \rightarrow CH_3CH{:}CH_2 + H_2O$$

deionization The removal of ions from a solution. The usual method is to use an ion-exchange resin. The term is commonly applied to the purification of tap water; deionized water is cheaper to produce than distilled water and is adequate for many applications.

deliquescent Describing a solid compound that absorbs water from the atmosphere, eventually forming a solution. *See also* hygroscopic.

delocalization A spreading out of bonding electrons in a molecule over the molecule. *See* delocalized bond.

delocalized bond A type of bonding in molecules that occurs in addition to sigma bonding. The electrons forming the delocalized bond are not localized between two atoms; i.e. the electron density of the delocalized electrons is spread over several atoms and may spread over the whole molecule.

The electron density of the delocalized bond is spread by means of a delocalized molecular orbital and may be regarded as a series of pi bonds extending over several atoms, for example the pi bonds in butadiene and the C–O pi bonds in the CARBOXYLATE ION.

denaturation The changes in structure that occur when a PROTEIN is heated. These changes are irreversible and affect the properties of the protein.

denatured alcohol Alcohol (ethanol) that has been contaminated by the addition of small amounts of substances to make it unfit for drinking. Ethanol treated in this way may still be useful for many purposes (e.g. as a solvent) but not have restrictions or taxes on its sale.

dendritic growth The growth of crystals with a branching habit.

dendritic polymer *See* supramolecular chemistry.

density Symbol: ρ A property of substances equal to the mass of substance per unit volume. The units are g dm^{-3}, etc. *See also* relative density.

deoxyribonucleic acid *See* DNA.

deoxyribose *See* ribose.

depression of freezing point A colligative property of solutions in which the freezing point of a given solvent is lowered by the presence of a solute. The amount of the reduction is proportional to the molal concentration of the solute. The depression depends only on the concentration and is independent of solute composition. The proportionality constant, K_f, is called the *freezing point constant* or sometimes the *cryoscopic constant*. $\Delta t = K_f C_M$, where Δt is the lowering of the temperature and C_M is the molal concentration; the unit of K_f is kelvin kilogram mole^{-1} (K kg mol^{-1}). Although closely related to the property of boiling-point elevation, the cryogenic method can be applied to measurement of relative molecular mass with considerable precision. A known weight of pure solvent is slowly frozen, with stirring, in a suitable cold bath and the freezing temperature measured using a Beckmann thermometer. A known weight of solute of known molecular mass is introduced, the solvent thawed out, and the cooling process and measurement repeated. The addition is repeated several times and an average value of K_f for the solvent obtained by plotting Δt against C_M. The whole process is then repeated using the unknown solute and its relative molecular mass determined using the value of K_f previously obtained.

The effect is applied to more precise measurement of relative molecular mass by using a pair of Dewar flasks (pure solvent and solution) and measuring Δt by means of thermocouples. The theoretical explanation is similar to that for lowering of vapor pressure. The freezing point of the solvent is that point at which the curve representing the vapor pressure above the liquid phase intersects the curve representing the vapor pressure above the frozen solvent. The addition of solute depresses the former curve but as the solid phase that separates is always pure solvent (above the eutectic point), there is no attendant depression of the latter curve. Consequently the point of intersection is depressed, resulting in a lowering of the freezing point. *See also* lowering of vapor pressure.

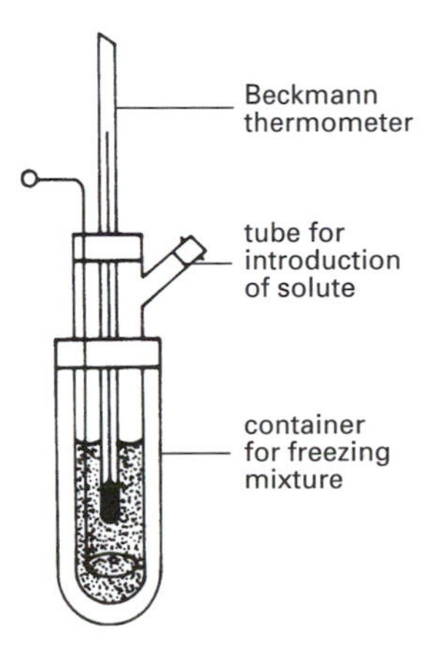

Depression of freezing point

derivative A compound that could be produced from another compound by chemical reaction. Usually, the term is applied to a compound that has a structural similarity to the parent compound; for example, chlorobenzene (C_6H_5Cl) is a derivative of benzene (C_6H_6).

derived unit A unit defined in terms of base units, and not directly from a standard value of the quantity it measures. For example, the newton is a unit of force defined as a kilogram meter second^{-2} (kg m s^{-2}). *See also* SI units.

desiccation Removal of moisture from a substance.

desiccator A piece of laboratory apparatus for drying solids or for keeping solids free of moisture. Typically, a dessicator is an air-tight container in which is kept a hygroscopic material (e.g. calcium chloride or silica gel) to absorb moisture from the atmosphere.

destructive distillation The process of heating an organic substance in the absence of air, so that it wholly or partially decomposes to produce volatile products, which are subsequently condensed. The destructive distillation of coal was the process for manufacturing coal gas and coal tar. At one time, methanol was made by the destructive distillation of wood.

detergent Any of a group of substances that improve the cleansing action of solvents, particularly water. The majority of detergents, including SOAP, have the same basic structure. Their molecules have a hydrocarbon chain (tail) that does not attract water molecules. The tail is said to be hydrophobic (water hating). Attached to this tail is a small group (head) that readily ionizes and attracts water molecules. It is said to be hydrophilic (water loving). Detergents reduce the surface tension of water and thus improve its wetting power. Because the detergent ions have their hydrophilic heads anchored in the water and their hydrophobic tails protruding above it, the water surface is broken up, enabling the water to spread over the material to be cleaned and penetrate between the material and the dirt. With the assistance of agitation, the dirt can be floated off. The hydrophobic tails of the detergent molecules 'dissolve' in grease and oils. The protruding hydrophilic heads repel each other causing the oil to roll up and form a drop, which floats off into the water as an emulsion. More recently synthetic detergents, often derived from petrochemicals, have been developed. Unlike soaps these detergents do not form insoluble scums with hard water.

Synthetic detergents are of three types. *Anionic detergents* form ions consisting of a hydrocarbon chain to which is attached either a sulfonate group, $-SO_2-O^-$, or a sulfate group, $-O-SO_2-O^-$. The corresponding metal salts are soluble in water. *Cationic detergents* have organic positive ions of the type RNH_3^+, in which R has a long hydrocarbon chain. Non-ionic detergents are complex chemical compounds called ethoxylates. They owe their detergent properties to the presence of a number of oxygen atoms in one part of the molecule, which are capable of forming hydrogen bonds with the surface water molecules, thus reducing the surface tension of the water. *See also* soap.

deuterated compound A compound in which one or more ^1H atoms have been replaced by deuterium (^2H) atoms.

deuterium Symbol: D, ^2H A naturally occurring stable isotope of hydrogen in which the nucleus contains one proton and one neutron. The atomic mass is thus approximately twice that of ^1H; deuterium is known as 'heavy hydrogen'. Chemically it behaves almost identically to hydrogen, forming analogous compounds, although reactions of deuterium compounds are often slower than those of the corresponding ^1H compounds. This is made use of in kinetic studies where the rate of a reaction may depend on transfer of a hydrogen atom.

deuterium oxide (D_2O) *See* heavy water.

deuteron The nucleus of a deuterium atom.

Dewar flask (vacuum flask) A double-walled container of thin glass with the space between the walls evacuated and sealed to stop conduction and convection of energy through it. The glass is often silvered to reduce radiation. It is named for the British chemist and physicist Sir James Dewar (1842–1923).

Dewar structure A representation of the structure of benzene in which there is a single bond between two opposite corners of the hexagonal ring and two double bonds at the sides of the ring. The Dewar structures contribute to the resonance hybrid of benzene. It is named for the British–American chemist Michael Dewar (1918–). The nonplanar compound with this structure, having two fused four-membered rings, was synthesized in 1963. *See* benzene.

dextrin A polysaccharide SUGAR produced from starch by the action of amylase enzymes or by chemical hydrolysis. Dextrins are used as adhesives.

dextro-form *See* optical activity.

dextronic acid *See* gluconic acid.

dextrorotatory *See* optical activity.

dextrose (grape-sugar) The dextrorotatory naturally occurring form of GLUCOSE, D-(+)-glucose. Because other stereochemical forms of glucose have no significance in biological systems the term 'glucose' is often used interchangeably with 'dextrose' in biology.

D-form *See* optical activity.

1,6-diaminohexane (hexamethylene diamine; $H_2N(CH_2)_6NH_2$) An organic compound used as a starting material in the production of nylon. It is manufactured from cyclohexane. *See* nylon.

diastereoisomer *See* isomerism.

diatomic molecule A molecule that consists of two atoms. Hydrogen (H_2), oxygen (O_2), nitrogen (N_2), and the halogens are examples of diatomic elements.

diazine *See* pyrazine.

diazole *See* pyrazole.

diazonium compound A compound of the type $ArN_2^+X^-$, where Ar is an aromatic group and X^- a negative ion. Diazonium salts are made by diazotization. They can be isolated but are very unstable, and are usually prepared in solution. The $-N_2^+$ group renders the benzene ring susceptible to nucleophilic substitution (rather than electrophilic substitution). Typical reactions are:

1. Reaction with water on warming the solution:

$$ArN_2^+ + H_2O \rightarrow ArOH + N_2 + H^+$$

2. Reaction with halogen ions (CuCl catalyst for chloride ions):

$$ArN_2^+ + I^- \rightarrow ArI + N_2$$

Diazonium ions can also act as electrophiles and undergo substitution reacting with other benzene rings (*diazo coupling*). *See also* azo compound.

diazotization The reaction of an aromatic amine (e.g. aniline) with nitrous acid at low temperatures (below 5°C).

$$C_6H_5NH_2 + HNO_2 \rightarrow C_6H_5N^+N + OH^- + H_2O$$

The acid is prepared *in situ* by reaction between nitric acid and sodium nitrite. The resulting *diazonium ion* is susceptible to attack by nucleophiles and provides a method of nucleophilic substitution onto the benzene ring.

dibasic acid An acid that has two acidic hydrogen atoms, such as sulfuric acid. Dibasic acids can give rise to two series of salts. For example, sulfuric acid (H_2SO_4) forms sulfates (SO_4^{2-}) and hydrogensulfates (HSO^-_4).

dibenzo-4-pyrone *See* xanthone.

1,2-dibromoethane (ethylene dibromide; $BrCH_2CH_2Br$) A colorless volatile organic liquid, made by reacting bromine with ethene. It is used as a fuel additive to remove lead (as lead bromide, which is also volatile).

dicarboxylic acid An organic acid that has two carboxyl groups (–COOH). An example is hexanedioic acid, $HOOC(CH_2)_4COOH$ (adipic acid).

dichloroacetic acid *See* chloroethanoic acid.

dichlorodiphenyltrichloroethane *See* DDT.

dichloroethanoic acid *See* chloroethanoic acid.

dichromate(VI) A salt containing the ion $Cr_2O_7^-$. Dichromates are strong oxidizing agents. *See* potassium dichromate.

Diels, Otto Paul Hermann (1876–1954) German organic chemist. The first major discovery which Diels made was carbon suboxide (C_3O_2). He discovered this compound in 1906 by dehydrating malonic acid with phosphorus pentoxide. His second major discovery was the process of removing hydrogen from steroids by heating them with selenium. He used this process on cholesterol. He was able to use the process he found to determine the structures of steroids. In 1928 Diels and his colleague Kurt ALDER discovered what came to be known as the Diels–Alder reaction for producing a ring compound from a diene. Diels and Alder shared the 1950

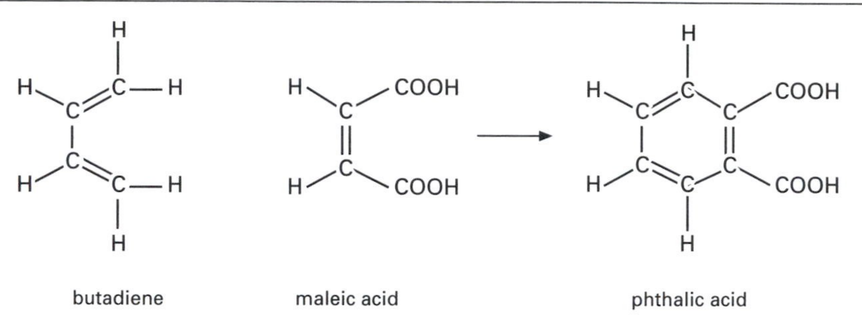

Diels–Alder reaction

Nobel Prize for chemistry for this discovery.

Diels–Alder reaction A type of reaction in which a conjugated DIENE adds to a compound containing a double C=C bond (called the *dienophile*) to give a ring compound. To be effective, the dienophile has to have electron-withdrawing groups on the double bond. The diene has to have a *cis*-conformation or to be able to adopt a *cis*-conformation. The reactants are mixed together and heated. The mechanism involves a single step in which electrons move to form different bonds. The reaction is an example of a cycloaddition reaction (*see* pericyclic reaction). It was named for the German chemists Otto Diels and Kurt Alder (1902–58), who described it in 1928.

diene An organic compound containing two carbon–carbon double bonds. In a conjugated diene the two double bonds are separated by a single C–C bond.

dienophile *See* Diels–Alder reaction.

diesel fuel A petroleum fraction consisting of various alkanes in the boiling range 200–350°C, used as a fuel for diesel (compression-ignition) engines.

diethylether *See* ethoxyethane.

diffusion Movement of a gas, liquid, or solid as a result of the random thermal motion of its particles (atoms or molecules). A drop of ink in water, for example, will slowly spread throughout the liquid. Diffusion in solids occurs very slowly at normal temperatures. *See also* Graham's law.

dihedral angle *See* conformation.

dihydrate A crystalline compound with two molecules of water of crystallization per molecule of compound.

dihydric alcohol *See* diol.

dihydroxypurine *See* xanthine.

diluent A solvent that is added to reduce the strength of a solution.

dilute Denoting a solution in which the amount of solute is low relative to that of the solvent. The term is always relative.

dimensionless units The radian and steradian in SI units. *See* SI units.

dimer A compound (or molecule) formed by combination or association of two molecules of a monomer. Cyclopentadiene, for example, exists as a dimer at room temperature. On heating it dissociates.

dimethylbenzene (xylene; $C_6H_4(CH_3)_2$) An organic hydrocarbon present in the light-oil fraction of crude oil. It is used extensively as a solvent. There are three isomeric compounds with this name and formula, distinguished as 1,2-, 1,3-, and 1,4-dimethylbenzene according to the positions of the methyl groups on the benzene ring.

2,4-dinitrophenylhydrazine (Brady's reagent) An orange solid commonly used in solution with methanol and sulfuric acid to produce crystalline derivatives by condensation with aldehydes and ketones. The derivatives, known as 2,4-dinitrophenylhydrazones, can easily be purified by recrystallization and have characteristic melting points, used to identify the original aldehyde or ketone.

dinucleotide A compound of two nucleotides linked by their phosphate groups. Important examples are the coenzymes NAD and FAD.

diol (dihydric alcohol; glycol) An alcohol that has two hydroxyl groups (–OH) per molecule of compound.

1,4-dioxan ($(CH_2)_2O_2$) A colorless liquid cyclic ether. It is an inert compound miscible with water used as a solvent.

dioxin Any of a related group of highly toxic chlorinated compounds. Particularly important is the compound 2,3,7,8-tetrachlorodibenzo-*p*-dioxin (*TCDD*), which is produced as a by-product in the manufacture of 2,4,5-T, and may consequently occur as an impurity in certain types of weedkiller. The defoliant known as AGENT ORANGE used in Vietnam contained significant amounts of TCDD. Dioxins cause a skin disease (chloracne) and birth defects. Dioxins have been released into the atmosphere as a result of explosions at herbicide manufacturing plants, most notably at Seveso, Italy, in 1976.

dipeptide *See* peptide.

diphosphane (diphosphine, P_2H_4) A yellow liquid that can be condensed out from phosphine in a freezing mixture. It ignites spontaneously in air.

diphosphine *See* diphosphane.

dipole A system in which two equal and opposite electric charges are separated by a finite distance. Polar molecules have permanent dipoles. Induced dipoles can also occur. *See also* van der Waals force.

dipole moment Symbol: μ A quantitative measure of polarity in either a bond (bond moment) or a molecule as a whole (molecular dipole moment). The unit is the debye (equivalent to 3.34×10^{-30} coulomb meter). Molecules such as HF, H_2O, NH_3, and $C_6H_5NH_2$ possess dipole moments; CCl_4, N_2, C_6H_6, and PF_5 do not.

The molecular dipole moment can be estimated by vector addition of individual bond moments if the bond angles are known. The possession of a dipole moment permits direct interaction with electric fields or interaction with the electric component of radiation.

dipyridyl (bipyridyl) A compound formed by linking two pyridine rings. There are various isomers, some of which are used in herbicides.

direct dyes A group of dyes that are mostly azo-compounds derived from benzidene or benzidene derivatives. They are used to dye cotton, viscose rayon, and other cellulose fibers directly, using a neutral bath containing sodium chloride or sodium sulfate as a mordant.

disaccharide A SUGAR with molecules composed of two monosaccharide units. These are linked by a –O– linkage (*glycosidic link*). Sucrose and maltose are examples.

disconnection *See* retrosynthetic analysis.

disperse dyes Water-insoluble dyes, which, when held in fine suspension, can be applied to acetate rayon fabrics. The dye, together with a dispersing agent, is warmed to a temperature of 45–50°C and the fabric added. By modifying the method of application it is possible to dye polyacrylic and polyester fibers. The yellow/orange shades are nitroarylamine derivatives and the green to bluish shades are derivatives of 1-amino anthraquinone.

Certain azo compounds are disperse dyes and these give a range of colors.

disperse phase *See* colloid.

dispersing agent A compound used to assist emulsification or dispersion.

dispersion force A weak type of intermolecular force. *See* van der Waals force.

displacement pump A commonly used device for transporting liquids and gases around chemical plants. It works on the principle of the bicycle pump: a piston raises the pressure of the fluid and, when it is high enough, a valve opens and the fluid is discharged through an outlet pipe. As the piston moves back the pressure falls and the cycle continues. Displacement pumps can be used to generate very high pressures but because of the system of valves, they are more expensive than other types of pump. *Compare* centrifugal pump.

displacement reaction A chemical reaction in which an atom or group displaces another atom or group from a molecule.

disproportionation A chemical reaction in which there is simultaneous oxidation and reduction of the same compound. The CANNIZZARRO REACTION is an example in organic chemistry.

dissociation Breakdown of a molecule into two molecules, atoms, radicals, or ions. Often the reaction is reversible, as in the ionic dissociation of weak acids in water:

$$CH_3COOH + H_2O \rightleftharpoons CH_3COO^- + H_3O^+$$

dissociation constant The equilibrium constant of a dissociation reaction. For example, the dissociation constant of a reaction:

$$AB \rightleftharpoons A + B$$

is given by:

$$K = [A][B]/[AB]$$

where the brackets denote concentration (activity).

Often the degree of dissociation is used – the fraction (α) of the original compound that has dissociated at equilibrium. For an original amount of AB of n moles in a volume V, the dissociation constant is given by:

$$K = \alpha^2 n/(1 - \alpha)V$$

Note that this expression is for dissociation into two molecules.

Acid dissociation constants (or *acidity constants*, symbol: K_a) are dissociation constants for the dissociation into ions in solution:

$$HA + H_2O \rightleftharpoons H_3O^+ + A^-$$

The concentration of water can be taken as unity, and the acidity constant is given by:

$$K_a = [H_3O^+][A^-]/[HA]$$

The acidity constant is a measure of the strength of the acid. Base dissociation constants (K_b) are similarly defined. The expression:

$$K = \alpha^2 n/(1 - \alpha)V$$

applied to an acid is known as *Ostwald's dilution law* (for the German chemist Friedrich Wilhelm Ostwald (1853–1932), who formulated it in 1888). In particular if α is small (a weak acid) then $K = \alpha^2 n/V$, or $\alpha = C\sqrt{V}$, where C is a constant. The degree of dissociation is then proportional to the square root of the dilution.

distillation The process of boiling a liquid and condensing the vapor. Distillation is used to purify liquids or to separate components of a liquid mixture. *See also* destructive distillation; fractional distillation; steam distillation; vacuum distillation.

distilled water Water that has been purified by distillation, perhaps several times.

diterpene *See* terpene.

divalent (bivalent) Having a valence of two.

Djerassi, Carl (1923–) Austrian-born American chemist. The first notable work which he and his colleagues at Syntex, in Mexico City, performed was to extract cortisone from a vegetable source. Djerassi and his colleagues then investigated the steroid hormone progesterone which acts as a natural contraceptive. They produced progesterone artificially in the early 1950s, thus reducing its price. Djerassi improved the power of progesterone by removing a particular methyl group. He used a similar trick with testosterone. This led to the development of the contraceptive pill. Djerassi published his autobiography *The Pill, Pigmy Chimps, and Degas Horse* in 1992.

D-L convention *See* optical activity.

DNA (deoxyribonucleic acid) A nucleic acid, mainly found in the chromosomes, that contains the hereditary information of organisms. The molecule is made up of two antiparallel helical polynucleotide chains coiled around each other to give a *double helix*. It is also known as the *Watson-Crick model* after James Watson and Francis Crick who first proposed this model in 1953. Phosphate molecules alternate with deoxyribose sugar molecules along both chains and each sugar molecule is also joined to one of four nitrogenous bases – adenine (A), guanine (G), cytosine (C), or thymine (T). The two chains are joined to each other by bonding between bases. The two purine bases (adenine and guanine) always bond with the pyrimidine bases (thymine and cytosine), and the pairing is quite specific: adenine with thymine and guanine with cytosine. The two chains are therefore complementary. The sequence of bases along the chain makes up a code – the genetic code – that determines the precise sequence of amino acids in proteins.

DNA is the hereditary material of all organisms with the exception of RNA viruses. Together with histones (and RNA in some instances) it makes up the chromosomes of eukaryotic cells. *See also* RNA. *See illustration overleaf.*

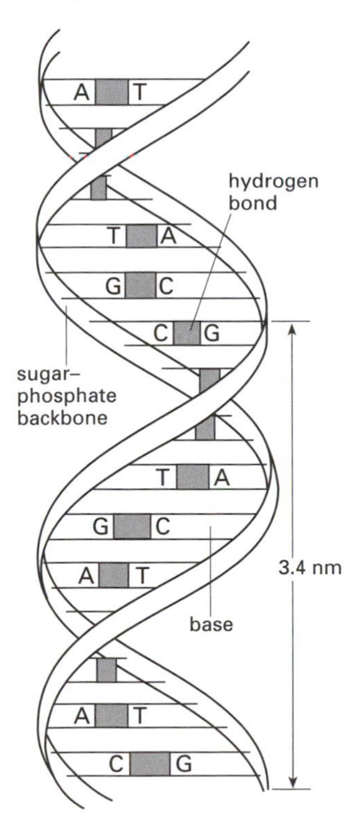

DNA: the double helix

dodecanoic acid (lauric acid; $CH_3(CH_2)_{10}COOH$) A white crystalline carboxylic acid, used as a plasticizer and for making detergents and soaps. Its glycerides occur naturally in coconut and palm oils.

donor **1.** The atom, ion, or molecule that provides the pair of electrons in forming a covalent bond.
2. The impurity atoms used in doping semiconductors.

dopamine A catecholamine precursor of epinephrine and norepinephrine. In mammals it is found in highest concentration in the corpus striatum of the brain, where it functions as an inhibitory neurotransmitter. High levels of dopamine are associated with Parkinson's disease in humans.

dormin A former name for *abscisic acid*.

double bond A covalent bond between two atoms that includes two pairs of electrons, one pair being the single bond equivalent (the sigma pair) and the other forming an additional bond, the pi bond (π bond). It is conventionally represented by two lines, for example $H_2C=O$. *See* multiple bond; orbital.

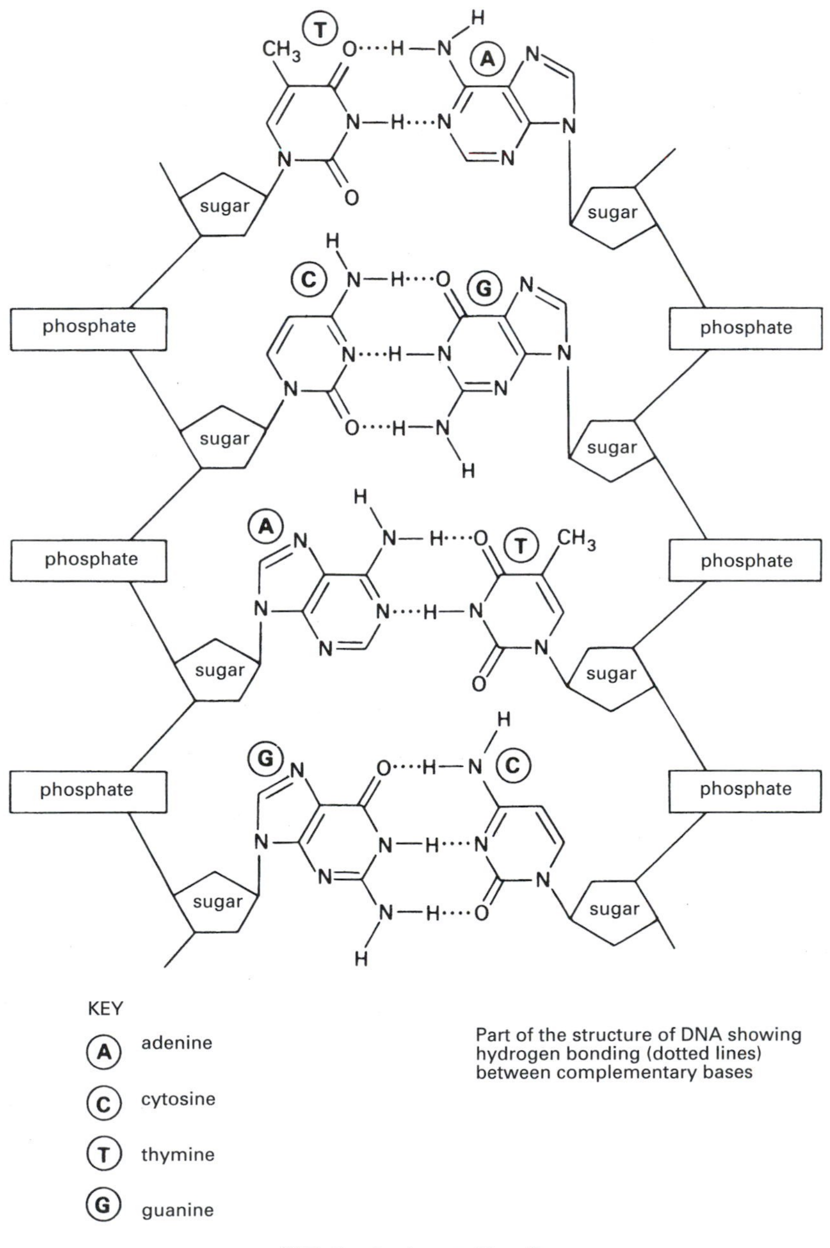

DNA: the structure and bonding

double helix *See* DNA.

double salt When equivalent quantities of certain salts are mixed in aqueous solution and the solution evaporated, a salt may form, e.g. $FeSO_4.(NH_4)_2SO_4.6H_2O$. In aqueous solution the salt behaves as a mixture of the two individuals. These salts are called double salts to distinguish them from complex salts, which yield complex ions in solution.

dryers Devices used in chemical processes to remove a liquid from a solid by evaporation. Drying equipment is classified by the method of transferring heat to a wet solid. This can be by direct contact between hot gases and the solid (direct dryers), heat transfer by conduction through a retaining metallic wall (indirect dryers), or infrared rays (infrared dryers).

dry ice Solid carbon dioxide, used as a refrigerant.

drying oil A natural oil, such as linseed oil, that hardens in air. Such oils contain unsaturated fatty acids, which polymerize on oxidation.

Dumas' method **1.** A method for determining the relative molecular mass of a volatile liquid. The method utilizes a glass bulb with a narrow entrance tube. The bulb is weighed 'empty' (i.e. full of air) then the sample is introduced and the bulb immersed in a heating bath so that the sample boils and expels all the air. When the surplus vapor has been expelled, the bulb is sealed off, cooled, dried, and weighed. The tip of the tube is then broken under water so that the water completely fills the tube and the whole weighed again. This enables the volume of the bulb to be calculated from the known density of water, and knowing the density of air one can compute the mass of vapor in a known volume of the sample.
2. A method of finding the amount of nitrogen in an organic compound by heating the compound with copper oxide to convert the nitrogen into nitrogen oxides. These are reduced by passing them over hot copper and the volume of nitrogen collected is measured.
Both methods are named for the French chemist Jean Baptiste André Dumas (1800–84).

dye A coloring material for fabric, leather, etc. Most dyes are now synthetic organic compounds (the first such was the dye *mauve* synthesized from aniline in 1856 by William Perkin). Dyes are often unsaturated organic compounds containing conjugated double bonds – the bond system responsible for the color is called the *chromophore*. *See also* azo compound.

dynamite A high explosive made by absorbing nitroglycerine into an earthy material such as diatomite (kieselguhr). Solid sticks of dynamite are much safer to handle than the highly sensitive liquid nitroglycerine.

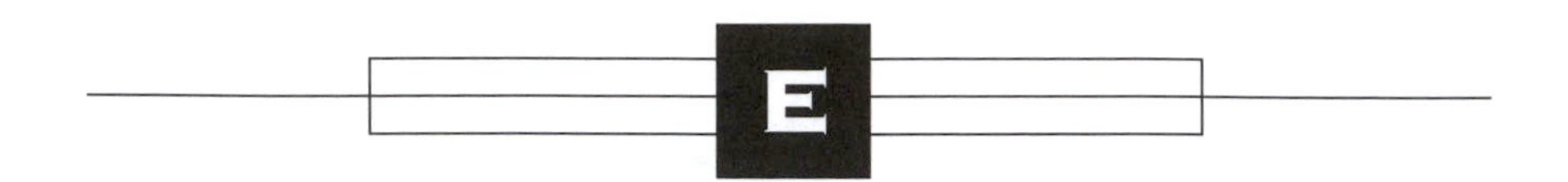

ebonite *See* vulcanite.

ebulioscopic constant *See* elevation of boiling point.

ebullition The boiling or bubbling of a liquid.

eclipsed conformation *See* conformation.

edta (ethylenediamine tetraacetic acid) A compound with the formula

$(HOOCCH_2)_2N(CH_2)_2N(CH_2COOH)_2$

It is used in forming chelates of transition metals. *See* chelate.

effervescence The evolution of gas in the form of bubbles in a liquid.

efflorescence The process in which a crystalline hydrated solid spontaneously loses water of crystallization to the air. A powdery deposit is gradually formed.

elastin A structural protein found in mammalian connective tissues, especially in elastic fibers. Glycine is the main component; proline, alanine, and valine are the other main residues.

elastomer An elastic substance, e.g. a natural or synthetic rubber.

electrochemical equivalent Symbol: z The mass of an element released from a solution of its ion when a current of one ampere flows for one second during ELECTROLYSIS.

electrochemical series (electromotive series) A series giving the activities of metals for reactions that involve ions in solution. In decreasing order of activity, the series is

K, Na, Ca, Mg, Al, Zn, Fe,
Pb, H, Cu, Hg, Ag, Pt, Au

Any member of the series will displace ions of a lower member from solution. For example, zinc metal will displace Cu^{2+} ions:

$$Zn(s) + Cu^{2+}(aq) \rightarrow Zn^{2+}(aq) + Cu(s)$$

Zinc has a greater tendency than copper to form positive ions in solution. Similarly, metals above hydrogen displace hydrogen from acids:

$$Zn + 2HCl \rightarrow ZnCl_2 + H_2$$

The series is based on electrode potentials, which measure the tendency to form positive ions. The series is one of increasing electrode potential for half cells of the type $M^{n+}|M$. Thus, copper ($E^\ominus$ for $Cu^{2+}|Cu$ = + 0.34 V) is lower than zinc ($E^\ominus$ for $Zn^{2+}|Zn$ = –0.76V). The hydrogen half cell has a value $E^\ominus = 0$.

electrochemistry The study of the formation and behavior of ions in solutions. It includes electrolysis and the generation of electricity by chemical reactions in cells.

electrochromatography *See* electrophoresis.

electrocyclic reaction *See* pericyclic reaction.

electrode Any part of an electrical device or system that emits or collects electrons or other charge carriers. An electrode may also be used to deflect charged particles by the action of the electrostatic field that it produces.

electrode potential Symbol: *E* A measure of the tendency of an element to form ions in solution. For example, a metal in a solution containing M^+ ions may dissolve in the solution as M^+ ions; the metal then has an excess of electrons and the solution an excess of positive ions – thus, the metal becomes negative with respect to the solution. Alternatively, the positive ions may gain electrons from the metal and be deposited as metal atoms. In this case, the metal becomes positively charged with respect to the solution. In either case, a potential difference is developed between solid and solution, and an equilibrium state will be reached at which further reaction is prevented. The equilibrium value of this potential difference would give an indication of the tendency to form aqueous ions.

It is not, however, possible to measure this for an isolated half cell – any measurement requires a circuit, which sets up another half cell in the solution. Therefore, electrode potentials (or *reduction potentials*) are defined by comparison with a hydrogen half cell, which is connected to the half cell under investigation by a salt bridge. The e.m.f. of the full cell can then be measured.

In referring to a given half cell the more reduced form is written on the right for a half-cell reaction. For the half cell $Cu^{2+}|Cu$, the half-cell reaction is a reduction:

$$Cu^{2+}(aq) + 2e \rightarrow Cu$$

The cell formed in comparison with a hydrogen electrode is:

$$Pt(s)H_2(g)|H^+(aq)|Cu^{2+}(aq)|Cu$$

The e.m.f. of this cell is +0.34 volt measured under standard conditions. Thus, the standard electrode potential (symbol: $E^{\ominus}$) is +0.34 V for the half cell $Cu^{2+}|Cu$. The standard conditions are 1.0 molar solutions of all ionic species, standard pressure, and a temperature of 298 K.

Half cells can also be formed by a solution of two different ions (e.g. Fe^{2+} and Fe^{3+}). In such cases, a platinum electrode is used under standard conditions.

electrolysis The production of chemical change by passing electric charge through certain conducting liquids (electrolytes). The current is conducted by migration of ions – positive ones (cations) to the cathode (negative electrode), and negative ones (anions) to the anode (positive electrode). Reactions take place at the electrodes by transfer of electrons to or from them.

In the electrolysis of water (containing a small amount of acid to make it conduct adequately) hydrogen gas is given off at the cathode and oxygen is evolved at the anode. At the cathode the reaction is:

$$H^+ + e^- \rightarrow H$$
$$2H \rightarrow H_2$$

At the anode:

$$OH^- \rightarrow e^- + OH$$
$$2OH \rightarrow H_2O + O$$
$$2O \rightarrow O_2$$

In certain cases the electrode material may dissolve. For instance, in the electrolysis of copper(II) sulfate solution with copper electrodes, copper atoms of the anode dissolve as copper ions

$$Cu \rightarrow 2e^- + Cu^{2+}$$

electrolyte A liquid containing positive and negative ions that conducts electricity by the flow of those charges. Electrolytes can be solutions of acids or metal salts ('ionic compounds'), usually in water. Alternatively they may be molten ionic compounds – again the ions can move freely through the substance. Liquid metals (in which conduction is by free electrons rather than ions) are not classified as electrolytes. *See also* electrolysis.

electrolytic Relating to the behavior or reactions of ions in solution.

electrolytic cell *See* cell; electrolysis.

electromagnetic radiation Energy propagated by vibrating electric and magnetic fields. Electromagnetic radiation forms a whole *electromagnetic spectrum*, depending on frequency and ranging from high-frequency radio waves to low-frequency gamma rays.

Electromagnetic radiation can be thought of as waves (*electromagnetic waves*) or as streams of photons. The frequency and wavelength are related by:

$$\lambda\nu = c$$

ELECTROMAGNETIC SPECTRUM
(*note: the figures are only approximate*)

Radiation	*Wavelength (m)*	*Frequency (Hz)*
gamma radiation	-10^{-10}	$10^{19}-$
x-rays	$10^{-12}-10^{-9}$	$10^{17}-10^{20}$
ultraviolet radiation	$10^{-9}-10^{-7}$	$10^{15}-10^{17}$
visible radiation	$10^{-7}-10^{-6}$	$10^{14}-10^{15}$
infrared radiation	$10^{-6}-10^{-4}$	$10^{12}-10^{14}$
microwaves	$10^{-4}-1$	$10^{9}-10^{13}$
radio waves	$1-$	-10^{9}

where c is the speed of light. The energy carried depends on the frequency.

electromagnetic spectrum *See* electromagnetic radiation.

electromotive series *See* electrochemical series.

electron An elementary particle of negative charge ($-1.602\ 192 \times 10^{-19}$C) and rest mass $9.109\ 558 \times 10^{-31}$ kg. Electrons are present in all atoms in shells around the nucleus.

electron affinity Symbol: A The energy released when an atom (or molecule or group) gains an electron in the gas phase to form a negative ion. It is thus the energy of:

$$A + e^- \rightarrow A^-$$

A positive value of A (often in electronvolts) indicates that heat is given out. Often the molar enthalpy is given for this process of electron attachment (ΔH). Here the units are joules per mole (J mol^{-1}), and, by the usual convention, a negative value indicates that energy is released.

electron-deficient compounds Compounds in which the number of electrons available for bonding is insufficient for the bonds to consist of conventional two-electron covalent bonds. Diborane, B_2H_6, is an example in which each boron atom has two terminal hydrogen atoms bound by conventional electron-pair bonds and in addition the molecule has two hydrogen atoms bridging the boron atoms (B–H–B). In each bridge there are only two electrons for the bonding orbital. *See also* multicenter bond.

electron diffraction A technique used to determine the structure of substances, principally the shapes of molecules in the gaseous phase. A beam of electrons directed through a gas at low pressure produces a series of concentric rings on a photographic plate. The dimensions of these rings are related to the interatomic distances in the molecules. *See also* x-ray diffraction.

electron donor *See* reduction.

electronegative Describing an atom or molecule that attracts electrons, forming negative ions. Examples of electronegative elements include the halogens (chlorine etc.), which readily form negative ions (F^-, Cl^-, etc.). *See also* electronegativity.

electronegativity A measure of the tendency of an atom in a molecule to attract electrons to itself. Elements to the right-hand side of the periodic table are strongly electronegative (values from 2.5 to 4); those on the left-hand side have low electronegativities (0.8–1.5) and are sometimes called electropositive elements. Different electronegativities of atoms in the same molecule give rise to polar bonds and sometimes to polar molecules.

As the concept of electronegativity is not precisely defined it cannot be precisely measured and several electronegativity scales exist. Although the actual values differ the scales are in good relative agreement. *See also* electron affinity; ionization potential.

electronic energy levels *See* atom; energy level.

electronic transition The demotion or promotion of an electron between electronic energy levels in an atom or molecule.

electron pair Two electrons in one orbital with opposing spins (*spin paired*), such as the electrons in a COVALENT BOND or LONE PAIR.

electron spin *See* atom.

electron spin resonance (ESR) A similar technique to nuclear magnetic resonance, but applied to unpaired electrons in a molecule (rather than to the nuclei). It is a powerful method of studying free radicals. ESR is also used in inorganic chemistry to study transition-metal complexes.

electron-transport chain (respiratory chain) A chain of chemical reactions involving proteins and enzymes, resulting in the formation of ATP and the transfer of hydrogen atoms to oxygen to form water. The enzymes and other proteins are, in eukaryotic cells, located in the inner membrane of the mitochondria and are grouped into discrete complexes. The reduced coenzyme NADH gives up two electrons to the first component in the chain, NADH dehydrogenase, and two hydrogen ions (H^+) are discharged from the matrix of the mitochondrion into the intermembrane space. The electrons are transferred along the chain to a carrier molecule (ubiquinone). Ubiquinone passes them to the next complex, which contains cytochromes b and c_1. Another carrier (cytochrome c) transfers the electrons to the final complex in the chain. There they act with the enzyme cytochrome oxidase to reduce an oxygen atom, which combines with two H^+ ions to form water. During this electron transfer, a further two pairs of H^+ ions are pumped into the intermembrane space by the complexes, making a total of six per molecule of NADH. If $FADH_2$ is the electron donor, only four H^+ ions are pumped across, as it donates electrons directly to ubiquinone.

The function of electron transport in the mitochondrion is to provide the energy required to phosphorylate ADP to ATP. According to the *chemiosmotic theory*, the H^+ ions in the intermembrane space diffuse back to the matrix through the inner mitochondrial membrane down a concentration gradient. As they do so they pass through the protein channel (the F_0 unit) of the enzyme ATP synthase. The energy released allows the catalytic F_1 unit of ATP synthase to synthesize ATP from ADP and inorganic phosphate. Each pair of H^+ ions catalyzes the formation of one molecule of ATP, so for each NADH molecule, three molecules of ATP may be synthesized (two molecules of ATP per molecule of $FADH_2$). A similar mechanism is involved in ATP formation by components of the light reaction in photosynthesis. *See* photosynthesis; oxidative phosphorylation.

electronvolt Symbol: eV A unit of energy equal to $1.602\ 191\ 7 \times 10^{-19}$ joule. It is defined as the energy required to move an electron charge across a potential difference of one volt. It has been used to measure the kinetic energies of elementary particles or ions, or the ionization potentials of molecules.

electrophile An electron-deficient ion or molecule that takes part in an organic reaction. The electrophile can be either a positive ion (H^+, NO_2^+), a molecule that can accept an electron pair (SO_3, O_3), or an electron-deficient group (e.g. a CARBENE). The electrophile attacks negatively charged areas of molecules, which usually arise from the presence in the molecule of a polar single bond or group or of pi-bonds. *Compare* nucleophile.

electrophilic addition A reaction involving the addition of a small molecule to an unsaturated organic compound, across the atoms joined by a double or triple bond with an ELECTROPHILE as the initial attacking species. The reaction is initiated by the attack of the electrophile on the electron-rich area of the molecule. The mechanism of electrophilic addition is thought to be ionic, as in the addition of HBr to ethene:

$$H_2C{:}CH_2 + H^+Br^- \rightarrow$$
$$H_3CCH_2^+ + Br^- \rightarrow$$
$$H_3CCH_2Br$$

In the case of higher alkenes (more than two carbon atoms) several isomeric products are possible. The particular isomer produced depends on the stability of the alternative intermediates and this is summarized empirically by MARKOVNIKOFF'S RULE. *See also* addition reaction.

electrophilic substitution A reaction involving substitution of an atom or group of atoms in an organic compound with an ELECTROPHILE as the attacking substituent. Electrophilic substitution is very common in aromatic compounds, in which electrophiles are substituted onto the ring. An example is the nitration of benzene:

$$C_6H_6 + NO_2^+ \rightarrow C_6H_5NO_2 + H^+$$

The nitronium ion (NO_2^+) is formed by mixing concentrated nitric and sulfuric acids:

$$HNO_3 + H_2SO_4 \rightarrow H_2NO_3^+ + HSO_4^-$$
$$H_2NO_3^+ \rightarrow NO_2^+ + H_2O$$

The accepted mechanism for a simple electrophilic substitution on benzene involves an intermediate of the form $C_6H_5HNO_2^+$. *See also* substitution reaction.

electrophoresis The use of an electric field (produced between two electrodes) to cause charged particles of a colloid to move through a solution. The technique is used to separate and identify colloidal substances such as carbohydrates, proteins, and nucleic acids. Various experimental arrangements are used. One simple technique uses a strip of adsorbent paper soaked in a buffer solution with electrodes placed at two points on the paper. This technique is sometimes called *electrochromatography*. In *gel electrophoresis*, used to separate DNA fragments, the medium is a layer of gel.

electropositive Describing an atom or molecule that tends to lose electrons, forming positive ions. Examples of electropositive elements include the alkali metals (lithium, sodium, etc.), which readily form positive ions (Li^+, Na^+, etc.).

electrovalent bond (ionic bond) A binding force between the ions in compounds in which the ions are formed by complete transfer of electrons from one element to another element or radical. For example, Na + Cl becomes $Na^+ + Cl^-$. The electrovalent bond arises from the excess of the net attractive force between the ions of opposite charge over the net repulsive force between ions of like charge. The magnitude of electrovalent interactions is of the order 10^2–10^3 kJ mol^{-1} and electrovalent compounds are generally solids with rigid lattices of closely packed ions.

element A substance that cannot be chemically decomposed into more simple substances. The atoms of an element all have the same proton number (and thus the same number of electrons, which determines the chemical activity).

At present there are 114 reported chemical elements, although research is continuing all the time to synthesize new ones. The elements from hydrogen (p.n. 1) to uranium (92) all occur naturally, with the exception of technetium (43), which is produced artificially by particle bombardment. Technetium and elements with proton numbers higher than 84 (polonium) are radioactive. Radioactive isotopes also exist for other elements, either naturally in small amounts or synthesized by particle bombardment. The elements with proton number higher than 92 are the *transuranic elements*. Neptunium (93) and plutonium (94) both occur naturally in small quantities in uranium ores, but the transuranics are all synthesized. Thus, neptunium and plutonium are made by neutron bombardment of uranium nuclei. Other transuranics are made by high-energy collision processes between nuclei. The higher proton number elements have been detected only in very small quantities – in some cases, only a few atoms have been produced.

elevation of boiling point A colligative property of solutions in which the boiling point of a solution is raised relative to that of the pure solvent. The elevation is directly proportional to the number of solute molecules introduced rather than to any specific aspect of the solute composition. The proportionality constant, k_B, is called the boiling-point elevation constant or sometimes the *ebulioscopic constant*. The relationship is

$$\Delta t = k_B C_M$$

where Δt is the rise in boiling point and C_M is the molal concentration; the units of k_B are kelvins kilograms moles^{-1} (K kg mol^{-1}). The property permits the measurement of relative molecular mass of involatile solutes. An accurately weighed amount of pure solvent is boiled until the temperature is steady, a known weight of solute of known molecular mass is quickly introduced, the boiling continued, and the elevation measured using a Beckmann thermometer. The process is repeated several times and the average value of k_B obtained by plotting Δt against C_M. The whole process is then repeated with the unknown material and its relative molecular mass obtained using the value of k_B previously obtained.

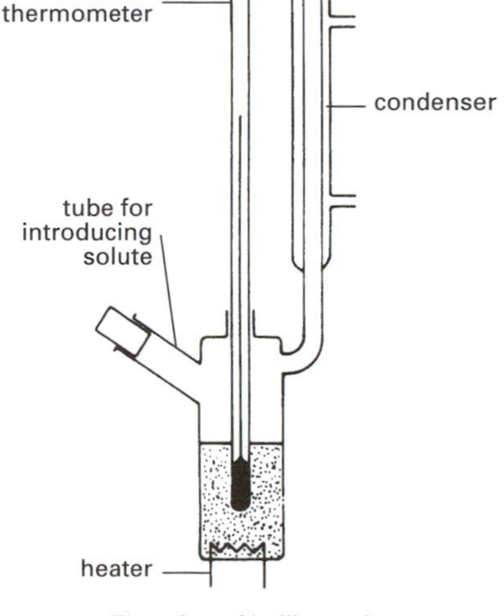

Elevation of boiling point

There are several disadvantages with this method and it is therefore used largely for demonstration purposes. The main problem is that the exact amount of solvent remaining in the liquid phase is unknown and varies with the rate of boiling. The theoretical explanation of the effect is identical to that for the lowering of vapor pressure; the boiling points are those temperatures at which the vapor pressure equals the atmospheric pressure.

elimination reaction A reaction involving the removal of a small molecule, e.g. water or hydrogen chloride, from an organic molecule to give an unsaturated compound. An example is the elimination of a water molecule from an alcohol to produce an alkene.

An elimination reaction is often in competition with a substitution reaction and the predominant product will depend on the reaction conditions. The reaction of bromoethane with sodium hydroxide could yield either ethene (by elimination of HBr) or ethanol (by substitution of the Br with OH). The former product predominates if the reaction is carried out in an alcoholic solution and the latter if the solution is aqueous.

eluate *See* elution.

eluent *See* elution.

elution The removal of an adsorbed substance in a CHROMATOGRAPHY column or ion-exchange column using a solvent (*eluent*), giving a solution called the *eluate*. The chromatography column can selectively adsorb one or more components from the mixture. To ensure efficient recovery of these components graded elution is used. The eluent is changed in a regular manner starting with a nonpolar solvent and gradually replacing it by a more polar one. This will wash the strongly polar components from the column.

Embden–Meyerhoff pathway *See* glycolysis.

emission spectrum *See* spectrum.

empirical formula *See* formula.

emulsion A colloid in which a liquid phase (small droplets with a diameter range 10^{-5}–10^{-7} cm) is dispersed or suspended in a liquid medium. Emulsions are classed as lyophobic (solvent-repelling and generally unstable) or lyophilic (solvent-attracting and generally stable).

enantiomer (enantiomorph) A compound whose structure is not superimposable on its mirror image; one of any pair of optical isomers. *See also* isomerism; optical activity.

enantiomorph *See* enantiomer.

encephalin *See* endorphin.

enclosure compound *See* clathrate.

endorphin (encephalin; enkephalin) One of a group of peptides produced in the brain and other tissues that are released after injury and have pain-relieving effects similar to those of opiate alkaloids, such as morphine. They include the *enkephalins*, which consist of just five amino acids. Other larger endorphins occur in the pituitary, while some are polypeptides, found mainly in pancreas, adrenal gland, and other tissues.

endothermic Describing a process in which heat is absorbed (i.e. heat flows from outside the system, or the temperature falls). The dissolving of a salt in water, for instance, is often an endothermic process. *Compare* exothermic.

end point *See* equivalence point; volumetric analysis.

energy Symbol: W A property of a system; a measure of its capacity to do work. Energy and work have the same unit: the joule (J). It is convenient to divide energy into *kinetic energy* (energy of motion) and *potential energy* ('stored' energy). Names are given to many different forms of energy (chemical, electrical, nuclear, etc.); the only real difference lies in the system under discussion. For example, chemical energy is the kinetic and potential energies of electrons in a chemical compound.

energy level One of the discrete energies that an atom, molecule, ion, etc., can have according to quantum theory. Thus in an atom there are certain definite orbits that the electrons can be in, corresponding to definite *electronic energy levels* of the atom. Similarly, a vibrating or rotating molecule can have discrete *vibrational* and *rotational energy levels*.

energy profile A diagram that traces the changes in the energy of a system during the course of a reaction. Energy profiles are obtained by plotting the potential energy of the reacting particles against the reaction coordinate. To obtain the reaction coordinate the energy of the total interacting system is plotted against position for the molecules. The *reaction coordinate* is the pathway for which the energy is a minimum.

enkephalin *See* endorphin.

enol An organic compound containing the C:CH(OH) group; i.e. one in which a hydroxyl group is attached to one of the carbon atoms of a double bond between two carbon atoms. *See* keto–enol tautomerism.

enthalpy Symbol: H The sum of the internal energy (U) and the product of pressure (p) and volume (V) of a system:

$$H = U + pV$$

In a chemical reaction carried out at constant pressure, the change in enthalpy measured is the internal energy change plus the work done by the volume change:

$$\Delta H = \Delta U + p\Delta V$$

entropy Symbol: S In any system that undergoes a reversible change, the change of entropy is defined as the heat absorbed divided by the thermodynamic temperature:

$$\delta S = \delta Q/T$$

A given system is said to have a certain entropy, although absolute entropies are seldom used: it is change in entropy that is important. The entropy of a system measures the availability of energy to do work.

In any real (irreversible) change in a closed system the entropy always increases. Although the total energy of the system has not changed (first law of thermodynamics) the available energy is less – a consequence of the second law of thermodynamics.

The concept of entropy has been widened to take in the general idea of disorder – the higher the entropy, the more disordered the system. For instance, a chemical reaction involving polymerization may well have a decrease in entropy because there is a change to a more ordered system. The 'thermal' definition of entropy is a special case of this idea of disorder – here the entropy measures how the energy transferred is distributed among the particles of matter.

enzyme A macromolecule that catalyzes biochemical reactions. Enzymes act with a given compound (the substrate) to produce a complex, which then forms the products of the reaction. The enzyme itself is unchanged in the reaction; its presence allows the reaction to take place. The names of most enzymes end in -ase, added to the substrate (e.g. lactase) or the reaction (e.g. hydrogenase).

Enzymes are extremely efficient catalysts for chemical reactions, and very specific to particular reactions. Most enzymes are proteins. They may have a nonprotein part (cofactor), which may be an inorganic ion or an organic constituent (coenzyme). The mechanism of action of most enzymes appears to be by *active sites* on the enzyme molecule. The substrate acting with the enzyme changes shape to fit the active site, and the reaction proceeds. Enzymes are very sensitive to their environment – e.g. temperature, pH, and the presence of other substances. Catalytic activity has also been found in some RNA molecules.

enzyme technology (enzyme engineering) A branch of biotechnology that utilizes enzymes for industrial purposes. For example rennet (impure rennin) is manufactured on a large scale to make cheese and junkets. Enzymes are also used to determine the concentration of reactants or products in specific reactions catalyzed by them.

epimerism A form of isomerism exhibited by carbohydrates in which the isomers (*epimers*) differ in the positions of –OH groups. The α- and β- forms of glucose are epimers. *See* sugar.

epinephrine (adrenaline) A hormone produced by the adrenal glands. The middle part of these glands, the adrenal medulla, secretes the hormone, which is chemically almost identical to the transmitter substance norepinephrine produced at the ends of sympathetic nerves. Epinephrine secretion into the bloodstream in stress causes acceleration of the heart, constriction of arterioles, and dilation of the pupils. In addition, epinephrine produces a marked increase in metabolic rate thus preparing the body for emergency.

epoxide A type of organic compound containing a three-membered ring containing two carbon atoms and one oxygen atom.

epoxyethane (ethylene oxide; C_2H_4O) A colorless gaseous cyclic ether. Epoxyethane is the simplest EPOXIDE. It is made by oxidation of ethene over a silver catalyst. The ring is strained and the compound is consequently highly reactive. It polymerizes to produce epoxy polymers (resins). The compound hydrolyzes to give 1,2-ethanediol ($CH_2(OH)CH_2(OH)$).

epoxy resin *See* epoxyethane.

equation *See* chemical equation.

equation of state An equation that interrelates the pressure, temperature, and volume of a system, such as a gas. The ideal gas equation (*see* gas laws) and the van der Waals equation are examples of equations of state.

equatorial conformation *See* cyclohexane.

equilibrium In a reversible chemical reaction:

$$A + B \rightleftharpoons C + D$$

The reactants are forming the products:

$$A + B \rightarrow C + D$$

which also react to give the original reactants:

$$C + D \rightarrow A + B$$

The concentrations of A, B, C, and D change with time until a state is reached at which both reactions are taking place at the same rate. The concentrations (or pressures) of the components are then constant – the system is said to be in a state of chemical equilibrium. Note that the equilibrium is a dynamic one; the reactions still take place but at equal rates. The relative proportions of the components determine the 'position' of the equilibrium, which may be *displaced* by changing the conditions (e.g. temperature or pressure).

equilibrium constant In a chemical equilibrium of the type

$$xA + yB \rightleftharpoons zC + wD$$

The expression:

$$[A]^x[B]^y/[C]^z[D]^w$$

where the square brackets indicate concentrations, is a constant (K_c) when the system is at equilibrium. K_c is the equilibrium constant of the given reaction; its units depend on the stoichiometry of the reaction. For gas reactions, pressures are often used instead of concentration. The equilibrium constant is then K_p, where $K_p = K_c^n$. Here n is the number of moles of product minus the number of moles of reactant; for instance, in

$$3H_2 + N_2 \rightleftharpoons 2NH_3$$

n is $2 - (1 + 3) = -2$.

equivalence point The point in a TITRATION at which the reactants have been added in equivalent proportions, so that there is no excess of either. It differs slightly from the *end point*, which is the observed point of complete reaction, because of the effect of the indicator, errors, etc.

equivalent proportions, law of (law of reciprocal proportions) The principle that when two chemical elements both form compounds with a third element, a compound of the first two elements contains them in the relative proportions that they have in compounds with the third element. For example, the mass ratio of carbon to hydrogen in methane (CH_4) is 12:4; the ratio of oxygen to hydrogen in water (H_2O) is 16:2. In carbon monoxide (CO), the ratio of carbon to oxygen is 12:16.

equivalent weight A measure of 'combining power' formerly used in calculations for chemical reactions. The equivalent weight of an element is the number of grams that could combine with or displace one gram of hydrogen (or 8 grams of oxygen or 35.5 grams of chlorine). It is the relative atomic mass (atomic weight) divided by the valence. For a compound the equivalent weight depends on the reaction considered. An acid, for instance, in acid–base reactions has an equivalent weight equal to its relative molecular mass (molecular weight) divided by the number of acidic hydrogen atoms.

ergosterol A sterol present in plants. It is converted, in animals, to vitamin D_2 by ultraviolet radiation, and is the most important of vitamin D's provitamins.

Erlenmeyer flask A glass laboratory flask with conical shape and a narrow neck. It is named for the German chemist Richard Erlenmeyer (1825–1909).

essential amino acid *See* amino acid.

essential fatty acid A polyunsaturated fatty acid (*see* carboxylic acid) required for growth and health that cannot be synthesized by the body and therefore must be included in the diet. Linoleic acid and (9,12,15)-linolenic acid are the only essential fatty acids in humans, being required for cell membrane synthesis and fat metabolism. Arachidonic acid is essential in some animals, such as the cat, but in humans it is synthesized from linoleic acid. Essential fatty acids occur mainly in vegetable-seed oils, e.g. safflower-seed and linseed oils.

essential oil Any pleasant-smelling volatile oil obtained from various plants, widely used in making flavorings and perfumes. Most consist of terpenes and they are obtained by steam distillation or solvent extraction.

ester A type of organic compound formed, or regarded as formed, by reaction between an alcohol and an acid. If the acid is a carboxylic acid, esters have the general formula R^1COOR^2, where R^1 and R^2 are alkyl or aryl groups. For example, ethanol (C_2H_5OH) reacts with ethanoic acid (acetic acid; CH_3COOH) to give the ester ethyl ethanoate ($C_2H_5OCOCH_3$) along with water:

$$C_2H_5OH + CH_3COOH \rightleftharpoons C_2H_5OCOCH_3 + H_2O.$$

Methanol reacts with propanoic acid to give methyl propanoate:

$$CH_3OH + C_2H_5COOH \rightleftharpoons CH_3OCOC_2H_5 + H_2O$$

This type of reaction, called *esterification*, is reversible and, in preparing esters, the equilibrium can be displaced toward the ester by using a large excess of alcohol or acid. It can also be displaced by distilling off the water or by removing it with a dehydrating agent (e.g. sulfuric acid). Esters can also be made from alcohols with ACYL HALIDES or ACID ANHYDRIDES.

The reverse reaction of esterification is hydrolysis. Both esterification and ester hydrolysis are acid-catalyzed. The mechanism involves protonation of the oxygen of the carbonyl group, allowing nucleophilic attack by water or alcohol at the carbon atom of the carbonyl group. Ester formation cannot be base-catalyzed but the hydrolysis can be catalyzed by OH^- ions, which attack the carbon atom of the carbonyl group. This type of hydrolysis is known as *saponification* (because it is the reaction used to make SOAP from fats and oils).

Simple esters are volatile compounds, often with pleasant odors. They are used as flavorings. Esters of triols occur as fats and oils. *See* glyceride.

esterification *See* ester.

ethanal (acetaldehyde; CH_3CHO) A water-soluble liquid aldehyde used as a starting material in the manufacture of several other compounds. Ethanal can be prepared by the oxidation of ethanol. It is manufactured by the catalytic oxidation of ethyne with oxygen using copper(II) chloride and palladium(II) chloride as catalysts. The mixture of gases is bubbled through an aqueous solution of the catalysts; the reaction involves formation of an intermediate organometallic complex with Pd^{2+} ions. With dilute acids ethanal polymerizes to *ethanal trimer* ($C_3O_3H_3(CH_3)_3$, formerly called *paraldehyde*), which is a sleep-inducing drug. Below 0°C *ethanal tetramer* is formed ($C_4O_4H_4(CH_3)_4$, formerly called *metaldehyde*), which is used as a slug poison and a fuel in small portable stoves.

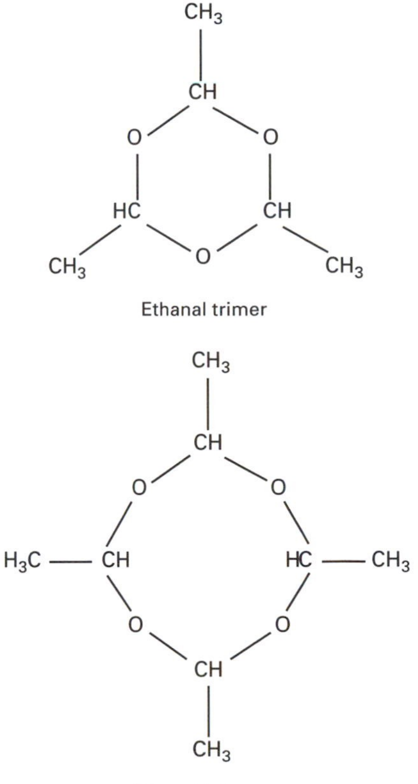

Ethanal: polymer forms of ethanal

ethanimide (acetamide; CH_3CONH_2) A colourless solid crystallizing in the form of long white crystals with a characteristic smell of mice. It is made by the dehydration of ammonium ethanoate or by the action of ammonia on ethanoyl chloride, ethanoic anhydride, or ethyl ethanoate.

ethane (C_2H_6) A gaseous alkane obtained either from the gaseous fraction of crude oil or by the 'cracking' of heavier fractions. Ethane is the second member of the homologous series of alkanes.

ethanedioic acid (oxalic acid; $(COOH)_2$) A white crystalline organic acid that occurs naturally in rhubarb, sorrel, and other plants of the genus *Oxalis*. It is slightly soluble in water, highly toxic, and used in dyeing and as a chemical reagent.

ethane-1,2-diol (ethylene glycol; glycol; $CH_2(OH)CH_2(OH)$) A syrupy organic liquid commonly used as antifreeze and as a starting material in the manufacture of Dacron. The compound is manufactured from ethene by oxidation over suitable catalysts to form epoxyethane, with subsequent hydrolysis to the diol.

ethanoate (acetate) A salt or ester of ethanoic acid (acetic acid). *See* ethanoic acid.

ethanoic acid (acetic acid; CH_3COOH) A colorless viscous liquid organic acid with a pungent odor (it is the acid in vinegar). Below 16.7°C it solidifies to a glassy solid (*glacial ethanoic acid*). It is made by the oxidation of ethanol or butane, or by the continued fermentation of beer or wine. It is made into ethenyl ethanoate (vinyl acetate) for making polymers. Cellulose ethanoate (acetate) is made from ethanoic anhydride. *See* cellulose acetate.

ethanol (ethyl alcohol; alcohol; C_2H_5OH) A colorless volatile liquid alcohol. Ethanol occurs in intoxicating drinks, in which it is produced by fermentation of a sugar:

$$C_6H_{12}O_6 \rightarrow 2C_2H_5OH + 2CO_2$$

Yeast is used to cause the reaction. At about 15% alcohol concentration (by volume) the reaction stops because the yeast is killed. Higher concentrations of alcohol are produced by distillation.

Apart from its use in drinks, alcohol is used as a solvent and to form ethanal. Formerly, the main source was by fermentation of molasses, but now catalytic hydration of ethene is used to manufacture industrial ethanol.

ethanoyl chloride (acetyl chloride; CH_3COCl) A liquid acyl chloride used as an acetylating agent.

ethanoyl group (acetyl group) The

ethene (ethylene; C_2H_4) A gaseous alkene. Ethene is not normally present in the gaseous fraction of crude oil but can be obtained from heavier fractions by catalytic cracking. This is the principal industrial source. The compound is important as a starting material in the organic-chemicals industry (e.g. in the manufacture of ethanol) and as the starting material for the production of polyethene. Ethene is the first member of the homologous series of alkenes.

ether A type of organic compound containing the group –O–. Simple ethers have the formula R^1–O–R^2, where R^1 and R^2 are alkyl or aryl groups, which may or may not be the same. They are either gases or very volatile liquids and are very flammable. The commonest example is ethoxyethane (diethylether; $C_2H_5OC_2H_5$) used formerly as an anesthetic. Ethers now find application as solvents. They are prepared in the laboratory by the dehydration of alcohols with concentrated sulfuric acid. An excess of alcohol is used to ensure that only one molecule of water is removed from each pair of alcohol molecules. They are generally unreactive, but the C–O bond can be cleaved by reaction with HI or PCl_5.

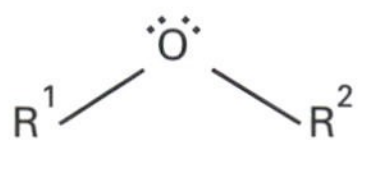

Ether

ethoxyethane (ether; diethylether; C_2H_5-OC_2H_5) A colorless volatile liquid. Ether is well known for its characteristic smell and anesthetic properties, also for its extreme flammability. It still finds some application as an anesthetic when more modern materials are unsuitable; it is also an excellent solvent. Its manufacture is an extension of the laboratory synthesis: ethanol vapor is passed into a mixture of excess ethanol and concentrated sulfuric acid at 140°C:

$$C_2H_5OH + H_2SO_4 \rightarrow C_2H_5.O.SO_2.OH + H_2O$$
$$C_2H_5O.SO_2.OH + C_2H_5OH \rightarrow C_2H_5OC_2H_5 + H_2SO_4$$

ethyl acetate *See* ethyl ethanoate.

ethyl alcohol *See* ethanol.

ethylamine (aminoethane; $C_2H_5NH_2$) A colorless liquid amine. It can be prepared from chloroethane heated with concentrated aqueous ammonia:

$$C_2H_5Cl + NH_3 \rightarrow C_2H_5NH_2 + HCl$$

It is used in manufacturing certain dyes.

ethyl bromide *See* bromoethane.

ethyl carbamate *See* urethane.

ethyl chloride *See* chloroethane.

ethylene *See* ethene.

ethylenediamine tetraacetic acid *See* edta.

ethylene glycol *See* ethane-1,2-diol.

ethylene oxide *See* epoxyethane.

ethyl ethanoate (ethyl acetate; C_2H_5-$OOCCH_3$) An ester formed from ethanol and ethanoic acid. It is a fragrant liquid used as a solvent for plastics and in flavoring and perfumery.

ethyl iodide *See* iodoethane.

ethyne (acetylene; C_2H_2) A gaseous alkyne. Traditionally ethyne has found use in oxy-acetylene welding torches, since its combustion with oxygen produces a flame of very high temperature. It is also important in the organic chemicals industry for the production of chloroethene (vinyl chloride), which is the starting material for the production of polyvinyl chloride (PVC), and for the production of other vinyl compounds. Formerly, ethyne was manufactured by the synthesis and subsequent hydrolysis of calcium dicarbide. Modern methods increasingly employ the cracking of alkanes.

ethynide *See* carbide.

eudiometer An apparatus for the volumetric analysis of gases.

evaporation 1. A change of state from liquid to gas (or vapor). Evaporation can take place at any temperature, the rate increasing with temperature. Some molecules in the liquid have enough energy to escape into the gas phase (if they are near the surface and moving in the right direction). Because these are the molecules with higher kinetic energies, evaporation results in a cooling of the liquid.
2. A change from solid to vapor, especially occurring at high temperatures close to the melting point of the solid. Thin films of metal can be evaporated onto a surface in this way.

exa- Symbol: E A prefix denoting 10^{18}.

excitation The process of producing an excited state of an atom, molecule, etc.

excitation energy The energy required to change an atom, molecule, etc. from one quantum state to a state with a higher energy. The excitation energy (sometimes called *excitation potential*) is the difference between two energy levels of the system.

excited state A state of an atom, molecule, or other system, with an energy greater than that of the ground state. *Compare* ground state.

exclusion principle The principle, enunciated by the Austrian–Swiss physicist Wolfgang Pauli (1900–58) in 1925, that no two electrons in an atom can have an identical set of quantum numbers.

exothermic Denoting a chemical reaction in which heat is evolved (i.e. heat flows from the system or the temperature rises). Combustion is an example of an exothermic process. *Compare* endothermic.

explosive A substance or mixture that can rapidly decompose upon detonation producing large amounts of heat and gases. The three most important classes of explosives are:

1. *Propellants* which burn steadily, and are used as rocket fuels.
2. *Initiators* which are very sensitive and are used in small amounts to detonate less sensitive explosives.
3. *High explosives* which need an initiator, but are very powerful.

***E–Z* convention** A convention for the description of a molecule showing cis-trans ISOMERISM. In a molecule ABC=CDE, where A, B, D, and E are different groups, the sequence rule (*see* CIP system) is applied to the pair A and B to find which has priority and it is similarly applied to the pair C and D. If the two groups of highest priority are on the same side of the bond then the isomer is designated *Z* (from German *zusammen*, together). If they are on opposite sides the isomer is designated *E* (German *entgegen*, opposite).

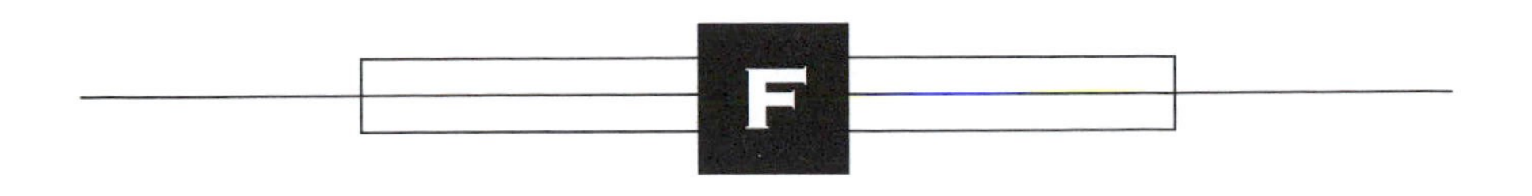

FAD (flavin adenine dinucleotide) A derivative of riboflavin that is a coenzyme in electron-transfer reactions. Its reduced form is written as $FADH_2$. *See also* flavoprotein.

Fahrenheit scale A temperature scale in which the ice temperature is taken as 32° and the steam temperature is taken as 212° (both at standard pressure). The scale is not used for scientific purposes. To convert between degrees Fahrenheit (F) and degrees Celsius (C) the formula $C/5 = (F - 32)/9$ is used. It is named for the German physicist Gabriel Daniel Fahrenheit (1686–1736) who proposed a scale of this type in 1714.

faraday Symbol: *F* A unit of electric charge equal to the charge required to discharge one mole of a singly-charged ion. One faraday is $9.648\ 670 \times 10^4$ coulombs. The unit is named for the British chemist and physicist Michael Faraday (1791–1867).

fat *See* glyceride.

fatty acid *See* carboxylic acid.

Fehling's solution A solution used to test for the aldehyde group (–CHO). It is a freshly made mixture of copper(II) sulfate solution with alkaline potassium sodium 2,3-dihydroxybutanedioate (tartrate). The aldehyde, when heated with the mixture, is oxidized to a carboxylic acid, and a red precipitate of copper(I) oxide and copper metal is produced. The tartrate is present to complex with the original copper(II) ions to prevent precipitation of copper(II) hydroxide. It is named for the German chemist H. C. von Fehling (1812–85).

femto- Symbol: f A prefix denoting 10^{-15}. For example, 1 femtometer (fm) = 10^{-15} meter (m).

fermentation A chemical reaction produced by microorganisms (molds, bacteria, or yeasts). A common example is the formation of ethanol from sugars:

$$C_6H_{12}O_6 \rightarrow 2C_2H_5OH + 2CO_2$$

ferredoxins A group of red-brown proteins found in green plants, many bacteria and certain animal tissues. They contain nonheme iron in association with sulfur at the active site. They are strong reducing agents (very negative redox potentials) and function as electron carriers, for example in photosynthesis and nitrogen fixation. They have also been isolated from mitochondria.

ferrocene ($Fe(C_5H_5)_2$) An orange crystalline solid. It is an example of a sandwich compound, in which an iron(II) ion is coordinated to two cyclopentadienyl ions. The bonding involves overlap of d orbitals on the iron with the pi electrons in the cyclopentadienyl ring. The compound can undergo substitution reactions on the rings, which have aromatic character. The systematic name is di-π-cyclopentadienyl iron(II).

filler A solid material used to modify the physical properties or reduce the cost of synthetic compounds, such as rubbers, plastics, paints, and resins. Slate powder, glass fiber, mica, and cotton are all used as fillers.

filter *See* filtration.

filter pump A type of vacuum pump in which a jet of water forced through a nozzle carries air molecules out of the system. Filter pumps cannot produce pressures below the vapor pressure of water. They are used in the laboratory for vacuum filtration, distillation, and similar techniques requiring a low-grade vacuum.

filtrate *See* filtration.

filtration The process of removing suspended particles from a fluid by passing or forcing the fluid through a porous material (the *filter*). The fluid that passes through the filter is the *filtrate*. In laboratory filtration, filter paper or sintered glass is commonly used.

fine organic chemicals Carbon compounds, such as dyes and drugs, that are produced only in small quantities. Their main requirement is that they must have a high degree of purity, often higher than 95%. They are manufactured for special purposes, e.g. for use in spectroscopy, pharmacology, and electronics.

fine structure Closely spaced lines seen at high resolution in a spectral line or band. Fine structure may be caused by vibration of the molecules or by electron spin. *Hyperfine structure*, seen at very high resolution, is caused by the atomic nucleus affecting the possible energy levels of the atom.

firedamp Methane present in coal mines.

first-order reaction A reaction in which the rate of reaction is proportional to the concentration of one of the reacting substances. The concentration of the reacting substance is raised to the power one; i.e. rate = k[A]. For example, the decomposition of hydrogen peroxide is a first-order reaction:

$$\text{rate} = k[H_2O_2]$$

Similarly the rate of decay of radioactive material is a first-order reaction:

$$\text{rate} = k[\text{radioactive material}]$$

For a first-order reaction, the time for a definite fraction of the reactant to be consumed is independent of the original concentration. The units of k, the RATE CONSTANT, are s^{-1}.

Fischer, Emil Hermann (1852–1919) German organic chemist. Fischer studied many compounds of biological interest. He is sometimes referred to as the father of biochemistry. In 1874 he discovered phenylhydrazine. He studied peptides, purines and sugars very thoroughly. His work on purines (a name he coined) led to the synthesis of many compounds such as caffeine and purine. In his early work he put forward incorrect structures but by 1897 he and his colleagues had established the correct structures. Fischer was awarded the 1902 Nobel Prize for chemistry for his work on purines and sugars.

Fischer, Hans (1881–1945) German organic chemist. Fischer devoted his career to the study of the molecular structures of the biologically significant molecules hemoglobin, chlorophyll and the bile pigment bilirubin. Fischer started investigating hemoglobin in 1921. He showed that the iron-containing nonprotein part consists of four pyrrole rings surrounding an iron atom. He synthesized this part by 1929 and thoroughly investigated the porphyrins. He won the 1930 Nobel prize for chemistry for this work. He then investigated the chlorophylls and demonstrated that they are substituted porphins surrounding a magnesium atom. He also demonstrated that bile acids are degraded porphins. In 1944 he synthesized bilirubin completely.

Fischer projection A way of representing the three-dimensional structure of a molecule in two dimensions. The molecule is drawn using vertical and horizontal lines. Horizontal lines represent bonds that come out of the paper. Vertical lines represent bonds that go into the paper (or are in the plane of the paper). Named for Emil Fischer, the convention was formerly used for representing the absolute configuration of SUGARS.

Fischer–Tropsch process A method of making a mixture of hydrocarbons using hydrogen and carbon monoxide (2:1 ratio) passed over a nickel or cobalt catalyst at a temperature of 200°C. The mixture, which also contains alcohols and carbonyl compounds, can be distilled to make fuels for diesel and gasoline engines. It was used for this by Germany in World War II. Now the process is one way of making SNG. It is named for the German chemist Franz Fischer (1852–1932) and the Czech chemist Hans Tropsch (1839–1935), who invented it in 1933.

Fittig reaction *See* Wurtz reaction.

flame-ionization detector *See* gas chromatography.

flare stack A chimney at the top of which unwanted gases are burnt in an oil refinery or other chemical plant.

flash photolysis A technique for investigating free radicals in gases. The gas is held at low pressure in a long glass or quartz tube, and an absorption spectrum taken using a beam of light passing down the tube. The gas is subjected to a very brief intense flash of light from a lamp outside the tube, producing free radicals, which are identified by their spectra. Measurements of the intensity of spectral lines can be made with time using an oscilloscope, and the kinetics of very fast reactions can thus be investigated.

flash point The lowest temperature at which sufficient vapor is given off by a flammable liquid to ignite in the presence of a spark. *See also* ignition temperature.

flavanone A type of flavonoid. Flavanone glycosides are found in flowering plants.

flavin A derivative of riboflavin occurring in the flavoproteins; i.e. FAD or FMN.

flavin adenine dinucleotide *See* FAD.

flavin mononucleotide *See* FMN.

flavone *See* flavonoid.

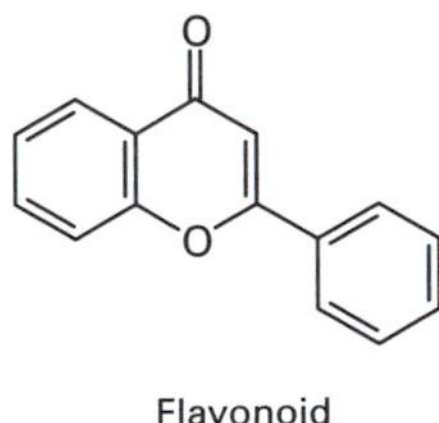

Flavonoid

flavonoid One of a common group of plant compounds having the C_6–C_3–C_6 chemical skeleton in which C_6 is a benzene ring. They are an important source of nonphotosynthetic pigments in plants. They are classified according to the C_3 portion and include the yellow chalcones and aurones; the pale yellow and ivory flavones and flavonols and their glycosides; the red, blue, and purple anthocyanins and anthocyanidins; and the colorless isoflavones, catechins, and leukoanthocyanidins. They are water soluble and usually located in the cell vacuole. *See* anthocyanin.

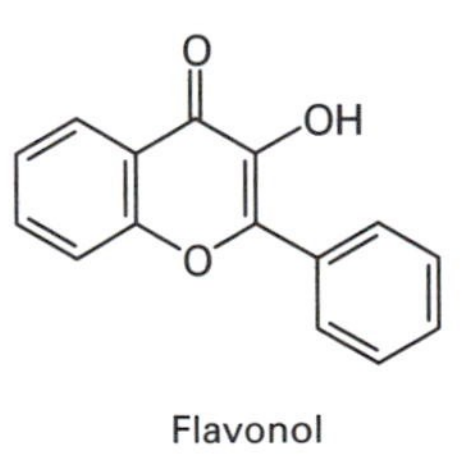

Flavonol

flavonol A plant pigment that modifies the effects of certain growth substances. *See* flavonoid.

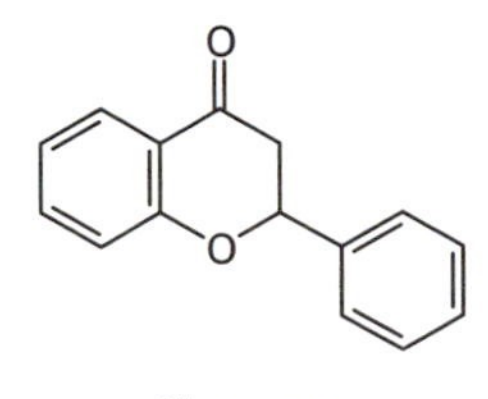

Flavanone

flavoprotein A conjugated protein in which a flavin (FAD or FMN) is joined to a protein component. Flavoproteins are enzymes in the electron-transport chain.

flocculation (coagulation) The combining of the particles of a finely divided precipitate, such as a colloid, into larger particles or clumps that sink and are easier to filter off.

flocculent Describing a precipitate that has aggregated in wooly masses.

Flory, Paul John (1910–85) American polymer chemist. Flory's early work consisted of helping Wallace CAROTHERS to develop nylon and neoprene. He began to investigate the properties of polymers in the 1930s. Flory solved the difficulty that a polymer molecule does not have a fixed size and structure by using statistical techniques to calculate a distribution of polymer chain lengths. Flory also worked on polymers in which there are links between chains. This led to work on the elasticity of rubber. Flory summarized his work in the classic books *Principles of Polymer Chemistry* (1953) and *Statistical Mechanics of Chain Molecules* (1969). Flory won the 1974 Nobel Prize for chemistry for his work on polymers.

fluid A state of matter that is not a solid – that is, a liquid or a gas. All fluids can flow, and the resistance to flow is the viscosity.

fluidization The suspension of a finely-divided solid in an upward-flowing liquid or gas. This suspension mimics many properties of liquids, such as allowing objects to 'float' in it. Fluidized beds so constructed are important in the chemical industry.

fluorescein A fluorescent dye used as an absorption indicator. It has a yellow solution with green fluorescence.

fluorescence The absorption of energy by atoms, molecules, etc., followed by immediate emission of electromagnetic radiation as the particles make transitions to lower energy states. *Compare* phosphorescence.

fluoridation The introduction of small quantities of fluoride compounds into the water supply as a public-health measure to reduce the incidence of tooth decay.

fluoride *See* halide.

fluorination *See* halogenation.

fluorine A slightly greenish-yellow highly reactive gaseous element belonging to the halogens (group 17 of the periodic table, formerly VIIA). It occurs notably as fluorite (CaF_2) and cryolite (Na_3AlF_3) but traces are also widely distributed with other minerals. It is slightly more abundant than chlorine, accounting for about 0.065% of the Earth's crust. The high reactivity of the element delayed its isolation. Fluorine is now prepared by electrolysis of molten KF/HF electrolytes, using copper or steel apparatus. Its preparation by conventional chemical methods is impossible.

Fluorine is strongly electronegative and exhibits a strong electron withdrawing effect on adjacent bonds, thus CF_3COOH is a strong acid (whereas CH_3COOH is not). Fluorine and hydrogen fluoride are extremely dangerous and should only be used in purpose-built apparatus; gloves and face shields should be used when working with hydrofluoric acid and accidental exposure should be treated as a hospital emergency.

Symbol: F; b.p. –188.14°C; m.p. –219.62°C; d. 1.696 kg m^{-3} (0°C); p.n. 9; r.a.m. 18.99840.32.

fluorocarbon A compound derived from a hydrocarbon by replacing hydrogen atoms with fluorine atoms. Fluorocarbons are unreactive and most are stable up to high temperatures. They have a variety of uses – in aerosol propellants, oils and greases, and synthetic polymers such as PTFE. *See also* halocarbon.

fluxional molecule A molecule in which the constituent atoms change their relative positions so quickly at room temperature that the normal concept of structure is inadequate; i.e. no specific structure exists for longer than about 10^{-2} second and the relative positions become indistinguishable. For example ClF_3 at –60°C has a distinct 'T' shape but at room temperature the fluorine atoms are visualized as moving rapidly over the surface of the chlorine atom in a state of exchange and are effectively identical.

FMN (flavin mononucleotide) A derivative of riboflavin that is a coenzyme in electron-transfer reactions. *See also* flavoprotein.

foam A dispersion of bubbles of gas in a liquid, usually stabilized by a SURFACTANT. Solid foams, such as expanded polystyrene or foam rubber, are made by allowing liquid foams to set.

folic acid (pteroylglutamic acid) One of the water-soluble B-group of vitamins. The principal dietary sources of folic acid are leafy vegetables, liver, and kidney. Deficiency of the vitamin exhibits itself in anemia in a similar manner to vitamin B_{12} deficiency, while deficiency during pregnancy increases the risk of birth defects in children.

Folic acid is important in metabolism in various coenzyme forms, all of which are specifically concerned with the transfer and utilization of the single carbon (C_1) group. Before functioning in this manner folic acid must be reduced to either dihydrofolic acid (FH_2) or tetrahydrofolic acid (FH_4). It is important in the growth and reproduction of cells, participating in the synthesis of purines and thymine. *See also* vitamin B complex.

formaldehyde *See* methanal.

formalin *See* methanal.

formate *See* methanoate.

formic acid *See* methanoic acid.

formula A representation of a molecule using symbols for the atoms. Subscripts indicate the numbers of atoms present. The *molecular formula* gives the numbers and types of atom present. For example, ethanoic acid (acetic acid) has the molecular formula $C_2H_4O_2$. The *empirical formula* gives the simplest ratios of atoms. Thus, the empirical formula of ethanoic acid is CH_2O. This is the formula that would be obtained by experimental determination of the amounts of each element present. The molecular formula can then

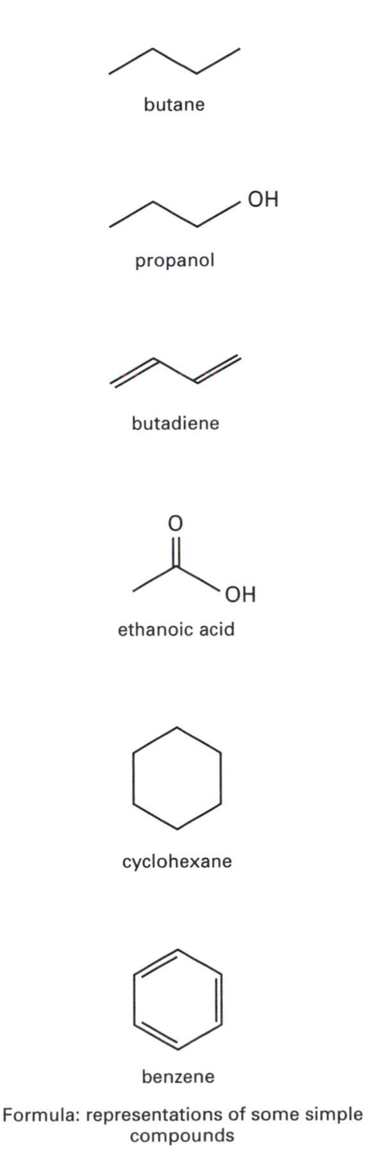

Formula: representations of some simple compounds

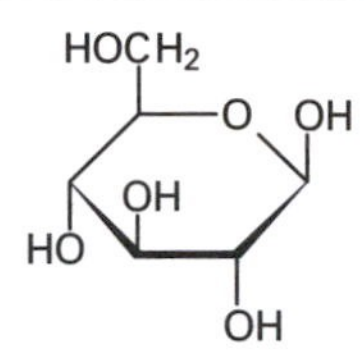

Formula: representation of the β-D-anomer of glucose

be obtained if the relative molecular mass is known.

More information is given by the *structural formula*, which shows how the atoms are joined together. The formula of ethanoic acid is usually written as CH_3-COOH, showing that it is formed from a methyl group (CH_3–) and a carboxylate group (–COOH). Sometimes full stops are used in such formulae to divide up the groups. Often it is necessary to show a ring compound or to show the disposition of the atoms or groups in space (*see* isomer; optical activity). In such cases a diagram of the structure has to be given.

In certain cases symbols are used for groups of atoms. Common ones are Me for methyl, Et for ethyl, Pr for propyl, Bu for butyl, and Ph for phenyl. These symbols are sometimes called *organic elements*. So, for example, ethanol is EtOH, phenol is PhOH, and ethanoic acid is MeCOOH.

In representing three-dimensional structures certain conventions are used. Particular types of *projection formulae* are used for certain types of compound. For example, the FISCHER PROJECTION has been extensively used for representing the open-chain form of sugars. The NEWMAN PROJECTION is used for discussions of rotation about C–C single bonds (*see also* conformation). More generally, it is conventional to use a straight line for a single bond in the plane of the paper. A bond coming out of the paper is represented as a solid narrow wedge, intended to give the impression of perspective. A bond into the paper is represented by a dotted or dashed line. Also, organic chemists commonly represent structures without the C or H atoms, except where these appear in functional groups. A hydrocarbon chain is drawn as a zig-zag line and the BENZENE ring is drawn as one of the Kekulé structures.

Often a *general formula* is used to represent a class of compounds. For instance C_nH_{2n} for alkenes. It is also common to use the symbol R for an organic group. So RCOOH is any carboxylic acid. When two different groups are needed, R and R′ are used (or R^1, R^2, R^3, etc.). Ar is sometimes used for any aryl group.

formyl group The group HCO–.

fossil fuel A mineral fuel that forms underground from the remains of living organisms. Fossil fuels include coal, natural gas, peat, and petroleum.

fraction A mixture of liquids with similar boiling points collected by fractional distillation.

fractional crystallization Crystallization of one component from a mixture in solution. When two or more substances are present in a liquid (or in solution), on cooling to a lower temperature one substance will preferentially form crystals, leaving the other substance in the liquid (or dissolved) state. Fractional crystallization can thus be used to purify or separate substances if the correct conditions are known.

fractional distillation (fractionation) A distillation carried out with partial reflux, using a long vertical column (fractionating column). It utilizes the fact that the vapor phase above a liquid mixture is generally richer in the more volatile component. If the region in which refluxing occurs is sufficiently long, fractionation permits the complete separation of two or more volatile liquids. Fractionation is the fundamental process for producing petroleum from crude oil.

Unlike normal reflux, the fractionating column may be insulated to reduce heat loss, and special designs are used to maximize the liquid-vapour interface.

fractionation *See* fractional distillation.

Frankland, Sir Edward (1825–99) British chemist. Frankland is best known for introducing the concept of what is now known as valence. In 1852 he noticed that nitrogen and phosphorus frequently form compounds in which there are either three or five atoms of the other elements. This suggested to Frankland that each element has a definite combining power which is satisfied by a certain number of atoms. He elaborated what came to be known as the theory of valence in 1866. In 1864 Frankland pointed out that the carboxyl group (CO_2H) is present in many organic acids. Frankland was also concerned with technological applications of chemistry, notably to coal gas and to water purification.

Franklin, Rosalind (1920–58) British x-ray crystallographer. Rosalind Franklin is best known for having played a key role in the discovery of the structure of DNA and for having been portrayed in an unflattering way by James WATSON in his book *The Double Helix* (1968). Her early work was on coal. Her x-ray photographs of DNA in 1952 led Francis CRICK and Watson to postulate the double helix structure of DNA. Franklin subsequently confirmed the double helix picture using x-ray crystallography. In her later years she worked on tobacco mosaic virus. Her early death prevented the possibility of her winning a Nobel Prize for her work on DNA.

free energy A measure of the ability of a system to do useful work. *See* Gibbs function; Helmholtz function.

free radical An atom or group of atoms with a single unpaired electron. Free radicals are produced by breaking a covalent bond; for example:

$$CH_3Cl \rightarrow CH_3\bullet + Cl\bullet$$

They are often formed in light-induced reactions. Most free radicals are extremely reactive and can be stabilized and isolated only under special conditions. They can be studied by electron spin resonance. *See also* carbene.

freezing The process by which a liquid is converted into a solid by cooling; the reverse of melting.

freezing mixtures Two or more substances mixed together to produce a low temperature. A mixture of sodium chloride and ice in water (–20°C) is a common example.

freezing point The temperature at which a liquid is in equilibrium with its solid phase at standard pressure and below which the liquid freezes or solidifies. This temperature is always the same for a particular liquid and is numerically equal to the melting point of the solid. *See also* depression of freezing point.

Freon (*Trademark*) Any of a number of chlorofluorocarbons (CFCs) and fluorocarbons used as refrigerants. *See* fluorocarbon; halocarbon.

Friedel–Crafts reaction A type of reaction in which an alkyl or acyl group is substituted on a benzene ring. In *Friedel–Crafts alkylation* the reactant is a haloalkane, and an alkylbenzene is produced:

$$CH_3Cl + C_6H_6 \rightarrow C_6H_5CH_3 + HCl$$

In *Friedel–Crafts acylation* the reactant is an acyl halide and the product is an aromatic ketone:

$$CH_3COCl + C_6H_6 \rightarrow C_6H_5COCH_3 + HCl$$

These reactions occur at about 100°C using aluminum chloride as a catalyst. This accepts a lone pair of electrons from the halogen atom on the haloalkane or acyl halide, which results in a positive charge on the adjoining carbon atom. The reaction is then ELECTROPHILIC SUBSTITUTION. It is also possible to use alkenes (for alkylation) and acid anhydrides (for acylation). It is named for the French chemist Charles Friedel (1832–99) and the US chemist James Crafts (1832–99).

frontier orbital Either of two orbitals in a molecule: the highest occupied molecular orbital (the *HOMO*) or the lowest unoccupied molecular orbital (the *LUMO*). The HOMO is the orbital with the highest energy level occupied at absolute zero temperature. The LUMO is the lowest-energy unoccupied orbital at absolute zero. For a

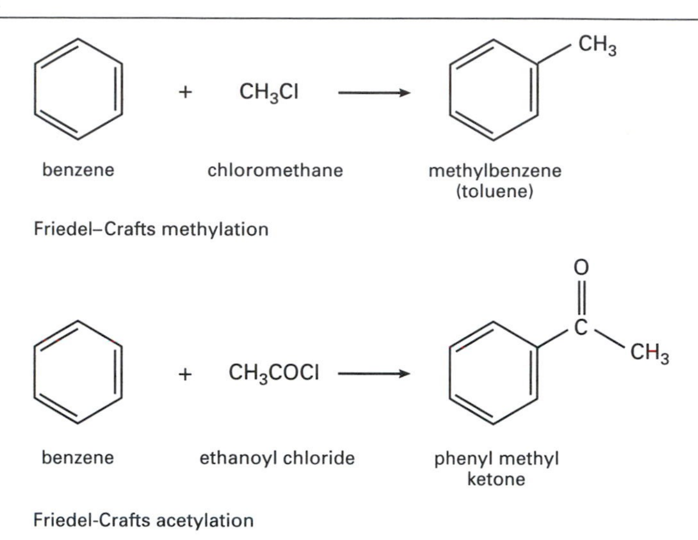

Friedel–Crafts reactions

particular molecule the nature of these two orbitals is very important in determining the chemical properties. *Frontier-orbital theory*, which considers the symmetry of these orbitals, has been very successful in explaining such reactions as the DIELS–ALDER REACTION. *See also* Woodward–Hoffmann rules.

fructan A polysaccharide made entirely of fructose residues. They are used as food stores in many plants.

fructose (fruit sugar; $C_6H_{12}O_6$) A SUGAR found in fruit juices, honey, and cane sugar. It is a ketohexose, existing in a pyranose form when free. In combination (e.g. in sucrose) it exists in the furanose form.

fruit sugar *See* fructose.

fucoxanthin A xanthophyll pigment of diatoms, brown algae, and golden brown algae. The light absorbed is used with high efficiency in photosynthesis, the energy first being transferred to chlorophyll *a*. It has three absorption peaks covering the blue and green parts of the spectrum.

fuel cell A type of cell in which fuel is converted directly into electricity. In one form, hydrogen gas and oxygen gas are fed to the surfaces of two porous nickel electrodes immersed in potassium hydroxide solution. The oxygen reacts to form hydroxyl (OH^-) ions, which it releases into the solution, leaving a positive charge on the electrode. The hydrogen reacts with the OH^- ions in the solution to form water, giving up electrons to leave a negative charge on the other electrode. Large fuel cells can generate tens of amperes. Usually the e.m.f. is about 0.9 volt and the efficiency around 60%.

Fukui, Kenichi (1918–98) Japanese physical and theoretical chemist. Fukui is best known for his work on *frontier orbital theory*, a theory which describes the changes in molecular orbitals during a chemical reaction. He was particularly interested in applying frontier orbital theory to the reactions of methyl radicals. He shared the 1981 Nobel Prize for chemistry with Roald HOFFMANN for his work on frontier orbital theory. He also studied the reaction between nitrogen molecules and transition metal complexes.

fullerene *See* buckminsterfullerene.

fullerite *See* buckminsterfullerene.

fuller's earth A natural clay used as an absorbent and industrial catalyst.

fumaric acid *See* butenedioic acid.

functional group A group of atoms in a compound that is responsible for the characteristic reactions of the type of compound. Examples are:

alcohol	–OH
alkoxide	–OR
aldehyde	–CHO
amide	–CO–NH_2
amine	–NH_2
ketone	=CO
carboxylic acid	–CO.OH
ester	–CO–OR
acyl halide	–CO.X (X = halogen)
nitro compound	–NO_2
sulfonic acid	–SO_2.OH
nitrile	–CN
diazonium salt	–N_2^+
diazo compound	–N=N–

fundamental units The units of length, mass, and time that form the basis of most systems of units. In SI, the fundamental units are the meter, the kilogram, and the second.

furan (furfuran; C_4H_4O) A heterocyclic liquid organic compound. Its five-membered ring contains four carbon atoms and one oxygen atom. The structure is characteristic of some monosaccharide sugars (furanoses).

furanose A SUGAR that has a five-membered ring (four carbon atoms and one oxygen atom).

furfuran *See* furan.

fused Describing a solid that has been melted and solidified into a single mass. Fused silica, for example, is produced by melting sand.

fused ring *See* ring.

fusel oil A mixture of high-molecular-weight alcohols together with some esters and fatty acids, formed from alcoholic fermentation and obtained during distillation. It is used as a source of higher alcohols.

fusion Melting.

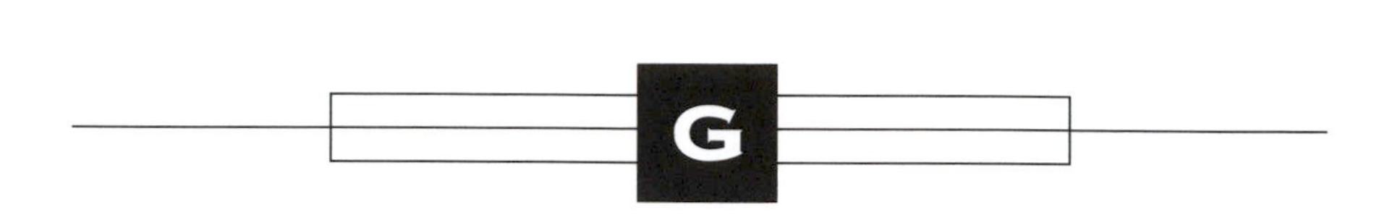

GA_3 *See* gibberellic acid

GAG *See* glycosaminoglycan.

galactose ($C_6H_{12}O_6$) A SUGAR found in lactose and many polysaccharides. It is an aldohexose, isomeric with glucose.

gas The state of matter in which forces of attraction between the particles of a substance are small. The particles have freedom of movement and gases and have no fixed shape or volume. The atoms or molecules of a gas are in a continual state of motion and are continually colliding with each other and with the walls of the containing vessel. These collisions with the walls create the pressure of a gas.

gas chromatography A type of CHROMATOGRAPHY widely used for the separation and analysis of mixtures. Gas chromatography employs a column packed with either a solid stationary phase (*gas–solid chromatography* or *GSC*) or a solid coated with a nonvolatile liquid (*gas–liquid chromatography* or *GLC*). The whole column is placed in a thermostatically controlled heating jacket. A volatile sample is introduced into the column using a syringe, and a *carrier gas*, such as hydrogen or nitrogen, is passed through it. The components of the sample will be carried along in this mobile phase. Some of the components will cling more readily to the stationary phase than others, either because they become attached to the solid surface or because they dissolve in the liquid. The time taken for different components to pass through the column is characteristic of a particular compound and can be used to identify it. The emergent sample is passed through a detector, which registers the presence of the different components in the carrier gas.

Two types of detector are in common use: the *katharometer*, which measures changes in thermal conductivity, and the *flame-ionization detector*, which turns the volatile components into ions and registers the change in electrical conductivity. Gas chromatography is also used in other techniques to identify the separated components, as in gas chromatography–mass spectroscopy (*GCMS*) and gas chromatography infrared (*GCIR*).

gas chromatography infrared *See* gas chromatography.

gas chromatography–mass spectroscopy *See* gas chromatography.

gas constant (universal gas constant) Symbol: *R* The universal constant 8.314 34 J mol^{-1} K^{-1} appearing in the equation of state for an ideal gas. *See* gas laws.

gas equation *See* gas laws.

gas laws Laws relating the temperature, pressure, and volume of a fixed mass of gas. The main gas laws are BOYLE'S LAW and CHARLES' LAW. The laws are not obeyed exactly by any real gas, but many common gases obey them under certain conditions, particularly at high temperatures and low pressures. A gas that would obey the laws over all pressures and temperatures is a *perfect* or *ideal gas*

Boyle's and Charles' laws can be combined into an equation of state for ideal gases:

$$pV_m = RT$$

where V_m is the molar volume and R the molar gas constant. For n moles of gas

$$pV = nRT$$

All real gases deviate to some extent from the gas laws, which are applicable only to idealized systems of particles of negligible volume with no intermolecular forces. There are several modified equations of state that give a better description of the behavior of real gases, the best known being the VAN DER WAALS EQUATION.

gas–liquid chromatography *See* gas chromatography.

gasohol Alcohol (ethanol) obtained by the industrial fermentation of sugar for use as a motor fuel. It has been produced on a large scale in Brazil.

gas oil One of the main fractions obtained from petroleum by distillation, used as a fuel for diesel engines. *See* diesel fuel; petroleum.

gasoline *See* petroleum.

gas–solid chromatography *See* gas chromatography.

Gatterman–Koch reaction A reaction for substituting a formyl group (HCO–) into a benzene ring of an aromatic hydrocarbon. It is used in the industrial production of benzaldehyde from benzene:

$$C_6H_6 \rightarrow C_6H_5CHO$$

The aromatic hydrocarbon is mixed with a Lewis acid, such as aluminum chloride, and a mixture of carbon monoxide and the hydrogen chloride is passed through. The first stage if the production of a $H\text{–}C\equiv O^+$ ion:

$$HCl + CO + AlCl_3 \rightarrow HCO^+ + AlCl_4^-$$

Copper(I) chloride (CuCl) is also added as a catalyst. The HCO^+ ion acts as an electrophile in electrophilic substitution on the benzene ring.

The Gatterman–Koch reaction is used for introducing the HCO– group into hydrocarbons. A variation of the reaction in which the HCO– group is substituted into the benzene ring of a phenol uses hydrogen cyanide rather than carbon monoxide. Typically, a mixture of zinc cyanide and hydrochloric acid is used, to give zinc chloride (which acts as a Lewis acid) and hydrogen cyanide. The electrophile in this case is protonated hydrogen cyanide:

$$HCN + HCl + ZnCl_2 \rightarrow HCNH^+ + ZnCl_3^-$$

The phenol is first substituted to give an IMINE:

$$C_6H_5OH \rightarrow HOC_6H_4CH{=}NH$$

This then hydrolyzes to the aromatic aldehyde:

$$HOC_6H_4CH{=}NH_2 \rightarrow HOC_6H_4CHO$$

Similar reactions using alkyl cyanides (nitriles) rather than hydrogen cyanide give aromatic ketones. This type of reaction, in which a cyanide is used to produce an aldehyde (or ketone) is often called the *Gatterman reaction.* The Gatterman–Koch reaction was reported by the German chemist Ludwig Gatterman (1860–1920) in 1897 (with J. C. Koch). The use of hydrogen cyanide was reported by Gatterman in 1907.

Gatterman reaction **1.** *See* Sandmeyer reaction.
2. *See* Gatterman–Koch reaction.

gauche conformation *See* conformation.

gauss Symbol: G The unit of magnetic flux density in the c.g.s. system. It is equal to 10^{-4} tesla.

Gay-Lussac's law **1.** The principle that gases react in volumes that are in simple ratios to each other and to the products if they are gases (all volumes measured at the same temperature and pressure). The law was first put forward in 1808 by the French chemist and physicist Joseph-Louis Gay Lussac (1778–1850).
2. *See* Charles' law.

GCIR (gas chromatography infrared) *See* gas chromatography.

GCMS (gas chromatography–mass spectroscopy) *See* gas chromatography.

gel A lyophilic colloid that is normally stable but may be induced to coagulate partially under certain conditions (e.g. lowering the temperature). This produces a pseudo-solid or easily deformable jellylike mass, called a gel, in which intertwining particles enclose the whole dispersing medium. Gels may be further subdivided into elastic gels (e.g. gelatin) and rigid gels (e.g. silica gel).

gelatin (gelatine) A pale yellow protein obtained from the bones, hides, and skins of animals. It forms a colloidal jelly when dissolved in hot water and is used in the food industry, to make capsules for various medicinal drugs, as an adhesive and sizing medium, and in photographic emulsions.

gelatine *See* gelatin.

gel electrophoresis *See* electrophoresis.

gel filtration A form of column chromatography in which a gel is used as the stationary medium. Components move through pores in the gel at a rate that depends on the size of their molecules. The technique is used to separate proteins.

gem positions Positions in a molecule on the same atom. For example, 1,1-dichloroethane (CH_3CHCl_2), in which both chlorine atoms are on the same carbon, is a gem dihalide.

general formula *See* formula.

geometrical isomerism *See* isomerism.

gibberellic acid (GA_3) A common GIBBERELLIN and one of the first to be discovered. Together with GA_1 and GA_2 it was isolated from *Gibberella fujikuroi*, a fungus that infects rice seedlings causing abnormally tall growth.

gibberellin A plant hormone involved chiefly in shoot extension. Gibberellins are diterpenoids; their molecules have the gibbane skeleton. More than thirty have been isolated, the first and one of the most common being gibberellic acid, GA_3.

Gibbs free energy *See* Gibbs function.

Gibbs function (Gibbs free energy) Symbol: *G* A thermodynamic function defined by

$$G = H - TS$$

where *H* is the enthalpy, *T* the thermodynamic temperature, and *S* the entropy. It is useful for specifying the conditions of chemical equilibrium for reactions for constant temperature and pressure (*G* is a minimum). It is named for the US mathematician and physicist Josiah Willard Gibbs (1839–1903), who first developed the theory of chemical thermodynamics. *See also* free energy.

giga- Symbol: G A prefix denoting 10^9. For example, 1 gigahertz (GHz) = 10^9 hertz (Hz).

glacial acetic acid *See* glacial ethanoic acid.

glacial ethanoic acid (glacial acetic acid) Pure water-free ethanoic acid.

GLC Gas–liquid chromatography. *See* gas chromatography.

globulin One of a group of proteins that are insoluble in water but will dissolve in neutral solutions of certain salts. They generally contain glycine and coagulate when heated. Three types of globulin are found in blood: *alpha* (α), *beta* (β), and *gamma* (γ). α and β globulins are made in the liver and are used to transport nonprotein material. γ globulins are made in reticuloendothelial tissues, lymphocytes, and plasma cells and most of them have antibody activity (*see* immunoglobulin).

glove box A sealed box with gloves fitted to ports in one side and having a transparent top, used for safety reasons or to handle materials in an inert or sterile atmosphere.

glucan *See* glycan.

gluconic acid (dextronic acid; $CH_2OH(CHOH)_4COOH$) A soluble crystalline organic acid made by the oxidation of glucose (using specific molds). It is used in paint strippers.

glucosan A POLYSACCHARIDE that is formed of glucose units. Cellulose and starch are examples.

glucose (dextrose; grape sugar; $C_6H_{12}O_6$) A monosaccharide occurring widely in nature as D-glucose. It occurs as glucose units in sucrose, starch, and cellulose. It is important to metabolism because it participates in energy-storage and energy-release systems. *See also* sugar.

glucoside *See* glycoside.

glue An adhesive, of which there are various types. Aqueous solutions of starch and ethyl cellulose are used as pastes for sticking paper; traditional wood glue is made by boiling animal bones (*see* gelatin); quick-drying adhesives are made by dissolving rubber or a synthetic polymer in a volatile solvent; and some polymers, such as epoxy resins and polyvinyl acetate (PVA), are themselves used as glues.

glutamic acid *See* amino acid.

glutamine *See* amino acid.

glutathione A tripeptide of cysteine, glutamic acid, and glycine, widely distributed in living tissues. It takes part in many oxidation–reduction reactions, due to the reactive thiol group (–SH) being easily oxidized to the disulfide (–S–S–), and acts as an antioxidant, as well as a coenzyme to several enzymes.

gluten A mixture of proteins found in wheat flour. It is composed mainly of two proteins (gliaden and glutelin), the proteins being present in almost equal quantities. Certain people are sensitive to gluten (celiac disease) and must have a gluten-free diet.

glycan A polysaccharide made of more than 10 monosaccharide residues. A *homoglycan* is made up of a single type of sugar unit (i.e. > 95%). As a class the glycans serve both as structural units (e.g. cellulose in plants and chitin in invertebrates) and energy stores (e.g. starch in plants and glycogen in animals). The most common homoglycans are made up of D-glucose units and called *glucans*.

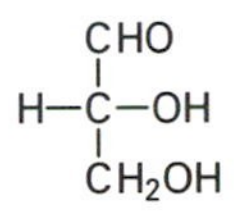

Glyceraldehyde: Fischer projection of D-glyceraldehyde

glyceraldehyde A simple triose sugar used in determining absolute configuration.

glyceride An ester formed between glycerol (propane-1,2,3-triol) and one or more carboxylic acids. Glycerol has three alcohol groups, and if all three groups have formed esters, the compound is a *triglyceride*. Naturally occurring fats and oils are triglycerides of long-chain carboxylic acids (hence the name 'fatty acid'). The main carboxylic acids forming glycerides in fats and oils are:

1. octadecanoic acid (stearic acid), a saturated acid $CH_3(CH_2)_{16}COOH$.
2. hexadecanoic acid (palmitic acid), a saturated acid $CH_3(CH_2)_{14}COOH$.
3. *cis*-9-octadecenoic acid (oleic acid), an unsaturated acid.
   $CH_3(CH_2)_7CH{:}CH(CH_2)_7COOH$

glycerin *See* propane-1,2,3-triol.

glycerine (glycerin) *See* propane-1,2,3-triol.

glycerol *See* propane-1,2,3-triol.

glyceryl trinitrate *See* nitroglycerine.

glycine *See* amino acid.

glycogen (animal starch) A polysaccharide that is the main carbohydrate store of animals. It is composed of many glucose units linked in a similar way to starch. Glycogen is readily hydrolyzed in a stepwise manner to glucose itself. It is stored largely in the liver and in muscle but is found widely distributed in the body.

glycol *See* diol.

glycolipid A sugar-containing lipid with one or more sugar residues attached to a lipid by a glycosidic link (or ester link in prokaryotes). Glycolipids play an important structural role in cell membranes, where the sugar residues are always extracellular. Animal glycolipids are derived from *sphingosine*, an amino alcohol with a long unsaturated hydrocarbon chain. In glycolipids, the amino group of sphingosine is joined to a fatty acid chain by an amide bond and the primary hydroxyl group is linked to a sugar residue. The simplest glycolipid (in animal cells) is *cerebroside*, which has one sugar residue (either glucose or galactose). Glycolipids with branched chains of sugar residues are known as gangliosides. The fatty acid chain and sphingosine chain are hydrophobic while the sugar residues are hydrophilic, making glycolipids amphiprotic.

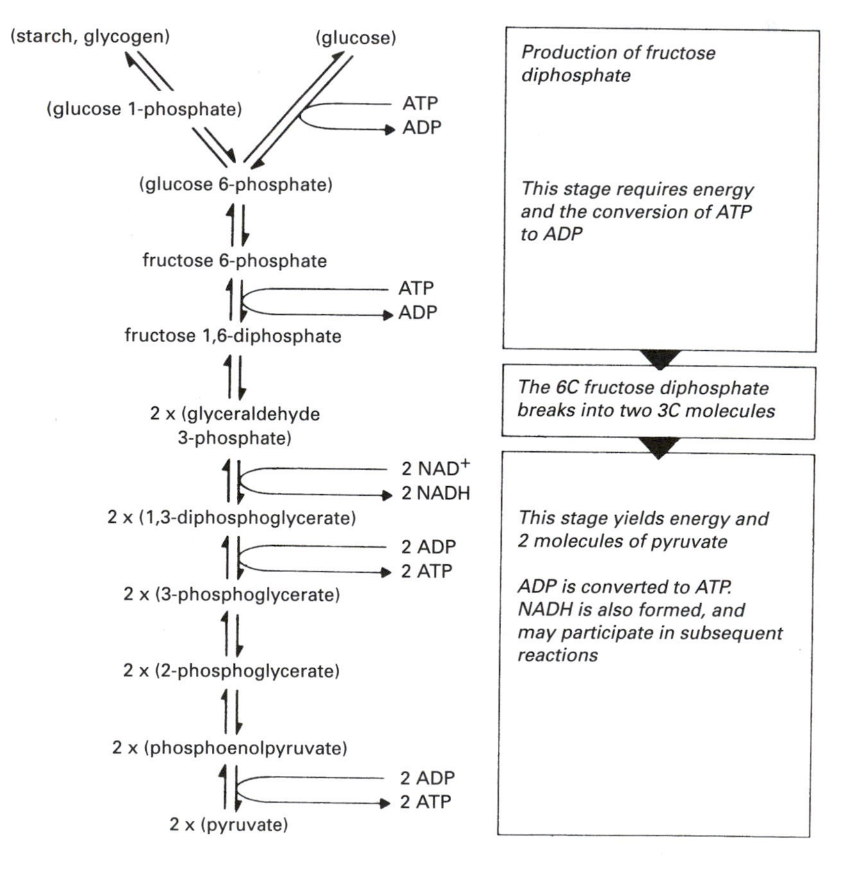

Glycolysis

glycolysis (Embden–Meyerhof pathway; glycolytic pathway) The conversion of glucose into pyruvate, with the release of some energy in the form of ATP. Glycolysis occurs in cell cytoplasm. It yields two molecules of ATP and two of $NADH_2$ per molecule of glucose. In anaerobic conditions, breakdown proceeds no further and pyruvate is converted into ethanol or lactic acid for storage or elimination. In aerobic conditions, glycolysis is followed by the KREBS CYCLE. The rate of glycolysis is controlled by the enzyme phosphofructokinase, which catalyzes an essentialy irreversible reaction. There are two other irreversible reactions catalyzed by hexokinase and pyruvate kinase.

glycolytic pathway *See* glycolysis.

glycoprotein A conjugated protein formed by the combination of a protein with carbohydrate side chains. Certain antigens, enzymes, and hormones are glycoproteins.

glycosaminoglycan (GAG) One of a group of compounds, sometimes called *mucopolysaccharides*, consisting of long unbranched chains of repeating disaccharide sugars, one of the two sugar residues being an amino sugar – either *N*-acetylglucosamine or *N*-acetylgalactosamine. These compounds are present in connective tissue; they include heparin and hyaluronic acid. Most glycosaminoglycans are linked to protein to form proteoglycans (sometimes called *mucoproteins*). *See also* glycoprotein.

glycoside A compound in which the ring form of a SUGAR is joined to some other organic group. The link in a glycoside occurs at the anomeric carbon atom. If the hydroxyl group at this carbon is removed, the result is a *glycosyl group*. If this group is joined to another organic group through an oxygen atom, then the resulting compound is an O-glycoside. The linkage ring–O–organic group is a *glycosidic link*. A simple example of a glycoside would be the compound formed by replacing the anomeric –OH group in a sugar by a methoxy group, $-OCH_3$. Such glycosides are described as alpha or beta according to whether the organic group is below or above the ring at the anomeric carbon. If the glycosyl group is attached through some other atom, it is often described as ppa *S*-glycoside, *C*-glycoside, etc. Compounds in which the glycosyl group is attached through nitrogen (i.e. *N*-glycosides) are also called *glycosylamines*. If the sugar is glucose, the compound is a *glucoside*. Glycosides occur in plants and include many useful substances. *See also* nucleoside.

glycosidic link *See* glycoside.

glycosylamine *See* glycoside.

glycosyl group *See* glycoside.

glyoxylate cycle A modification of the Krebs cycle occurring in some microorganisms, algae, and higher plants in regions where fats are being rapidly metabolized, e.g. in germinating fat-rich seeds. Acetyl groups formed from the fatty acids are passed into the glyoxylate cycle, with the eventual formation of mainly carbohydrates.

graft copolymer *See* polymerization.

Graham's law (of diffusion) The principle that gases diffuse at a rate that is inversely proportional to the square root of their density. Light molecules diffuse faster than heavy molecules. It is named for the Scottish chemist Thomas Graham (1805–69), who reported it in 1829.

gram (gramme; symbol: g) A unit of mass defined as 10^{-3} kilogram.

gram-atom *See* mole.

gram-equivalent The equivalent weight of a substance in grams.

gramme An alternative spelling of *gram*.

gram-molecule *See* mole.

granulation A process for enlarging particles to improve the flow properties of solid reactants and products in industrial chemical processes. The larger a particle, and the freer from fine materials in a solid, the more easily it will flow. Dry granulation produces pellets from dry materials, which are crushed into the desired size. Wet granulation involves the addition of a liquid to the material, and the resulting paste is extruded and dried before cutting to the required size.

grape sugar *See* glucose.

gravimetric analysis A method of quantitative analysis in which the final analytical measurement is made by weighing. There are many variations in the method but in essence they all consist of:

1. taking an accurately weighed sample into solution;
2. precipitation as a known compound by a quantitative reaction;
3. digestion and coagulation procedures;
4. filtration and washing;
5. drying and weighing as a pure compound.

Filtration is a key element in the method and a variety of special filter papers and sinter-glass filters are available.

gray Symbol: Gy The SI unit of absorbed energy dose per unit mass resulting from the passage of ionizing radiation through living tissue. One gray is an energy absorption of one joule per kilogram of mass. The unit is named for the British radiobiologist L. H. Gray (1905–65).

Grignard, François Auguste Victor (1871–1935) Franch organic chemist. Grignard is best remembered for the organomagnesium compounds known as *Grignard reagents* which he discovered in 1901. He found that these compounds, which have the general formular RMgX, where R is an organic group and X is a halogen, can be used in the synthesis of many types of compounds, including alcohols, hydrocarbons and carboxylic acids. Grignard shared the 1912 Nobel Prize for chemistry for this work. He started compiling his *Treatise on Organic Chemistry* after World War I. The first volumes appeared in 1935, with others helping to complete this multi-volume treatise after his death.

Grignard reagent A type of organometallic compound with the general formula RMgX, where R is an alkyl or aryl group and X is a halogen (e.g. CH_3MgCl). Grignard reagents are prepared by reacting the haloalkane or haloaryl compound with magnesium in dry ether:

$$CH_3Cl + Mg \rightarrow CH_3MgCl$$

Grignard reagents probably have the form $R_2Mg.MgCl_2$. They are used extensively in organic chemistry. With methanal a primary alcohol is produced:

$$RMgX + HCHO \rightarrow RCH_2OH + Mg(OH)X$$

Other aldehydes give secondary alcohols:

$$RMgX + R'CHO \rightarrow RR'CHOH + Mg(OH)X$$

Alcohols and carboxylic acids give hydrocarbons:

$$RMgX + R'OH \rightarrow RR' + Mg(OH)X$$

Water also gives a hydrocarbon:

$$RMgX + H_2O \rightarrow RH + Mg(OH)X$$

Solid carbon dioxide in acid solution gives a carboxylic acid:

$$RMgX + CO_2 + H_2O \rightarrow RCOOH + Mg(OH)X$$

They are named for Victor Grignard.

ground state The lowest energy state of an atom, molecule, or other system. *Compare* excited state.

group **1.** In the periodic table, a series of chemically similar elements that have similar electronic configurations. A group is thus a column of the periodic table. For example, the alkali metals, all of which have outer s^1 configurations, belong to group 1. *See also* periodic table.

2. (functional group) In organic chemistry, an arrangement of atoms that bestows a particular type of property on a molecule and enables it to be placed in a particular class, e.g. the aldehyde group –CHO. *See also* functional group.

GSC Gas–solid chromatography. *See* gas chromatography.

GTP (guanosine triphosphate) A nucleoside triphosphate occurring in all cells as a coenzyme for various key processes. Often it provides energy by undergoing hydrolysis to GDP (guanosine diphosphate) and a phosphate group, a reaction catalyzed by an enzyme or other component having *GTPase* activity. In protein synthesis, GTP is essential for the assembly of ribosomes and elongation of the polypeptide chain. It is also required for the assembly of microtubules, for protein transport within cells, and for the relaying of messages to various cell components in signal transduction.

guanidine (iminourea; $HN{:}C(NH_2)_2$) A strongly basic crystalline organic compound which can be nitrated to make a powerful explosive. It is also used in making dyestuffs, medicines and polymer resins.

guanine A nitrogenous base found in DNA and RNA. Guanine has a purine ring structure.

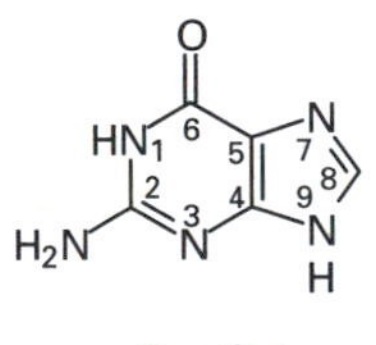

Guanine

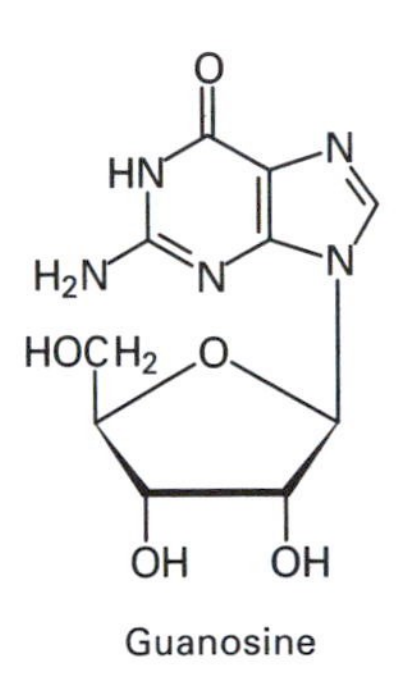

Guanosine

guanosine (guanine nucleoside) A nucleoside present in DNA and RNA and consisting of guanine linked to D-ribose via a β-glycosidic bond.

guanosine triphosphate *See* GTP.

gum One of a group of substances that swell in water to form gels or sticky solutions. Similar compounds that produce slimy solutions are called *mucilages*. Gums and mucilages are not distinguishable chemically. Most are heterosaccharides, being large, complex, flexible, and often highly-branched molecules.

guncotton *See* cellulose trinitrate.

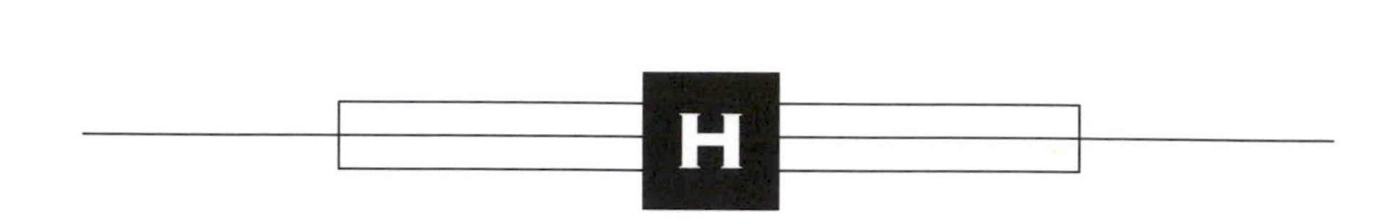

habit *See* crystal habit.

half-chair conformation *See* cyclohexane.

half life For a certain radioactive nucleus, the time taken for half the original nuclei in a sample to decay.

halide A compound containing a halogen. The HALOALKANES are examples.

haloalkane (alkyl halide) A type of organic compound in which one or more hydrogen atoms of an alkane have been replaced by halogen atoms. Haloalkanes can be made by direct reaction of the alkane with a halogen. Other methods are:

1. Reaction of an alcohol with the halogen acid (e.g. from NaBr + H_2SO_4) or with phosphorus halides (red phosphorus and iodine can be used):
$$ROH + HBr \rightarrow RBr + H_2O$$
$$ROH + PCl_5 \rightarrow RCl + POCl_3 + HCl$$
2. Addition of an acid to an alkene:
$$RCH{:}CH_2 + HBr \rightarrow RCH_2CH_2Br$$

The haloalkanes are much more reactive than the alkanes, and are useful starting compounds for preparing a wide range of organic chemicals. In particular, they undergo nucleophilic substitutions in which the halogen atom is replaced by some other group (iodine compounds are the most reactive). Some reactions of haloalkanes are:

1. Refluxing with aqueous potassium hydroxide to give an alcohol:
$$RI + OH^- \rightarrow ROH + I^-$$
2. Refluxing with potassium cyanide in alcoholic solution to give a nitrile:
$$RI + CN^- \rightarrow RCN + I^-$$
3. Refluxing with an alkoxide to give an ether:
$$RI + {}^-OR' \rightarrow ROR' + I^-$$
4. Reaction with alcoholic ammonia solution (100°C in a sealed tube) to give an amine:
$$RI + NH_3 \rightarrow RNH_2 + HI$$
5. Boiling with alcoholic potassium hydroxide, to eliminate an acid and produce an alkene:
$$RCH_2CH_2I + KOH \rightarrow KI + H_2O + RCH{:}CH_2$$

See also Grignard reagent; Wurtz reaction.

halocarbon A chemical compound that contains carbon atoms bound to halogen atoms and (sometimes) hydrogen atoms. The halocarbons include haloalkanes such as tetrachloromethane (CCl_4) and the haloforms ($CHCl_3$, $CHBr_3$, etc.). There are various types of halocarbon that are useful but are also significant pollutants. For example, the *chlorofluorocarbons* (CFCs) contain carbon, fluorine, and chlorine. They are useful as refrigerants, aerosol propellants, and in making rigid plastic foams. However, they are also thought to damage the ozone layer and an international agreement exists to phase out their use. Similar compounds are the *hydrochlorofluorocarbons* (HCFCs), which contain hydrogen as well as chlorine and fluorine, and the *hydrofluorocarbons* (HFCs), which contain hydrogen and fluorine.

The *halons* are a class of halocarbons that contain bromine as well as hydrogen and other halogens. Their main use is in fire extinguishers. They are, however, significantly more active than CFCs in their effect on the ozone layer. The halocarbons are also thought to contribute to global warming.

haloform Any of the four compounds CHX_3, where X is a halogen atom (F, fluoroform; Cl, chloroform; Br, bromoform; I, iodoform). The systematic names are trifluoromethane, trichloromethane, tribromomethane, and triiodomethane.

haloform reaction A reaction of a methyl ketone with NaOX, where X is Cl, Br, or I, to give a haloform. With sodium chlorate(I), for example:

$$RCOCH_3 + 3NaOCl \rightarrow RCOCCl_3 + 3NaOH$$

$$RCOCl_3 + NaOH \rightarrow NaOCOR + CHCl_3$$

The reaction can be used to make carboxylic acids (from the NaOCOR), and is especially useful when R is an aromatic group because the starting ketone, $RCOCH_3$, can be produced by Friedel–Crafts acetylation. *See also* triiodomethane.

halogenating agent A compound used to introduce halogen atoms into a molecule. Examples are phosphorus trichloride (PCl_3) and aluminum trichloride ($AlCl_3$).

halogenation A reaction in which a halogen atom is introduced into a molecule. Halogenations are specified as *chlorinations*, *brominations*, *fluorinations*, etc., according to the element involved. There are several methods.

1. Direct reaction with the element using high temperature or ultraviolet radiation:
$$CH_4 + Cl_2 \rightarrow CH_3Cl + HCl$$
2. Addition to a double bond:
$$H_2C{:}CH_2 + HCl \rightarrow C_2H_5Cl$$
3. Reaction of a hydroxyl group with a halogenating agent, such as PCl_3:
$$C_2H_5OH \rightarrow C_2H_5Cl + OH^-$$
4. In aromatic compounds direct substitution can occur using aluminum chloride as a catalyst:
$$2C_6H_6 + Cl_2 \rightarrow 2C_6H_5Cl$$
5. Alternatively in aromatic compounds, the chlorine can be introduced by reacting the diazonium ion with copper(I) chloride:
$$C_6H_5N_2^+ + Cl^- \rightarrow C_6H_5Cl + N_2$$

halogens A group of elements (group 17, formerly VIIA, of the periodic table) consisting of fluorine, chlorine, bromine, iodine, and the short-lived radioactive element astatine. The halogens all have outer valence shells that are one electron short of a rare-gas configuration. Because of this, the halogens are characterized by high electron affinities and high electronegativities, fluorine being the most electronegative element known.

A wide range of organic halides is formed in which the C–F bond is characteristically resistant to chemical attack; the C–Cl bond is also fairly stable, particularly in aryl compounds, but the alkyl halogen compounds become increasingly susceptible to nucleophilic attack and generally more reactive.

halon *See* halocarbon.

halothane ($CHBrClCF_3$) A colorless nonflammable liquid halocarbon used as a general anesthetic. The systematic name is 1-chloro-1-bromo-2,2,2-trifluoroethane.

hammer mill A device used in the chemical industry for crushing and grinding solid materials at high speeds to a specified size. The impact between the particles, grinding plates, and grinding hammers pulverizes the particles. Hammer mills can be used for a greater variety of soft material than other types of grinding equipment. *Compare* ball mill.

hardening (of oils) The conversion of liquid plant oils into a more solid form for use in margarine by hydrogenation using a nickel catalyst. In vegetable oils the fatty acids present (as glycerides) contain double bonds (i.e. they are unsaturated). The hydrogenation process increases the amount of unsaturated material, increasing the melting point, but still leaves unsaturated fatty acids. For this reason it is claimed that margarines are healthier than animal fats (e.g. butter), which contains saturated fats, because the unsaturated fats are less likely to lead to cholesterol build-up in the body, and consequent risk of coronary heart disease. However, the hydrogenation process may also affect the nature of the double bonds. Natural unsaturated fatty acids mostly have a *cis* configuration about the double bonds. In the hydrogenation process, a proportion of these are converted into fatty acids with a *trans* configuration. Glycerides of these are known as *trans-fats*. It has been claimed that there is also a link between trans-fats and coronary heart disease. *See* Sabatier–Senderens process.

Haworth, Sir Walter Norman (1883–1950) British organic chemist. Haworth was a pioneer of the study of carbohydrates, particularly sugars. He showed that sugar molecules are ring molecules, with a puranose ring consisting of five carbon atoms and an oxygen atom and a furanose ring consisting of four carbon atoms and an oxygen atom. He and his colleagues subsequently investigated the chain structures of various polysaccharides. This established the structures of cellulose, starch and glycogen. In 1929 he published the book *The Constitution of the Sugars* which soon became the standard work on this topic. In 1933, together with his colleague Edmund Hirst, he synthesized vitamin C (ascorbic acid), having previously established its structure. Haworth shared the 1937 Nobel Prize for chemistry with Paul KARRER.

HCFC Hydrochlorofluorocarbon. *See* halocarbon.

heat Energy transferred as a result of a temperature difference. The term is often loosely used to mean internal energy (i.e. the total kinetic and potential energy of the particles). It is common in chemistry to define such quantities as *heat of combustion*, *heat of neutralization*, etc. These are in fact molar enthalpies for the change, given the symbol $\Delta H_M^{\ominus}$. The superscript symbol denotes standard conditions, while the subscript M indicates that the enthalpy change is for one mole. The unit is usually the kilojoule per mole (kJ mol^{-1}). By convention, ΔH is negative for an exothermic reaction. Molar enthalpy changes stated for chemical reactions are changes for standard conditions, which are defined as 298 K (25°C) and 101 325 Pa (1 atmosphere). Thus, the standard molar enthalpy of reaction is the enthalpy change for reaction of substances under these conditions producing reactants under the same conditions. The substances involved must be in their normal equilibrium physical states under these conditions (e.g. carbon as graphite, water as the liquid, etc.). Note that the measured enthalpy change will not usually be the standard change. In addition, it is common to specify the entity involved. For instance $\Delta H_f^{\ominus}(H_2O)$ is the standard molar enthalpy of formation for one mole of H_2O species.

heat exchanger A device that enables the heat from a hot fluid to be transferred to a cool fluid without allowing the fluids to come into contact. The normal arrangement is for one of the fluids to flow in a coiled tube through a jacket containing the second fluid. Both the cooling and heating effect may be of benefit in conserving the energy used in a chemical plant and in controlling the process.

heat of atomization The energy required in dissociating one mole of a substance into atoms. *See* heat.

heat of combustion The energy liberated when one mole of a substance burns in excess oxygen. *See* heat.

heat of crystallization The energy liberated when one mole of a substance crystallizes from a saturated solution of this substance.

heat of dissociation The energy required to dissociate one mole of a substance into its constituent elements.

heat of formation The energy change when one mole of a substance is formed from its elements. *See* heat.

heat of neutralization The energy liberated when one mole of an acid or base is neutralized.

heat of reaction The energy change when molar amounts of given substances react completely. *See* heat.

heat of solution The energy change when one mole of a substance is dissolved in a given solvent to infinite dilution (in practice, to form a dilute solution).

heavy hydrogen *See* deuterium.

heavy water Deuterium oxide, D_2O.

hecto- Symbol: h A prefix denoting 10^2. For example, 1 hectometer (hm) = 10^2 meters (m).

helicate *See* supramolecular chemistry.

Hell–Volard–Zelinsky reaction A method for the preparation of halogenated carboxylic acids using free halogen in the presence of a phosphorus halide. The halogenation occurs at the carbon atom adjacent to the –COOH group. With Br_2 and PBr_3:

$$RCH_2COOH \rightarrow RCHBrCOOH \rightarrow RCBr_2COOH$$

Helmholtz free energy *See* Helmholtz function.

Helmholtz function (Helmholtz free energy) Symbol: F A thermodynamic function defined by

$$F = U - TS$$

where U is the internal energy, T the thermodynamic temperature, and S the entropy. It is a measure of the ability of a system to do useful work in an isothermal process. The function is named for the German physiologist and physicist Hermann Ludwig Ferdinand von Helmholtz (1821–94). *See also* free energy.

heme (haeme) An iron-containing porphyrin that is the prosthetic group in HEMOGLOBIN, myoglobin, and some cytochromes.

hemiacetal *See* acetal.

hemicellulose One of a group of substances that make up the amorphous matrix of plant cell walls together with pectic substances (and occasionally, in mature cells, with lignin, gums, and mucilages). They are *heteropolysaccharides*, i.e. polysaccharides built from more than one type of sugar, mainly the hexoses (mannose and galactose) and the pentoses (xylose and arabinose). They vary greatly in composition between species. In some seeds (e.g. the endosperm of dates) hemicelluloses are a food reserve.

hemiketal *See* acetal.

hemin The hydrochloride form of heme. Hemin is the crystalline form in which heme can be isolated and studied in the laboratory. The iron present is the trivalent state (iron(III)). Hemin can be made to crystallize by heating hemoglobin gently with acetic acid and sodium chloride. A variety of crystal forms are known.

hemocyanin A blue copper-containing blood pigment found in many mollusks and arthropods. Hemocyanin is the second most abundant blood pigment after hemoglobin and functions similarly in acting as an oxygen-carrier in the blood.

hemoerythrin A pigment occurring in the blood of certain invertebrates, similar in structure to HEMOGLOBIN.

hemoglobin The pigment of the red blood cells in humans and other verte-

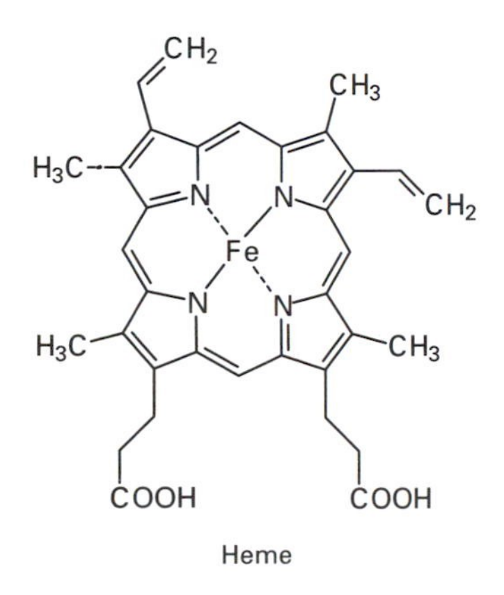

Heme

brates that is responsible for the transport of oxygen from the lungs to the tissues. It consists of a basic protein, globin, linked with four *heme* groups. Heme is a complex compound containing an iron atom. The most important property of hemoglobin is its ability to combine reversibly with one molecule of oxygen per iron atom to form *oxyhemoglobin*, which has a bright red color. The iron is present in the divalent state (iron(II)) and this remains unchanged with the binding of oxygen. There are variations in the polypeptide chains, giving rise to different types of hemoglobins in different species. The binding of oxygen depends on the oxygen partial pressure; high pressure favors formation of oxyhemoglobin and low pressure favors release of oxygen.

henry Symbol: H The SI unit of inductance, equal to the inductance of a closed circuit that has a magnetic flux of one weber per ampere of current in the circuit. $1\ \mathrm{H} = 1\ \mathrm{Wb\ A^{-1}}$. It is named for the US physicist Joseph Henry (1797–1828).

Henry's law The concentration (C) of a gas in solution is proportional to the partial pressure (p) of that gas in equilibrium with the solution, i.e. $p = kC$, where k is a proportionality constant. The relationship is similar in form to that for RAOULT'S LAW, which deals with ideal solutions. A consequence of Henry's law is that the 'volume solubility' of a gas is independent of pressure. The law is named for the British physician and chemist William Henry (1755–1836), who formulated it in 1801.

heparin A polysaccharide that inhibits the formation of thrombin from prothrombin and thereby prevents the clotting of blood. It is used in medicine as an anticoagulant. *See* polysaccharide.

heptane (C_7H_{16}) A colorless liquid alkane obtained from petroleum refining. It is used in gasoline and as a solvent.

hertz Symbol: Hz The SI unit of frequency, defined as one cycle per second (s^{-1}). Note that the hertz is used for regularly repeated processes, such as vibration or wave motion. It is named for the German physicist Heinrich Hertz (1857–94).

Hess's law A derivative of the first law of thermodynamics. It states that the total heat change for a given chemical reaction involving alternative series of steps is independent of the route taken. It is named for the Russian chemist Germain Henri Hess (1802–50), who proposed it in 1840.

hetero atom *See* heterocyclic compound.

heterocyclic compound A compound that has a ring containing more than one type of atom. Commonly, heterocyclic compounds are organic compounds with at least one atom in the ring that is not a carbon atom. Pyridine and glucose are examples. The noncarbon atom is called a *hetero atom*. *Compare* homocyclic compound.

heterogeneous Relating to more than one phase. A heterogeneous mixture, for instance, contains two or more distinct phases. Heterogeneous catalysis involves a catalyst that is a different phase than that of the reactants (usually gaseous reactants passed over a solid catalyst).

heterolysis *See* heterolytic fission.

heterolytic fission (heterolysis) The breaking of a covalent bond so that both electrons of the bond remain with one fragment. A positive ion and a negative ion are produced:

$$RX \rightarrow R^+ + X^-$$

Compare homolytic fission.

heteropolymer *See* polymerization.

heteropolysaccharide *See* hemicellulose.

hexadecanoate (palmitate) A salt or ester of hexadecanoic acid.

hexadecanoic acid (palmitic acid) A crystalline carboxylic acid:

$$CH_3(CH_2)_{14}COOH$$

It is present as glycerides in fats and oils. *See* glyceride.

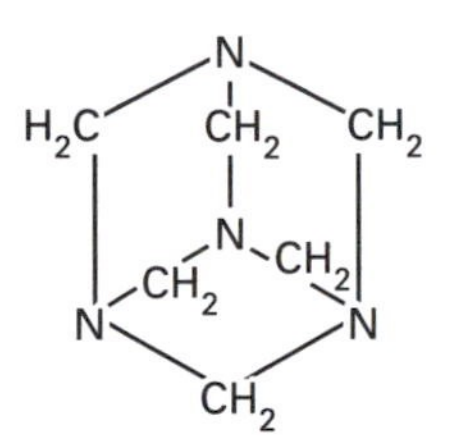

Hexamethylenetetramine

hexamethylenetetramine (hexamine; $C_6H_{12}N_4$) A white crystalline organic compound made by condensing methanal with ammonia. It is used as a fuel for camping stoves, in vulcanizing rubber, and as a urinary disinfectant. It can be nitrated to make the high explosive cyclonite.

hexamine *See* hexamethylenetetramine.

hexane (C_6H_{14}) A liquid alkane obtained from the light fraction of crude oil. The principal use of hexane is in gasoline and as a solvent.

hexanedioic acid (adipic acid; HOOC-$(CH_2)_4COOH$) A colorless crystalline organic dicarboxylic acid that occurs in rosin. It is used in the manufacture of NYLON.

hexanoate A salt or ester of hexanoic acid.

hexanoic acid (caproic acid; CH_3-$(CH_2)_4COOH$) An oily carboxylic acid found (as glycerides) in cow's milk and some vegetable oils.

hexose A SUGAR that has six carbon atoms in its molecules.

hexose monophosphate shunt *See* pentose phosphate pathway.

hexyl group The group $C_5H_{11}CH_2$–, having a straight chain of carbon atoms.

HFC Hydrofluorocarbon. *See* halocarbon.

high-performance liquid chromatography *See* HPLC.

histamine An amine formed from the amino acid histidine by decarboxylation and produced mainly in connective tissue as a response to injury or allergic reaction. It causes contraction of smooth muscle, stimulates gastric secretion of hydrochloric acid and pepsin, and dilates blood vessels, which lowers blood pressure and produces inflammation, itching, or allergic symptoms.

histidine *See* amino acid.

Hodgkin, Dorothy Mary Crowfoot (1910–94) British x-ray crystallographer Dorothy Hodgkin determined the structure of many complex organic molecules. She won the 1964 Nobel Prize for chemistry for this work. Together with Charles Bunn, she published the structure of penicillin in 1949. She then started work on the structure of vitamin B_{12} and found its structure in 1956. This work made use of early electronic computers. In 1969 she determined the structure of insulin.

Hoffmann, Roald (1937–) Polish-born American chemist, Hoffmann is best known for his collaboration with Robert WOODWARD in the mid 1960s which led to the formulation of the Woodward–Hoffmann rules. These rules stated which chemical reactions can take place in terms of molecular orbitals. Woodward and Hoffmann were largely concerned with organic reactions but their rules apply more generally. They summarized their results in the book *Conservation of Orbital Symmetry* (1970). Hoffmann shared the 1979 Nobel Prize for chemistry with Kenichi FUKUI for this work. Hoffmann has also done a great deal to popularize chemistry in books and television programmes.

Hofmann, August Wilhelm von (1818–92) German organic chemist. Hofmann was one of the most influential organic chemists of the 19th century. He was particularly influential in Britain and Germany. Much of his work was concerned with the constituents of coal tar, particularly aniline and phenol. He discovered or investigated a number of compounds, including quaternary ammonium salts. He also discovered the reaction known as *Hofmann degradation*, which consists of treating an amide with bromine and alkali to give an amine with one fewer carbon atom. He was one of the first people to investigate formaldehyde. In 1858 he obtained the dye magenta by reacting carbon tetrachloride with aniline.

Hofmann degradation A method of preparing primary amines from acid amides. The amide is refluxed with aqueous sodium hydroxide and bromine:

$$RCONH_2 + NaOH + Br_2 \rightarrow RCONHBr + NaBr + H_2O$$

$$RCONHBr + OH^- \rightarrow RCON^-Br + H_2O$$

$$RCON^-Br \rightarrow R{-}N = C = O + Br^-$$

$$RNCO + 2OH^- \rightarrow RNH_2 + CO_2^{2-}$$

The reaction is a 'degradation' in the sense that a carbon atom is removed from the amide chain. It is named for the German chemist August Wilhelm von Hofmann.

Hofmann's method A method formerly used for determining the vapor density of volatile liquids. A known weight of sample is introduced into a mercury barometer tube, which is surrounded by a heating jacket. The volume of vapor can thus be read off directly, the temperature is known, and the pressure is obtained by taking the atmospheric pressure minus the mercury height in the barometer (with corrections for the density of mercury at higher temperatures). The method's only advantage is that it may be used for samples that decompose at their normal boiling point.

holoenzyme A catalytically active complex made up of an apoenzyme and a coenzyme. The former is responsible for the specificity of the holoenzyme whilst the latter determines the nature of the reaction.

HOMO *See* frontier orbital.

homocyclic compound A compound containing a ring made up of the same atoms. Benzene is an example of a homocyclic compound. *Compare* heterocyclic compound.

homogeneous Relating to a single phase. A homogeneous mixture, for instance, consists of only one phase. In homogeneous catalysis, the catalyst has the same phase as the reactants.

homologous series A group of organic compounds possessing the same functional group and having a regular structural pattern so that each member of the series differs from the next one by a fixed number of atoms. The members of a homologous series can be represented by a general formula. For example, the homologous series of alkane alcohols CH_3OH, C_2H_5OH, C_3H_7OH, ..., has a general formula $C_nH_{2n+1}OH$. Each member differs by CH_2 from the next. Any two successive members of a series are called *homologs*.

homologs *See* homologous series.

homolysis *See* homolytic fission.

homolytic fission (homolysis) The breaking of a covalent bond so that one electron from the bond is left on each fragment. Two free radicals result:

$$RR' \rightarrow R\bullet + R'\bullet$$

Compare heterolytic fission.

homopolymer *See* polymerization.

host–guest chemistry A branch of supramolecular chemistry in which a molecular structure acts as a 'host' to hold an ion or molecule (the 'guest'). The guest may be coordinated to the host or may be trapped by its structure. For example, *calixarenes* are compounds with cup-shaped molecules that may accept guest molecules. *See also* crown ether; supramolecular chemistry.

HPLC High-performance liquid chromatography; a sensitive analytical technique, similar to gas-liquid chromatography but using a liquid carrier. The carrier is specifically choosen for the particular substance to be detected.

Hückel, Erich Armand Arthur Joseph (1896–1980) German physical and theoretical chemist. Hückel has two main claims to fame. His first is the theory of electrolytes which Peter Debye and he produced in 1923 and which gives a good description of the electrical and thermodynamic properties of dilue electrolytes. His second major work was the application of quantum mechanics to aromatic molecules such as benzene. He used molecular orbital theory to show that in the benzene molecule the electrons in the pi orbitals are spread out directly above and below the ring of carbon atoms, thus making the molecule more stable than it would be if one had alternating double and single bonds. Hückel started this work in 1930 and soon extended it to predict the *Hückel rule* which states that a molecule is aromatic if it has $(4n + 2)$ pi electrons.

Hückel rule *See* aromatic compound.

humectant A hygroscopic substance used to maintain moisture levels. Glycerol, mannitol, and sorbitol are commonly used in foodstuffs, tobacco, etc.

Hund's rule A rule that states that the electronic configuration in degenerate orbitals will have the minimum number of paired electrons.

hybrid orbital *See* orbital.

hydrate A compound coordinated with water molecules. When water is bound up in a compound it is known as *water of crystallization*.

hydration The solvation of such species as ions in water.

hydrazine (N_2H_4) A colorless liquid that can be prepared by the oxidation of ammonia with sodium chlorate(I) or by the gas phase reaction of ammonia with chlorine. Hydrazine is a weak base, forming salts (e.g. $N_2H_4.HCl$) with strong acids and is also a powerful reducing agent. With aldehydes and ketones it forms HYDRAZONES.

hydrazone A type of organic compound containing the $C:NNH_2$ group, formed by the reaction between an aldehyde or ketone and hydrazine (N_2H_4). Derivatives of hydrazine were formerly used to produce crystalline products, which have sharp melting points that can be used to characterize the original aldehyde or ketone. Phenylhydrazine ($C_6H_5NH.NH_2$), for instance, produces *phenylhydrazones*.

hydride A compound of hydrogen. Ionic hydrides are formed with highly electropositive elements and contain the H^- ion (hydride ion). Non-metals form covalent hydrides, as in methane (CH_4) or silane (SiH_4). The boron hydrides are electron-deficient covalent compounds. Many transition metals absorb hydrogen to form interstitial hydrides.

hydrobromic acid (HBr) A colorless liquid produced by adding hydrogen bromide to water. It shows the typical properties of a strong acid and it is a strong reducing agent.

hydrocarbon Any compound containing only the elements carbon and hydrogen. Examples are the alkanes, alkenes, alkynes, and aromatics such as benzene and naphthalene.

hydrochloric acid (HCl) A colorless fuming liquid made by adding hydrogen chloride to water. Dissociation into ions is extensive and hydrochloric acid shows the typical properties of a strong acid. Hydrochloric acid is used in making dyes, drugs, and photographic materials.

hydrochlorofluorocarbon *See* halocarbon.

hydrocyanic acid (prussic acid; HCN) A highly poisonous weak acid formed when hydrogen cyanide gas dissolves in water. Its salts are cyanides. Hydrogen cyanide is used in making acrylic plastics.

hydrofluoric acid (HF) A colorless liquid produced by dissolving hydrogen fluoride in water. It is a weak acid, but will dissolve most silicates and hence can be used to etch glass. As the interatomic distance in HF is relatively small, the H–F bond energy is very high and hydrogen fluoride is not a good proton donor. It does, however, form hydrogen bonds.

hydrofluorocarbon *See* halocarbon.

hydrogen A colorless gaseous element. Hydrogen has some similarities to both the alkali metals (group 1) and the halogens (group 17), but is not normally classified in any particular group of the periodic table. It is the most abundant element in the Universe and the ninth most abundant element in the Earth's crust and atmosphere (by mass). It occurs principally in the form of water and petroleum products; traces of molecular hydrogen are found in some natural gases and in the upper atmosphere.

Symbol: H; m.p. 14.01 K; b.p. 20.28 K; d. 0.089 88 kg m^{-3} (0°C); p.n. 1; r.a.m. 1.0079.

hydrogenation The reaction of a compound with hydrogen. In organic chemistry, hydrogenation usually refers to the addition of hydrogen to multiple bonds, often with the aid of a catalyst. Unsaturated natural liquid vegetable oils can be hydrogenated to form saturated semisolid fats – a reaction used in making types of margarine. *See* Bergius process.

hydrogen bond An intermolecular bond between molecules in which hydrogen is bound to a strongly electronegative element. Bond polarization by the electronegative element X leads to a positive charge on hydrogen $X^{\delta-}–H^{\delta+}$; this hydrogen can then interact directly with electronegative elements of adjacent molecules. The hydrogen bond is represented as a dotted line:

$$X^{\delta-} - H^{\delta+} \ldots\ldots X^{\delta-} - H^{\delta+} \ldots$$

The length of a hydrogen bond is characteristically 0.15–0.2 nm. Hydrogen bonding may lead to the formation of dimers (for example, in carboxylic acids) and is used to explain the anomalously high boiling points of H_2O and HF. Hydrogen bonding is important in many biochemical systems. It occurs between bases in the chains of DNA. It also occurs between C=O and N–H groups in PROTEINS, where it is responsible for maintaining the secondary structure.

hydrogen bromide (HBr) A colorless sharp-smelling gas that is very soluble in water. It is produced by direct combination of hydrogen and bromine in the presence of a platinum catalyst or by the reaction of phosphorus tribromide with water. It dissolves in water to give HYDROBROMIC ACID.

hydrogencarbonate (bicarbonate) A salt containing the ion $^{-}HCO_3$.

hydrogen chloride (HCl) A colorless gas that has a strong irritating odor and fumes strongly in moist air. It is prepared by the action of concentrated sulfuric acid on sodium chloride. The gas is made industrially by burning a stream of hydrogen in chlorine. It is not particularly reactive but will form dense white clouds of ammonium chloride when mixed with ammonia. It is very soluble in water and ionizes almost completely to give HYDROCHLORIC ACID. Hydrogen chloride is used in the manufacture of organic chlorine compounds, such as polyvinyl chloride (PVC).

hydrogen cyanide *See* hydrocyanic acid.

hydrogen fluoride (HF) A colorless liquid produced by the reaction of concentrated sulfuric acid on calcium fluoride. It produces toxic corrosive fumes and dissolves readily in water to give HYDROFLUORIC ACID. Hydrogen fluoride is atypical of the hydrogen halides as the individual H–F units are associated into much larger units, forming zigzag chains and rings. This is caused by hydrogen bonds that form between the hydrogen and the highly electronegative fluoride ions. Hydrogen fluoride is used extensively as a catalyst in the petroleum industry.

hydrogen ion A positively charged hydrogen atom, H^+, i.e. a proton. Hydrogen ions are produced by all acids in water, in which they are hydrated to hydroxonium (hydronium) ions, H_3O^+. *See* acid; pH.

hydrogen peroxide (H_2O_2) A colorless syrupy liquid, usually used in solution in water. Although it is stable when pure, on contact with bases such as manganese(IV) oxide it gives off oxygen, the manganese(IV) oxide acting as a catalyst:

$$2H_2O_2 \rightarrow 2H_2O + O_2$$

Hydrogen peroxide can act as an oxidizing agent, converting iron(II) ions to iron(III) ions, or as a reducing agent with potassium manganate(VII). It is used as a bleach and in rocket fuel. The strength of solutions is usually given as *volume strength* – the volume of oxygen (dm^3) at STP given by decomposition of 1 dm^3 of the solution.

hydrogensulfate (bisulfate, HSO_4^-) An acidic salt or ester of sulfuric acid (H_2SO_4), in which only one of the acid's hydrogen atoms has been replaced by a metal or organic radical. An example is sodium hydrogensulfate, $NaHSO_4$.

hydrogen sulfide (sulfuretted hydrogen, H_2S) A colorless very poisonous gas with an odor of bad eggs. Hydrogen sulfide is prepared by reacting hydrochloric acid with iron(II) sulfide. It is tested for by mixing with lead nitrate, with which it gives a black precipitate. Its aqueous solution is weakly acidic. Hydrogen sulfide reduces iron(III) chloride to iron(II) chloride, forming hydrochloric acid and a yellow precipitate of sulfur. Hydrogen sulfide precipitates insoluble sulfides, and is used in qualitative analysis. It burns with a blue flame in oxygen to form sulfur(IV) oxide and water. Natural gas contains some hydrogen sulfide, which is removed before supply to the consumer.

hydrogensulfite (bisulfite, HSO_3^-) An acidic salt or ester of sulfurous acid (H_2SO_3), in which only one of the acid's hydrogen atoms has been replaced by a metal or organic radical. An example is sodium hydrogensulfite, $NaHSO_3$.

hydrolysis A reaction between a compound and water. An example is the hydrolysis of an ESTER to give a carboxylic acid and an alcohol:

$$CH_3COOC_2H_5 + H_2O \rightleftharpoons CH_3COOH + C_2H_5OH$$

hydron The positive ion H^+. The name is used when the isotope is not relevant, i.e. a hydron could be a proton, deuteron, or triton.

hydrophilic Water attracting. *See* lyophilic.

hydrophobic Water repelling. *See* lyophobic.

hydroquinone *See* benzene-1,4-diol.

hydrosol A colloid in aqueous solution.

hydroxide A compound containing the ion OH^- or the group –OH.

hydroxonium ion *See* hydrogen ion.

hydroxybenzene *See* phenol.

hydroxybenzoate *See* salicylate.

hydroxybenzoic acid *See* salicylic acid.

2-hydroxypropanoic acid *See* lactic acid.

hydroxyl group A group (–OH) containing hydrogen and oxygen, characteristic of alcohols and phenols, and some hydroxides. It should not be confused with the hydroxide ion (OH^-).

hygroscopic Describing a substance that absorbs moisture from the atmosphere. *See also* deliquescent.

hyperconjugation The interaction of sigma bonds with pi bonds. It is sometimes described in terms of resonance structures of the type:

$$C_6H_5CH_3 \rightarrow C_6H_5CH_2^-H^+$$

to explain the interaction of the methyl group with the pi electrons of the benzene ring in methylbenzene (toluene).

hyperfine structure *See* fine structure.

hypertonic solution A solution that has a higher osmotic pressure than some other solution. *Compare* hypotonic solution.

hypotonic solution A solution that has a lower osmotic pressure than some other solution. *Compare* hypertonic solution.

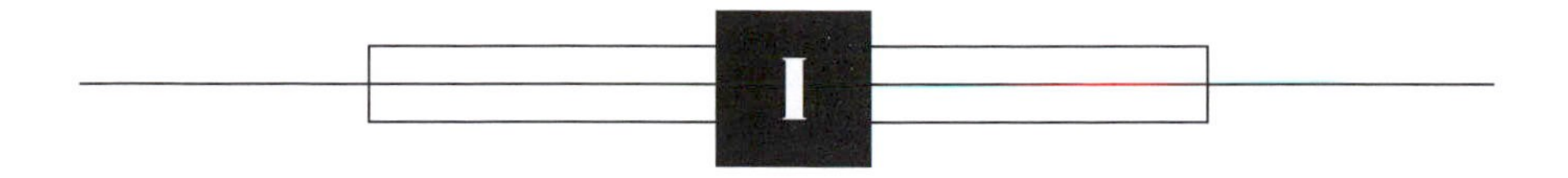

I

IAA (indole acetic acid) A naturally occurring auxin. *See* auxin.

ideal gas (perfect gas) *See* gas laws.

ideal solution A hypothetical solution that obeys RAOULT'S LAW.

ignis fatuus (will-o'-the-wisp) A light sometimes seen over marshy ground. It is caused by methane produced by rotting vegetation, which is ignited by the presence of small amounts of spontaneously flammable phosphine (PH_3).

ignition temperature The lowest temperature to which a given substance can be heated before it ignites (without the application of a spark or flame). It is often called the *autoignition temperature*. *Compare* flash point.

imide An organic compound containing the group –CO.NH.CO–, i.e. a –NH group attached to two carbonyl groups. Simple imides have the general formula R^1.CO.NH.CO.R^2, where R^1 and R^2 are alkyl or aryl groups. The group is known

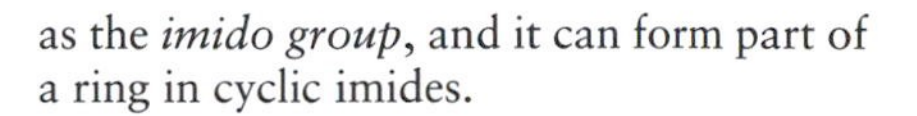

as the *imido group*, and it can form part of a ring in cyclic imides.

imido group *See* imide.

imine An organic compound containing the group C=N–, in which there is a double bond between the carbon and the nitrogen. A general formula for imines is $R^1R^2C{=}N{-}R^3$, where R^1, R^2, and R^3 are hydrocarbon groups or hydrogen. They can be made by the reaction of aldehydes and ketones with primary amines. For example, propanone (acetone; CH_3COCH_3) with ethylamine ($C_2H_6NH_2$):

$$CH_3COCH_3 + C_2H_6NH_2 \rightarrow (CH_3)_2C{=}N{-}C_2H_6 + H_2O$$

The reaction is acid-catalyzed. When ammonia is used the imine contains the C=N–H group:

$$CH_3COCH_3 + NH_3 \rightarrow (CH_3)_2C{=}N{-}H + H_2O$$

Most imines are unstable and can be detected only in solution unless R^1, R^2, or R^3 are aryl groups. Related compounds such as OXIMES, HYDRAZONES, and SEMICARBAZONES, in which the nitrogen is attached to an electronegative group, are more stable and can be isolated. Intermediates in

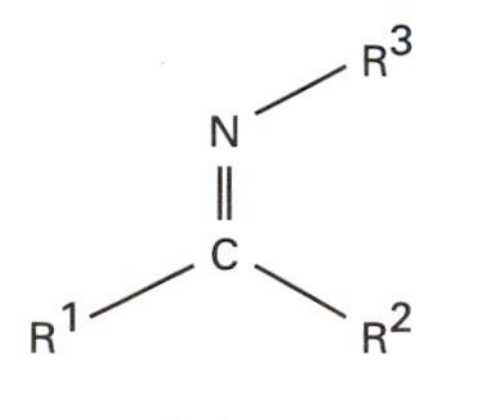

imine

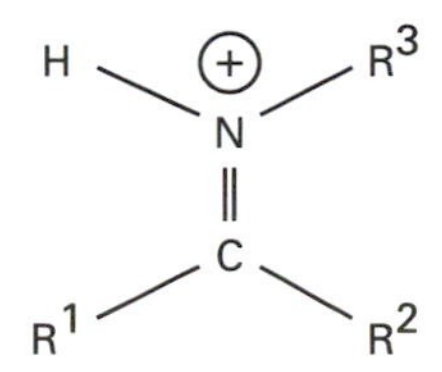

iminium ion

Imine

which an imine adds a proton to form a positive ion (e.g. $R^1R^2C{=}NR^3H^+$) are known as *iminium ions*.

iminium ion *See* imine.

imino group *See* imine.

iminourea *See* guanidine.

immiscible Describing two or more liquids that will not mix, such as oil and water. After being shaken together and left to stand immiscible liquids form separate layers.

indene (C_9H_8) A colorless flammable hydrocarbon. It has a benzene ring fused to a five-membered ring.

indicator A compound that reversibly changes color depending on the presence or absence of a chemical substance. In acid–base indicators the color depends on the pH of the solution in which it is dissolved. Methyl orange and phenolphthalein are examples of acid–base indicators. Redox titrations require either specific indicators, which detect one of the components of the reaction (e.g. starch for iodine, potassium thiocyanate for Fe^{3+}) or true redox indicators in which the transition potential of the indicator between oxidized and reduced forms is important. The transition potential of a redox indicator is analogous to the transition pH in acid–base systems. Complexometric titrations require indicators that complex with metal ions and change color between the free state and the complex state.

indigo ($C_{16}H_{10}N_2O_2$) A blue organic dye that occurs (as a glucoside) in plants of the genus *Indigofera*. It is a derivative of indole, and is now made synthetically.

indole (benzpyrrole; C_8H_7N) A colorless solid organic compound that occurs in coal tar and various plants, and is the basis of indigo and of several plant hormones. The indole molecule has a benzene ring fused to a pyrrole ring.

indole acetic acid (IAA) A naturally occurring auxin. *See* auxin.

inductive effect The effect in which substituent atoms or groups in an organic compound can attract (–I) or push away electrons (+I), forming polar bonds. Electron-attracting groups include $-NO_2$, –CN, –COOH, and the halogens. Electron-releasing groups include –OH, $-NH_2$, –OR, and R, where R is an alkyl group. Inductive effects can influence the reactivity of other parts of a molecule. For example, an electron-attracting group substituted on a benzene ring withdraws electrons and makes the ring less susceptible to electrophilic substitution. An electron-releasing group makes the ring more susceptible.

infrared (IR) Electromagnetic radiation with longer wavelengths than visible radiation. The wavelength range is approximately 0.7 μm to 1 mm. Many materials transparent to visible light are opaque to infrared, including glass. Rock salt, quartz, germanium, or polyethene prisms and lenses are suitable for use with infrared. Infrared radiation is produced by movement of charges on the molecular scale; i.e. by vibrational or rotational motion of molecules. *Infrared spectroscopy* is of particular importance in organic chemistry and absorption spectra are used extensively in identifying compounds. Certain bonds between pairs of atoms (C–C, C=C, C=O, etc.) have characteristic vibrational frequencies, which correspond to bands in the infrared spectrum. Infrared spectra are thus used in finding the structures of new organic compounds by indicating the presence of certain groups. They are also used to 'fingerprint' and thus identify known compounds. At shorter wavelengths, infrared absorption corresponds to transitions between rotational energy levels, and

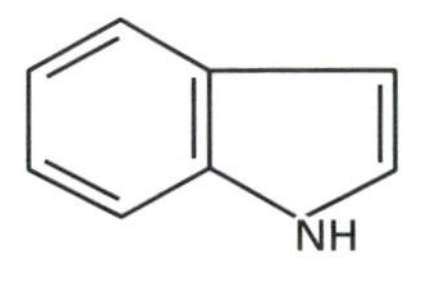

Indole

can be used to find the dimensions of molecules (by their moment of inertia).

Ingold, Sir Christopher Kelk (1893–1970) British organic chemist. Ingold devoted his career to understanding the mechanisms of organic reactions in terms of the electrons of the molecules concerned. For example, he postulated the concept of mesomerism in 1926 to explain how a molecule could exist as a hybrid of two possible structures. Ingold was particularly concerned with the mechanisms of elimination and substitution reactions. Ingold summarized his work in the classic book *Structure and Mechanisms in Organic Chemistry*, the second edition of which was published in 1969.

inhibitor A substance that slows down the rate of a chemical reaction.

inner Describing a ring compound that is formed, or regarded as formed, by one part of a molecule reacting with another. For example, a LACTAM is an inner amide and a LACTONE is an inner ester.

inorganic chemistry The branch of chemistry concerned with elements other than carbon and with the preparation, properties, and reactions of their compounds. Certain simple carbon compounds are treated in inorganic chemistry, including carbon oxides, carbon disulfide, carbon halides, hydrogen cyanide, and certain simple salts, such as the cyanides, cyanates, carbonates, and hydrogencarbonates.

insertion reaction A reaction in which an atom or group is inserted between two other groups. *See* carbene.

insoluble Describing a compound that has a very low solubility (in a specified solvent).

intermediate **1.** A compound that requires further chemical treatment to produce a finished industrial product.
2. A transient chemical entity in a complex reaction.

See also precursor.

intermediate bond A form of covalent bond that also has an ionic or electrovalent character. *See* polar bond.

intermolecular force A force of attraction between molecules, as distinguished from a force within the molecule (a chemical bond). Forces of attraction between molecules are the result of interactions between dipoles. *See* hydrogen bond; Van der Waals force.

internal energy Symbol: U The energy of a system that is the total of the kinetic and potential energies of its constituent particles (e.g. atoms and molecules). If the temperature of a substance is raised, by transferring energy to it, the internal energy increases (the particles move faster). Similarly, work done on or by a system results in an increase or decrease in the internal energy. The relationship between heat, work, and internal energy is given by the first law of thermodynamics. Sometimes the internal energy of a system is loosely spoken of as 'heat' or 'heat energy'. Strictly, this is incorrect; heat is the transfer of energy as a result of a temperature difference.

inversion A change in which a compound is converted from one optical isomer to the other. *See* Walden inversion.

invert sugar *See* sucrose.

in vitro Literally 'in glass'; describing experiments or techniques performed in laboratory apparatus rather than in the living organism. Cell tissue cultures and *in vitro* fertilization (to produce 'test-tube babies') are examples. *Compare in vivo.*

in vivo Literally 'in life'; describing processes that occur within the living organism. *Compare in vitro.*

iodide *See* halide.

iodine A dark-violet volatile solid element belonging to the HALOGENS (group 17, formerly VIIA, of the periodic table). It occurs in seawater and is concentrated by various marine organisms in the form of iodides. Significant deposits also occur in the form of iodates. A large number of organic iodine compounds are known.

iodoethane (ethyl iodide; C_2H_5I) A colorless liquid haloalkane made by reaction of ethanol with iodine in the presence of red phosphorus.

iodoform *See* triiodomethane.

iodoform reaction *See* triiodomethane.

iodomethane (methyl iodide; CH_3I) A liquid haloalkane made by reaction of methanol with iodine in the presence of red phosphorus.

ion An atom or molecule that has a negative or positive charge as a result of losing or gaining one or more electrons. *See also* ionization.

ion exchange A process that takes place in certain insoluble materials that contain ions capable of exchanging with ions in the surrounding medium. Zeolites, the first ion-exchange materials, were used for water softening. These have largely been replaced by synthetic resins made of an inert backbone material, such as polyphenylethene, to which ionic groups are weakly attached. If the ions exchanged are positive, the resin is a cationic resin. An anionic resin exchanges negative ions. When all available ions have been exchanged (e.g. sodium ions replacing calcium ions) the material can be regenerated by passing concentrated solutions (e.g. sodium chloride) through it. The calcium ions are then replaced by sodium ions. Ion-exchange techniques are used for a vast range of purification and analytical purposes.

ionic bond *See* electrovalent bond.

ionic crystal A crystal composed of ions of two or more elements. The positive and negative ions are arranged in definite patterns and are held together by electrostatic attraction. Sodium chloride is a typical example.

ionic product The product of concentrations:

$$K_W = [H^+][OH^-]$$

in water as a result of a small amount of self-ionization:

$$H_2O \rightleftharpoons H^+ + OH^-$$

ionic radius A measure of the effective radius of an ion in a compound. For an isolated ion, the concept is not very meaningful, since the ion is a nucleus surrounded by an 'electron cloud'. Values of ionic radii can be assigned, however, based on the distances between ions in crystals.

ionic strength For an ionic solution a quantity can be introduced that emphasizes the charges of the ions present:

$$I = \tfrac{1}{2}\Sigma_i m_i z^2{}_i$$

where m is the molality and z the ionic charge. The summation is continued over all the different ions in the solution, i.

ionization The process of producing ions. There are several ways in which ions may be formed from atoms or molecules. In certain chemical reactions ionization occurs by transfer of electrons; for example, sodium atoms and chlorine atoms react to form sodium chloride, which consists of sodium ions (Na^+) and chloride ions (Cl^-). Certain molecules can ionize in solution; acids, for example, form hydrogen ions as in the reaction

$$H_2SO_4 \rightarrow 2H^+ + SO_4{}^{2-}$$

The 'driving force' for ionization in a solution is solvation of the ions by molecules of the solvent. H^+, for example, is solvated as a hydroxonium (hydronium) ion, H_3O^+.

Ions can also be produced by ionizing radiation; i.e. by the impact of particles or photons with sufficient energy to break up molecules or detach electrons from atoms: $A \rightarrow A^+ + e^-$. Negative ions can be formed by capture of electrons by atoms or molecules: $A + e^- \rightarrow A^-$.

ionization energy *See* ionization potential.

ionization potential (IP; Symbol: I) The energy required to remove an electron from an atom (or molecule or group) in the gas phase, i.e. the energy required for the process:

$$M \rightarrow M^+ + e^-$$

It gives a measure of the ability of metals to form positive ions. The second ionization potential is the energy required to remove two electrons and form a doubly charged ion:

$$M \rightarrow M^{2+} + 2e^-$$

Ionization potentials stated in this way are positive; often they are given in electronvolts. *Ionization energy* is the energy required to ionize one mole of the substance, and is usually stated in kilojoules per mole (kJ mol^{-1}).

In chemistry, the terms 'second', 'third', etc., ionization potentials are usually used for the formation of doubly, triply, etc., charged ions. However, in spectroscopy and physics, they are often used with a different meaning. The second ionization potential is the energy to remove the second least strongly bound electron in forming a singly charge ion. For lithium ($ls^2 2s^1$) it would refer to removal of a 1s electron to produce an excited ion with the configuration $1s^1 2s^1$.

ionizing radiation Radiation of sufficiently high energy to cause IONIZATION. It may be short-wavelength electromagnetic radiation (ultraviolet, x-rays, or gamma rays) or streams of particles.

ion pair A positive ion and a negative ion in close proximity in solution, held by the attractive force between their charges.

IP *See* ionization potential.

IR *See* infrared.

iron A transition element occurring in many ores, especially the oxides (hematite and magnetite) and carbonate. Iron is present in a number of bioinorganic compounds, notably HEMOGLOBIN.

irreversible change *See* reversible change.

irreversible reaction A reaction in which conversion to products is complete; i.e. there is little or no back reaction.

isoenzyme (isozyme) An ENZYME that occurs in different structural forms within a single species. The isomeric forms all have the same molecular weight but differing structural configurations and properties. Large numbers of different enzymes are known to have isomeric forms; for example, lactate dehydrogenase has five forms. Variations in the isoenzyme constitution of individuals can be distinguished by electrophoresis.

isocyanide *See* isonitrile.

isocyanide test (carbylamine reaction) A test for the primary amine group in organic compounds. The sample is warmed with trichloromethane in an alcoholic solution of potassium hydroxide. If a primary amine is present the resulting isocyanide (RNC) has a characteristic smell of bad onions (and is very toxic):

$$CHCl_3 + 3KOH + RNH_2 \rightarrow RNC + 3KCl + 3H_2O$$

isoelectronic Describing compounds that have the same number of electrons. For example, carbon monoxide (CO) and nitrogen (N_2) are isoelectronic.

isoleucine *See* amino acid.

isomer *See* isomerism.

isomerism The existence of two or more chemical compounds with the same molecular formulae but different structural formulae or different spatial arrangements of atoms. The different forms are known as *isomers*. For example, the compound C_4H_{10} may be butane (with a straight chain of carbon atoms) or 2-methyl propane ($CH_3CH(CH_3)CH_3$, with a branched chain).

Structural isomerism is the type of isomerism in which the structural formulae of the compounds differ. There are two main types. In one the isomers are different types of compound. An example is the com-

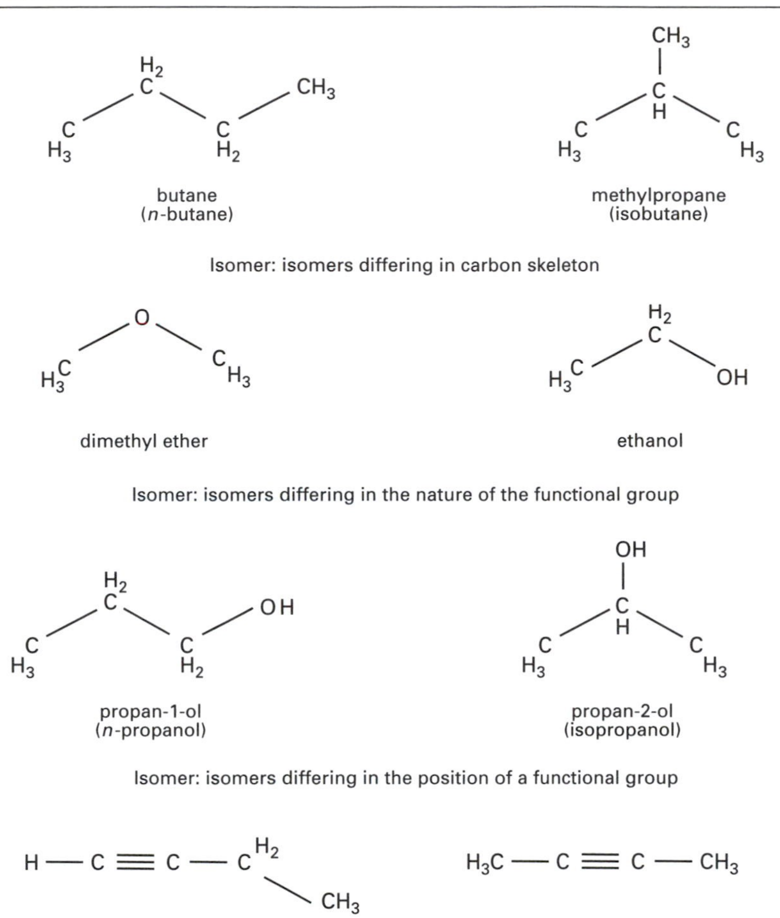

Isomer: isomers differing in carbon skeleton

Isomer: isomers differing in the nature of the functional group

Isomer: isomers differing in the position of a functional group

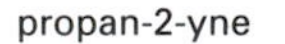

Isomer: isomers differing in the position of a multiple bond. See also illustration at alkene.

pounds ethanol (C_2H_5OH) and methoxymethane (CH_3OCH_3), both having the formula C_2H_6O but different functional groups. In the other type of structural isomerism, the isomers differ because of the position of a functional group in the molecule. For example, the primary alcohol propan-1-ol ($CH_3CH_2CH_2OH$) and the secondary alcohol propan-2-ol ($CH_3CH(OH)CH_3$) are isomers; both have the molecular formula C_3H_7OH.

Stereoisomerism occurs when two compounds with the same molecular formulae and the same groups differ only in the arrangement of the groups in space. There are two types of stereoisomerism.

Cis-trans isomerism (or *geometrical isomerism*) occurs when there is restricted rotation about a bond between two atoms. Groups attached to each atom may be on the same side of the bond (the *cis isomer*) or opposite sides (the *trans isomer*). *See also* E–Z convention.

Optical isomerism occurs when the compound has no plane of symmetry and can exist in left- and right-handed forms that are mirror images of each other. Such molecules have an asymmetric atom – i.e.

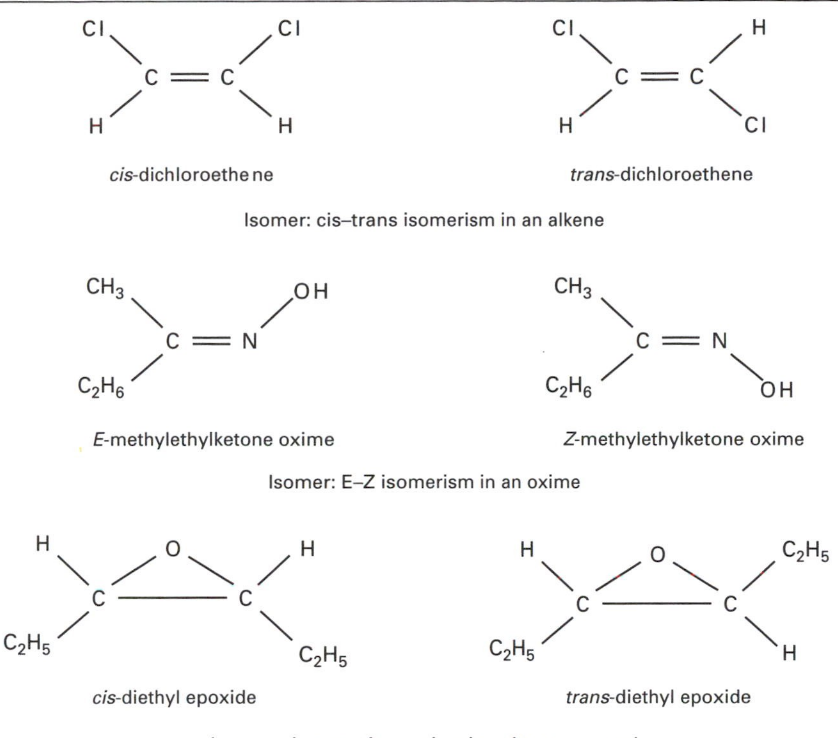

Isomer: cis–trans isomerism in an alkene

Isomer: E–Z isomerism in an oxime

Isomer: cis–trans isomerism in a ring compound

one attached to four different groups – called a *chiral center*. Molecules that are mirror images of each other are more properly called *enantiomers*. Stereoisomers that are not mirror images are called *diastereoisomers*. ANOMERS are examples of diastereoisomers. *See also* optical activity.

isonitrile (isocyanide) An organic compound of the formula R–NC.

isoprene *See* methylbuta-1,3-diene.

isopropanol Propan-2-ol. *See* propanol.

isotactic polymer *See* polymerization.

isotherm A line on a chart or graph joining points of equal temperature. *See also* isothermal change.

isothermal change A process that takes place at a constant temperature. Throughout an isothermal process, the system is in thermal equilibrium with its surroundings. For example, a cylinder of gas in contact with a constant-temperature box may be compressed slowly by a piston. The work done appears as energy, which flows into the reservoir to keep the gas at the same temperature. Isothermal changes are contrasted with *adiabatic changes*, in which no energy enters or leaves the system, and the temperature of the system changes. In practice no process is perfectly isothermal and none is perfectly adiabatic, although some can approximate in behavior to one of these ideals.

isotones Two or more nuclides that have the same neutron numbers but different proton numbers.

isotonic Describing solutions that have the same osmotic pressure.

isotope One of two or more species of the same element differing in their mass numbers because of differing numbers of neutrons in their nuclei. The nuclei must have the same number of protons (an element is characterized by its proton number). Isotopes of the same element have very similar properties because they have the same electron configuration, but differ slightly in their physical properties. An unstable isotope is termed a *radioactive isotope* or *radioisotope*

Isotopes of elements are useful in chemistry for studies of the mechanisms of chemical reactions. A standard technique is to *label* one of the atoms in a molecule by using an isotope of the element. It is then possible to trace the way in which this atom behaves throughout the course of the reaction. For example, in the esterification reaction:

$$ROH + R'COOH \rightleftharpoons H_2O + R'COOR$$

it is possible to find which bonds are broken by using a labeled oxygen atom. If the reaction is performed using ^{18}O in the alcohol it is found that this nuclide appears in the ester, showing that the C–OH bond of the acid is broken in the reaction. In labeling, radioisotopes are detected by counters; stable isotopes can also be used, and detected by a mass spectrum.

Isotopes are also used in kinetic studies to investigate the mechanism of a particular reaction. For example, if the bond between two atoms X–Y is broken in the rate-determining step, and Y is replaced by a heavier isotope of the element, Y*, then the reaction rate will be slightly lower with the Y* present. This difference in rate, known as a *kinetic isotope effect*, is significant only for reactions in which the rate-determining step involves breaking a bond to hydrogen (or deuterium) because the deuterium atom has twice the mass of the hydrogen atom.

isotopic mass (isotopic weight) The mass number of a given isotope of an element.

isotopic number The difference between the number of neutrons in an atom and the number of protons.

isotopic weight *See* isotopic mass.

isozyme *See* isoenzyme.

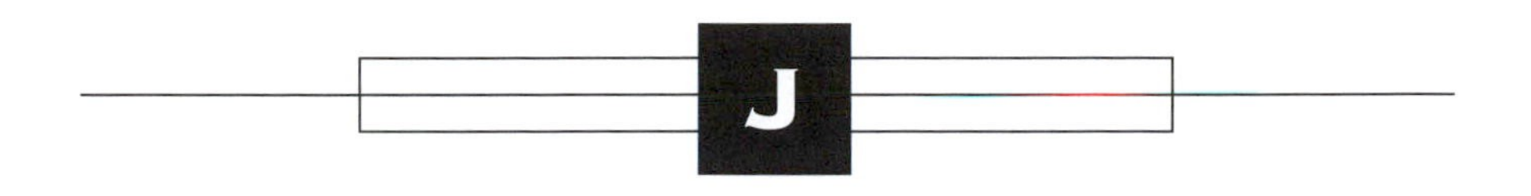

Jones oxidation The reaction in which a secondary alcohol is oxidized to a ketone using sodium chromate(VII) ($Na_2Cr_2O_7$) in dilute sulfuric acid. The reaction involves the formation of an intermediate chromate ester of the type $R_2HC–O–CrO_3H$.

joule Symbol: J The SI unit of energy and work, equal to the work done when the point of application of a force of one newton moves one meter in the direction of action of the force. 1 J = 1 N m. The joule is the unit of all forms of energy.

kairomone A chemical messenger emitted by an individual of a species and causing a response in an individual of another species. This may be detrimental to the producer of the kairomone, for example many parasites are attracted to their hosts by an excreted kairomone. *See also* pheromone.

Karrer, Paul (1889–1971) Swiss organic chemist. Karrer was famous for his work on vitamins and vegetable pigments. In 1930 he determined the structure of carotene. In 1931 he synthesized vitamin A, having worked out the structure. Karrer was aware of the similarity between the two molecules. He subsequently determined the structures of each and synthesized riboflavin (vitamin B_2) in 1937 and tocopherol (vitamin E) in 1938. Karrer shared the 1937 Nobel Prize for chemistry with Norman HAWORTH for 'the constitution of carotenoids, flavins, and vitamins A and B'. Karrer wrote an influential textbook entitled *Textbook of Organic Chemistry* (1927).

katharometer *See* gas chromatography.

Kekulé, Friedrich August von Stradonitz (1829–96) German organic chemist. Kekulé was the founder of the concept of structure in organic chemistry but is probably best remembered for the ring structure of benzene. In 1858, independently of Archibald Couper, Kekulé put forward the idea that carbon is a quadrivalent element which can combine with other carbon atoms. In 1865 he proposed the idea that the structure of benzene is a hexagonal ring of carbon atoms. In response to objections that benzene does not behave like a substance with double bonds Kekulé put forward the idea in 1872 that there is oscillation between two structures that are isomers. This view was justified about 60 years later by Linus PAULING using quantum mechanics.

kelvin Symbol: K The SI base unit of thermodynamic temperature. It is defined as the fraction 1/273.16 of the thermodynamic temperature of the triple point of water. Zero kelvin (0 K) is absolute zero. One kelvin is the same as one degree on the Celsius scale of temperature. The unit is named for the British physicist Lord Kelvin (William Thomson; 1824–1907).

Kendrew, Sir John Cowdery (1917–97) British biochemist. Kendrew is renowned for having determined the structure of the protein molecule myoglobin. He did so using x-ray crystallography. He was one of the first people to use electronic computers as an aid in analyzing the data produced by x-ray diffraction. He was able to determine the structure of myoglobin by 1960. He shared the 1962 Nobel Prize for chemistry with Max PERUTZ, his colleague at the Laboratory for Molecular Biology, Cambridge.

keratin One of a group of fibrous insoluble sulfur-containing proteins (scleroproteins) found in ectodermal cells of animals, as in hair, horns, and nails. Leather is almost pure keratin. There are two types: α keratins and β keratins. The former have a coiled structure, whereas the latter have a beta pleated sheet structure.

kerosene *See* petroleum.

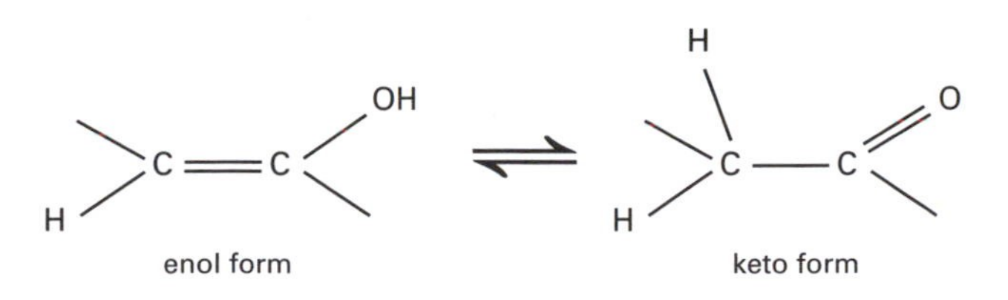

Keto–enol tautomerization

ketal *See* acetal.

keten *See* ketene.

ketene (keten) A member of a group of organic compounds, general formula $R_2C:CO$, where R is hydrogen or an organic radical. The simplest member is the colorless gas ketene, $CH_2:CO$; the other ketenes are colored because of the presence of the double bonds. Ketenes are unstable and react with other unsaturated compounds to give cyclic compounds.

keto–enol tautomerism A type of tautomerism in which a compound containing the $–CH_2–CO–$ group (the *keto form*) interconverts with one containing $–CH{=}CH(OH)–$ (the *enol*) by hydrogen migration.

keto form *See* keto–enol tautomerism.

ketohexose A ketose sugar with six carbon atoms. *See* sugar.

ketone A type of organic compound with the general formula RCOR, having two alkyl or aryl groups bound to a carbonyl group. They are made by oxidizing secondary alcohols (just as aldehydes are made from primary alcohols). Simple examples are propanone (acetone, CH_3COCH_3) and butanone (methyl ethyl ketone, $CH_3COC_2H_5$).

The chemical reactions of ketones are similar in many ways to those of ALDEHYDES. The carbonyl group is polarized, with positive charge on the carbon and negative charge on the oxygen. Thus nucleophilic addition can occur at the carbonyl group. Ketones thus:

1. Undergo addition reactions with hydrogen cyanide and hydrogensulfite (bisulfite) ions.
2. Undergo condensation reactions with hydroxylamine, hydrazine, and their derivatives.
3. Are reduced to (secondary) alcohols.

They are not, however, easily oxidized. Strong oxidizing agents give a mixture of carboxylic acids. They do not react with Fehling's solution or Tollen's reagent, and do not easily polymerize.

ketone body One of a group of organic substances formed in fat metabolism, mainly in the liver. Examples are acetoacetic acid and acetone. Ketone bodies are the major fuel source for resting skeletal muscle. If the body has little or no carbohydrate as a respiratory substrate, *ketosis* occurs, in which more ketone bodies are produced than the body can use.

ketopentose A ketose SUGAR with five carbon atoms.

ketose A SUGAR containing a keto- (=CO) or potential keto- group.

kieselguhr A siliceous deposit formed by diatoms, used as an absorbent, filter, and filler.

kilo- Symbol: k A prefix denoting 10^3. For example, 1 kilometer (km) = 10^3 meters (m).

kilogram (kilogramme; symbol: kg) The SI base unit of mass, equal to the mass of the international prototype of the kilogram, which is a piece of platinum–iridium kept at Sèvres in France.

kilogramme An alternative spelling of *kilogram.*

kilowatt-hour Symbol: kWh A unit of energy, usually electrical, equal to the energy transferred by one kilowatt of power in one hour. It has a value of 3.6×10^6 joules.

kinetic energy *See* energy.

kinetic isotope effect *See* isotope.

kinetics A branch of physical chemistry concerned with the study of rates of chemical reactions and the effect of physical conditions that influence the rate of reaction, e.g. temperature, light, concentration, etc. The measurement of these rates under different conditions gives information on the mechanism of the reaction, i.e. on the sequence of processes by which reactants are converted into products.

Kipp's apparatus An apparatus for the production of a gas from the reaction of a liquid on a solid. It consists of three globes, the upper globe being connected via a wide tube to the lower globe. The upper globe is the liquid reservoir. The middle globe contains the solid and also has a tap at which the gas may be drawn off. When the gas is drawn off the liquid rises from the lower globe to enter the middle globe and reacts with the solid, thereby releasing more gas. When turned off the gas released forces the liquid back down into the lower globe and up into the reservoir, thus stopping the reaction.

Kjeldahl's method A method used for the determination of nitrogen in organic compounds. The nitrogenous substance is converted to ammonium sulfate by boiling with concentrated sulfuric acid (often with a catalyst such as $CuSO_4$) in a specially designed long-necked *Kjeldahl flask.* The mixture is then made alkaline and the ammonia distilled off into standard acid for measurement by titration. It is named for the Danish chemist Johan Kjeldahl (1849–1900).

Klug, (Sir) Aaron (1926–) Lithuanian-born British molecular biologist distinguished for his determination of the structure of transfer RNA and for his work on three-dimensional structure of complexes of proteins and nucleic acids. He was awarded the Nobel Prize for chemistry in 1982 for his development of crystallographic electron microscopy and his work on the structures of biologically important nucleic acid–protein complexes.

Kolbe, Adolph Wilhelm Hermann (1818–84) German organic chemist. Kolbe made many important contributions to the development of organic chemistry. These included the synthesis of ethanoic acid from completely inorganic materials, the formation and hydrolysis of nitriles, the electrolysis of solutions of fatty acid salts and the synthesis of salicylic acid from phenol and carbon dioxide. In his later years his influence had a negative effect on the development of chemistry because he opposed concepts such as structure.

Kolbe electrolysis The electrolysis of sodium salts of carboxylic acids to prepare alkanes. The alkane is produced at the anode after discharge of the carboxylate anion and decomposition of the radical:

$$RCOO^- \rightarrow RCOO\bullet + e^-$$
$$RCOO\bullet \rightarrow R\bullet + CO_2$$
$$\rightarrow R - R + CO_2$$

and

$$2R\bullet \rightarrow R - R$$

As the reaction is a coupling reaction, only alkanes with an even number of carbon atoms in the chain can be prepared in this way. It is named for the German chemist Adolph Kolbe.

Kornberg, Arthur (1918–) American biochemist. Kornberg conducted research on enzymes at the start of his career. In this period he clarified the chemical reactions leading to certain enzymes. In 1956 he discovered an enzyme, which he called DNA polymerase, that catalyses the process of forming polynucleotides from nucleoside triphosphates. This enabled him to synthesize short DNA molecules, starting from a DNA template and triphosphate bases. He shared the 1959 Nobel Prize for medicine for this work with Severo Ochoa, who found the enzyme that catalyses the formation of RNA.

Kossel, Karl Martin Leonhard Albrecht (1853–1927) German chemist noted for his discovery of adenine, cytosine, thymine, and uracil as breakdown products of nucleic acids. He also discovered histone and histidine. He was awarded the Nobel Prize for physiology or medicine in 1910 in recognition of his work on cell chemistry.

Krebs, (Sir) Hans Adolf (1900–81) German-born British biochemist renowned for his work on metabolism and, in particular, for the discovery of the cyclic metabolic pathways named after him. He was awarded the Nobel Prize for physiology or medicine in 1953 for his discovery of the TCA cycle. The prize was shared with F. A. Lipmann.

Krebs cycle (tricarboxylic acid cycle; TCA cycle; citric acid cycle) A complex and almost universal cycle of reactions in which the acetyl group of acetyl CoA is oxidized to carbon dioxide and water, with the production of large amounts of energy. It is the final common pathway for the oxidation of carbohydrates, fatty acids, and amino acids. It requires oxygen, and in eukaryotes occurs in the mitochondrial matrix.

2-carbon acetate reacts with 4-carbon oxaloacetate to form 6-carbon citrate, which is then decarboxylated to reconstitute oxaloacetate. Some ATP is produced by direct coupling with cycle reactions, but most production is coupled to the electron-transport chain via the generation of reduced coenzymes, NADH and $FADH_2$. *See* electron-transport chain.

Kroto, Sir Harold Walter (1939–) British chemist. In the mid-1980s Harold Kroto heard that the American chemist

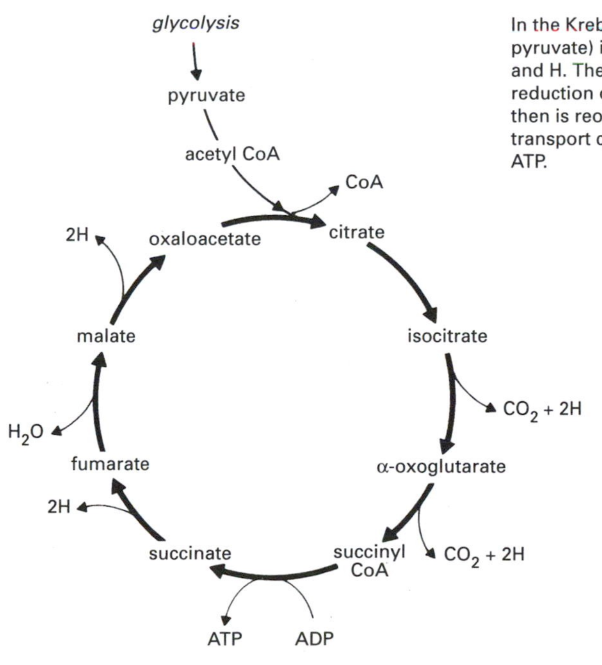

In the Krebs cycle acetyl (from pyruvate) is broken down to CO_2 and H. The H is held in NADH (from reduction of NAD^+). The NADH then is reoxidized in an electron-transport chain, with production of ATP.

Krebs cycle

Richard Smalley was using laser bombardment as a new technique to produce clusters of atoms. When Kroto visited Smalley he persuaded him to bombard graphite with a laser beam. When Smalley did so he found not only the chains that Kroto had been expecting but a molecule with 60 carbon atoms. Kroto and Smalley correctly postulated that this molecule is a polyhedron in which the faces are pentagons or hexagons. Kroto called this molecule *buckminsterfullerene* after its resemblance to the designs of the architect Buckminster Fuller. The name of the molecule is usually abbreviated to *fullerene*. Such molecular structures are frequently called 'bucky balls'. Kroto and Smalley shared the 1996 Nobel Prize for chemistry with Robert Curl for their parts in the discovery of C_{60} molecules.

label A stable or radioactive ISOTOPE used to investigate a chemical reaction. Labeling is a common method of investigating chemical reaction mechanisms. A classic example is the hydrolysis of an ester to give a carboxylic acid and an alcohol, as in:

$$H_2O + CH_3COOCH_3 \rightarrow CH_3COOH + CH_3OH$$

It is possible to investigate the mechanism by using an ester enriched with the isotope ^{18}O. If the ^{18}O is on the oxygen attached to the carbonyl, it is found that it ends up in the alcohol rather than the acid:

$$H_2O + CH_3CO^{18}OH \rightarrow CH_3COOH + CH_3{}^{18}OH$$

This, and similar experiments using labeled water and labeling of the carbonyl oxygen, help to establish the mechanism.

Certain radioisotopes such as tritium (^3H) and ^{14}C have been used as labels but it is now more usual to use deuterium (^2H), ^{13}C, and ^{18}O, which can be detected by mass spectroscopy.

lactam A type of organic compound containing the –NH.CO– group as part of a ring in the molecule. Lactams can be regarded as formed from a straight-chain compound that has an amino group ($-NH_2$) at one end of the molecule and a carboxylic acid group (–COOH) at the other; i.e. from an amino acid. The reaction of the amine group with the carboxylic acid group, with elimination of water, leads to the cyclic lactam, which is thus an internal (or inner) amide. Lactams can exist in an alternative tautomeric form in which the hydrogen atom has migrated from the nitrogen onto the O of the carbonyl. This, the *lactim* form, contains the group –N=C(OH)–.

lactate A salt or ester of lactic acid.

lactic acid (2-hydroxypropanoic acid) A colorless liquid carboxylic acid:

$$CH_3CH(OH)COOH$$

See optical activity.

lactim *See* lactam.

lactone A type of organic compound containing the group –O.CO– as part of a ring in the molecule. A lactone can be regarded as formed from a compound with an alcohol (–OH) group on one end of the chain and a carboxylic acid (–COOH) group on the other. The lactone then results from reaction of the –OH group with the –COOH group; i.e. it is an internal ester.

lactose (milk sugar; $C_{12}H_{22}O_{11}$) A SUGAR found in milk. It is a disaccharide composed of glucose and galactose units.

Ladenburg benzene An incorrect structure for BENZENE in which the six carbon atoms are at the corners of a triangular prism. It is named for Albert Ladenburg (1842–1911). The actual compound, known as *prismane*, was synthesized in 1973.

lake A pigment formed by absorbing an organic dyestuff on an inorganic oxide, hydroxide, or salt.

lamellar compound A compound with a crystal structure composed of thin plates or layers. Silicates form many compounds with distinct layers. Typical examples are talc ($Mg_3(OH)_2Si_4O_{10}$) and pyrophyllite ($Al_2(OH)_2Si_4O_{10}$).

lanolin A yellowish viscous substance obtained from wool fat. It contains cholesterol and terpene compounds, and is used in cosmetics, in ointments, and in treating leather.

Lapworth, Arthur (1872–1941) British organic chemist. In the early part of his career Lapworth investigated the structure of camphor and related compounds. However, his most significant contribution to chemistry was his work, starting in 1920, on the mechanisms of organic molecules in terms of the electrons in the molecules. For example, he was one of the first people to emphasize that ionization can occur in organic molecules and that parts of these molecules can have electrical charges, either on a permanent basis or when the reaction is occurring. Similar views on describing organic reactions in terms of electrons were developed by Sir Christopher INGOLD and Sir Robert ROBINSON.

laser An acronym for Light Amplification by Stimulated Emission of Radiation. A laser device produces high-intensity, monochromatic, coherent beams of light. In the laser process the molecules of a sample (such as ruby doped with Cr^{3+} ions) are promoted to an excited state. As the sample is in a cavity between two reflective surfaces, when a molecule emits spontaneously, the photon so generated ricochets backward and forward. In this way other molecules are stimulated to emit photons of the same energy. If one of the reflective surfaces is partially transmitting this radiation can be tapped.

latent heat The heat evolved or absorbed when a substance changes its physical state, e.g. the latent heat of fusion is the heat absorbed when a substance changes from a solid to a liquid.

latex A liquid found in some flowering plants contained in special cells or vessels called laticifers (or laticiferous vessels). It is a complex variable substance that may contain terpenes (e.g. rubber), resins, tannins, waxes, alkaloids, sugar, starch, enzymes, crystals, etc. It is often milky in appearance but may be colorless, orange, or brown. Its function is obscure, but may be involved in wound healing as well as a repository for excretory substances. Commercial rubber comes from the latex of the rubber plants *Ficus elastica* and *Hevea brasiliensis*.Opium comes from alkaloids found in the latex of the opium poppy.

lattice A regular three-dimensional arrangement of points. A lattice is used to describe the positions of the particles (atoms, ions, or molecules) in a crystalline solid. The lattice structure can be examined by x-ray diffraction techniques.

lauric acid *See* dodecanoic acid.

law of conservation of energy *See* conservation of energy; law of.

law of conservation of mass *See* conservation of mass; law of.

law of constant composition *See* constant proportions; law of.

law of constant proportions *See* constant proportions; law of.

law of definite proportions *See* definite proportions; law of.

law of equivalent proportions *See* equivalent proportions, law of.

law of mass action *See* mass action; law of.

law of reciprocal proportions *See* equivalent proportions, law of.

laws of chemical combination *See* chemical combination; laws of.

LDL *See* low-density lipoprotein.

leaching The washing out of a soluble material from an insoluble solid using a solvent. This is often carried out in batch tanks or by dispersing the crushed solid in a liquid.

lead-free fuel Vehicle fuel that contains none of the anti-knock agent LEAD TETRAETHYL.

lead tetraethyl (tetraethyl lead; $Pb(CH_2CH_3)_4$) A poisonous liquid that is insoluble in water but soluble in organic solvents. It is manufactured by the reaction of an alloy of sodium and lead with 1-chloroethane. The product is obtained by steam distillation. Lead tetraethyl was formerly extensively used as an additive in internal-combustion engine fuel to increase its octane number and thus prevent preignition (knocking).

leaving group The group leaving a molecule in a substitution or elimination reaction. The nature of the leaving group is an important factor in the progress of the reaction.

Le Chatelier's principle If a system is at equilibrium and a change is made in the conditions, the equilibrium adjusts so as to oppose the change. The principle can be applied to the effect of temperature and pressure on chemical reactions. A good example is the Haber process for synthesis of ammonia:

$$N_2 + 3H_2 \rightleftharpoons 2NH_3$$

The 'forward' reaction

$$N_2 + 3H_2 \rightarrow 2NH_3$$

is exothermic. Thus, reducing the temperature displaces the equilibrium towards production of NH_3 (as this tends to increase temperature). Increasing the pressure also favors formation of NH_3, because this leads to a reduction in the total number of molecules (and hence pressure). The principle is named for the French chemist Henri Louis Le Chatalier (1850–1936), who discovered it in 1884.

leucine *See* amino acid.

levo-form *See* optical activity.

levorotatory *See* optical activity.

Lewis acid *See* acid.

Lewis base *See* acid.

L-form *See* optical activity.

Liebig, Justus von (1803–73) German organic chemist. Lieberg was one of the most influential chemists of the 19th century. The early part of his career was devoted to classical organic chemistry. After 1840 he mostly devoted himself to biochemistry and the application of chemistry to agriculture. In 1826 he made his first significant discovery when, together with his long-standing colleague Friedrich WÖHLER, he discovered that silver fulminate and silver cyanate have the same chemical formula. In this way, Liebig and Wöhler discovered *isomerism*. Liebig and Wöhler discovered the benzoyl radical in 1832. When Liebig moved into biochemistry his tendency towards dogmatism became even more pronounced than it had been previously. As a result, he became involved in many acrimonious controversies. He was a prolific writer of books and papers.

Liebig condenser A simple type of laboratory condenser. It consists of a straight glass tube, in which the vapor is condensed, with a surrounding glass jacket through which cooling water flows. It is named for the German chemist Justus von Liebig.

ligand A molecule or ion that forms a coordinate bond to a metal atom or ion in a complex.

light (visible radiation) A form of electromagnetic radiation able to be detected by the human eye. Its wavelength range is between about 400 nm (far red) and about 700 nm (far violet). The boundaries are not precise because individuals vary in their ability to detect extreme wavelengths; this ability also declines with age.

lignin One of the main structural materials of vascular plants. With cellulose it is one of the main constituents of wood. Lignified tissues include sclerenchyma and xylem. Lignin is deposited during secondary thickening of cell walls. The degree of lignification varies but values of 25–30% lignin and 50% cellulose are average.

It is a complex variable polymer, derived from sugars via aromatic alcohols. Phenylpropane (C_6–C_3) units are linked in various ways by oxidation reactions during polymerization.

lignite (brown coal) The poorest grade of coal, containing up to 45% carbon and with a high moisture content.

ligroin A mixture of hydrocarbons obtained from petroleum, used as a general solvent. It has a boiling range of 80 to 120°C. *See* petroleum.

limiting step *See* rate-determining step.

line spectrum A SPECTRUM composed of a number of discrete lines corresponding to single wavelengths of emitted or absorbed radiation. Line spectra are produced by atoms or simple (monatomic) ions in gases. Each line corresponds to a change in electron orbit, with emission or absorption of radiation.

linoleic acid ($C_{17}H_{31}COOH$) An unsaturated carboxylic acid that occurs in LINSEED OIL and other plant oils. It contains two double bonds.

linolenic acid ($C_{17}H_{29}COOH$) A liquid unsaturated carboxylic acid that occurs in LINSEED OIL and other plant oils. It contains three double bonds.

linseed oil An oil extracted from the seeds of flax (linseed). It hardens on exposure to air (it is a drying oil) because it contains linoleic acid, and is used in enamels, paints, putty, and varnishes. *See* linoleic acid.

lipid A collective term used to describe a group of substances in cells characterized by their solubility in organic solvents such as ether and benzene, and their absence of solubility in water. The group is rather heterogeneous in terms of both function and structure. They encompass the following broad bands of biological roles: (1) basic structural units of cellular membranes and cytologically distinct subcellular bodies such as chloroplasts and mitochondria; (2) compartmentalizing units for metabolically active proteins localized in membranes; (3) a store of chemical energy and carbon skeletons; and (4) primary transport systems of nonpolar material through biological fluids. There are also the more physiologically specific lipid hormones, e.g. the steroid hormones and lipid vitamins.

The simple lipids include *neutral lipids* or glycerides, which are esters of glycerol and fatty acids, and the waxes, which are esters of long-chain monohydric alcohols and fatty acids.

Compound lipids have one of the fatty acid parts replaced, such that complete hydrolysis gives only two fatty acids; the phospholipids, which are particularly important examples.

lipoic acid A sulfur-containing fatty acid found in a wide variety of natural materials. It is an essential as a coenzyme for certain dehydrogenase enzymes, notably pyruvate dehydrogenase, which catalyzes the dehydrogenation of pyruvic acid to form acetyl-CoA. Lipoic acid is classified with the water-soluble B vitamins and has not yet been demonstrated to be required in the diet of higher animals.

lipopolysaccharide A conjugated polysaccharide in which the noncarbohydrate part is a lipid. Lipopolysaccharides are a constituent of the cell walls of certain bacteria.

lipoprotein Any conjugated protein formed by the combination of a protein with a lipid. In the blood of humans and other mammals, cholesterol, triglycerides, and phospholipids associate with various plasma proteins to form lipoproteins. These are particles with diameters in the region of 7.5–70 nm, and are placed in several classes. The largest lipoproteins in this range are the *very low-density lipoproteins* (*VLDLs*), which are formed in the liver and contain up to about 20% cholesterol. *Low-density lipoproteins* (*LDLs*) are formed in plasma from VLDLs and contain over 50% cholesterol. LDLs transport cholesterol from the liver to peripheral tissues, and an excess of LDLs in the blood is a factor in the development of fatty arterial deposits and cardiovascular disease. *High-density lipoproteins* (HDLs), with about 20% cholesterol, are the smallest of the plasma lipoproteins, and apparently function in transporting cholesterol from tissues to the liver.

liquefied natural gas (LNG) Liquid methane, obtained from natural gas and used as a fuel. *See* methane; natural gas.

liquefied petroleum gas (LPG) A mixture of liquefied hydrocarbon gases (mainly propane) extracted from petroleum, used as a fuel for internal combustion engines and for heating. *See* petroleum; propane.

liquid The state of matter in which the particles of a substance are loosely bound by intermolecular forces. The weakness of these forces permits movement of the particles and consequently liquids can change their shape within a fixed volume. The liquid state lacks the order of the solid state. Thus, amorphous materials, such as glass, in which the particles are disordered and can move relative to each other, can be classed as liquids.

liquid–liquid extraction *See* solvent extraction.

liter Symbol: l A unit of volume now defined as 10^{-3} meter^3; i.e. 1000 cm^3. The milliliter (ml) is thus the same as the cubic centimeter (cm^3). However, the name is not recommended for precise measurements because the liter was formerly defined as the volume of one kilogram of pure water at 4°C and standard pressure. On this definition, one liter is the same as 1000.028 cubic centimeters.

lithium aluminum hydride *See* lithium tetrahydridoaluminate(III).

lithium tetrahydridoaluminate(III) (lithium aluminum hydride; $LiAlH_4$) A white solid produced by action of lithium hydride on aluminum chloride, the hydride being in excess. Lithium tetrahydridoaluminate reacts violently with water. It is a powerful reducing agent, reducing ketones and carboxylic acids to their corresponding alcohols. In inorganic chemistry it is used in the preparation of hydrides.

litmus A natural pigment that changes color when in contact with acids and alkalis; above a pH of 8.3 it is blue and below a pH of 4.5 it is red. Thus it gives a rough indication of the acidity or basicity of a solution; because of its rather broad range of color change it is not used for precise work. Litmus is used both in solution and as litmus paper.

lixiviation The process of separating soluble components from a mixture by washing them out with water.

LNG *See* liquefied natural gas.

localized bond A bond in which the electrons contributing to the bond remain between the two atoms concerned, i.e. the bonding orbital is localized. The majority of bonds are of this type. *Compare* delocalized bond.

lone pair A pair of valence electrons having opposite spin that are located together on one atom, i.e. are not shared as in a covalent bond. Lone pairs occupy similar positions in space to bond pairs and account for the shapes of molecules. A molecule with a lone pair can donate the pair of electrons to an electron acceptor, such as H^+ or a metal ion, to form coordinate bonds.

long period *See* period.

Lonsdale, Dame Kathleen (1903–71) British x-ray crystallographer. Dame Kathleen Lonsdale was an early pioneer of x-ray crystallography. In 1929 she showed the hexagonal ring nature of the hexamethylbenzene molecule. In 1931 she investigated the structure of hexachlorobenzene. This was the first investigation in which Fourier analysis had been used for an organic molecule. She subsequently investigated many problems including thermal motion in crystals, the magnetic susceptibility of crystals and the composition of bladder stones. She reviewed her research in her book *Crystals and X-Rays* (1948). She was also the editor of the first three volumes of the *International Tables for X-Ray Crystallography* (1952, 1959, 1962).

low-density lipoprotein (LDL) A spherical particle, typically about 20–25 nm in diameter, that is found in blood plasma and transports cholesterol to tissue cells. It is bounded by a single layer of phospholipid and free cholesterol, which encloses a core of cholesterol esterified to a long-chain fatty acid. Embedded in the surface layer is a single large protein, called apo-B, which assists in binding of the LDL to cell-surface receptors. LDLs are taken into cells by receptor-mediated endocytosis. The cholesterol is incorporated into cell membranes or stored as lipid droplets. High concentrations of LDLs in the blood have been associated with an increased risk of atherosclerosis ('hardening of the arteries').

lowering of vapor pressure A colligative property of solutions in which the vapor pressure of a solvent is lowered as a solute is introduced. When both solvent and solute are volatile the effect of increasing the solute concentration is to lower the partial vapor pressure of each component. When the solute is a solid of negligible vapor pressure the lowering of the vapor pressure of the solution is directly proportional to the number of species introduced rather than to their nature and the proportionality constant is regarded as a general solvent property. Thus the introduction of the same number of moles of any solute causes the same lowering of vapor pressure, if dissociation does not occur. If the solute dissociates into two species on dissolution the effect is doubled. The kinetic model for the lowering of vapor pressure treats the solute molecules as occupying part of the surface of the liquid phase and thereby restricting the escape of solvent molecules. The effect can be used in the measurement of relative molecular masses, particularly for large molecules, such as polymers. *See also* Raoult's law.

Lowry–Brønsted theory *See* acid.

LPG *See* liquefied petroleum gas.

LSD *See* lysergic acid diethylamide.

lumen Symbol: lm The SI unit of luminous flux, equal to the luminous flux emitted by a point source of one candela in a solid angle of one steradian. 1 lm = 1 cd sr.

luminescence The emission of radiation from a substance in which the particles have absorbed energy and gone into excited states. They then return to lower energy states with the emission of electromagnetic radiation. If the luminescence persists after the source of excitation is removed it is called *phosphorescence*: if not, it is called *fluorescence*.

lutein The commonest of the xanthophyll pigments. It is found in green leaves and certain algae, e.g. the Rhodophyceae. *See* photosynthetic pigments.

lux Symbol: lx The SI unit of illumination, equal to the illumination produced by a luminous flux of one lumen falling on a surface of one square meter. 1 lx = 1 lm m^{-2}.

lyophilic Solvent attracting. When the solvent is water, the word *hydrophilic* is often used. The terms are applied to:

1. Ions or groups on a molecule. In aqueous or other polar solutions ions or polar groups are lyophilic. For example, the $-COO^-$ group on a soap is the lyophilic (hydrophilic) part of the molecule.
2. The disperse phase in colloids. In lyophilic colloids the dispersed particles have an affinity for the solvent, and the colloids are generally stable. *Compare* lyophobic.

lyophobic Solvent repelling. When the solvent is water, the word *hydrophobic* is used. The terms are applied to:

1. Ions or groups on a molecule. In aqueous or other polar solvents, the lyophobic group will be nonpolar. For example, the hydrocarbon group on a soap molecule is the lyophobic (hydrophobic) part.
2. The disperse phase in colloids. In lyophobic colloids the dispersed particles are not solvated and the colloid is easily solvated. Gold and sulfur sols are examples. *Compare* lyophilic.

lysergic acid diethylamide (LSD) A synthetic organic compound that has physiological effects similar to those produced by alkaloids in certain fungi. Even small quantities, if ingested, produce hallucinations and extreme mental disturbances. The initials LSD come from the German form of the chemical's name, *Lysergic-Saure-Diathylamide*.

lysine *See* amino acid.

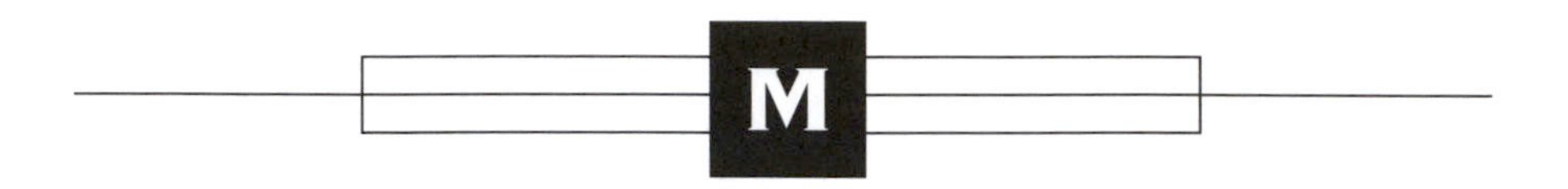

macromolecular crystal A crystal composed of atoms joined together by covalent bonds that form giant three-dimensional or two-dimensional networks. Diamond is an example of a macromolecular crystal.

macromolecule A large molecule; e.g. a natural or synthetic polymer.

magic acid *See* superacid.

magnetic quantum number *See* atom.

magnetism The study of the nature and cause of magnetic force fields, and how different substances are affected by them. Magnetic fields are produced by moving charge – on a large scale (as with a current in a coil, forming an *electromagnet*), or on the small scale of the moving charges in the atoms. It is generally assumed that the Earth's magnetism and that of other planets, stars, and galaxies have the same cause.

Substances may be classified on the basis of how samples interact with fields. Different types of magnetic behavior result from the type of atom. *Diamagnetism*, which is common to all substances, is due to the orbital motion of electrons. *Paramagnetism* is due to electron spin, and a property of materials containing unpaired electrons. It is particularly important in transition-metal chemistry, in which the complexes often contain unpaired electrons. Magnetic measurements can give information about the bonding in these complexes. *Ferromagnetism*, the strongest effect, also involves electron spin and the alignment of magnetic moments in domains.

maleic acid *See* butenedioic acid.

malic acid (2-hydroxybutanedioic acid; $HCOOCH_2CH(OH)COOH$) A colorless crystalline CARBOXYLIC ACID found in unripe fruits. It tastes of apples and is used in food flavorings.

malonic acid *See* propanedioic acid.

maltose ($C_{12}H_{22}O_{11}$) A SUGAR found in germinating cereal seeds. It is a disaccharide composed of two glucose units. Maltose is an important intermediate in the enzyme hydrolysis of starch. It is further hydrolyzed to glucose.

mannitol ($HOCH_2(CHOH)_4CH_2OH$) A soluble hexahydric alcohol that occurs in many plants and fungi. It is used in medicines and as a sweetener (particularly in foods for diabetics). It is an isomer of sorbitol.

mannose ($C_6H_{12}O_6$) A simple SUGAR found in many polysaccharides. It is an aldohexose, isomeric with glucose.

manometer A device for measuring pressure. A simple type is a U-shaped glass tube containing mercury or other liquid. The pressure difference between the arms of the tube is indicated by the difference in heights of the liquid.

Markovnikoff's rule A rule that predicts the quantities of the products formed when an acid (HA) adds to the double bond in an alkene. If the alkene is not symmetrical two products may result; for instance $(CH_3)_2C{:}CH_2$ can yield either $(CH_3)_2HCCH_2A$ or $(CH_3)_2ACCH_3$. The rule states that the major product will be the one in which the hydrogen atom attaches itself to the carbon atom with the larger number of hydrogen atoms. In the example above, therefore, the major product is $(CH_3)_2ACCH_3$.

The Markovnikoff rule is explainable if the mechanism is ionic. The first step is addition of H^+ to one side of the double bond, forming a carbonium ion. The more stable form of carbonium ion will be the form in which the positive charge appears on the carbon atom with the largest number of alkyl groups – thus the hydrogen tends to attach itself to the other carbon. The positive charge is partially stabilized by the electron-releasing (inductive) effect of the alkyl groups.

Additions of this type do not always follow the Markovnikoff rule. Under certain conditions the reaction may involve the free radicals H• and A•, in which case the opposite (*anti-Markovnikoff*) effect occurs.

marsh gas Methane produced in marshes by decomposing vegetation.

Martin, Archer John Porter (1910–) British biochemist. Martin is best known for the development of paper chromatography with Richard SYNGE, starting in 1941. Martin and Synge won the 1952 Nobel Prize for chemistry for this work. In the method they used a drop of the mixture which is being analyzed is placed at one of the ends of a strip of filter paper. This is then soaked in a solvent which carries the components of the mixture at different rates as it moves along the strip. The positions of the components are revealed by spraying the strip with a reagent which causes the components to change color. Martin and Synge were able to identify amino acids in a protein this way, using ninhydrin to record the positions of the amino acid on the strip.

mass Symbol: m A measure of the quantity of matter in an object. Mass is determined in two ways: the *inertial mass* of a body determines its tendency to resist change in motion; the *gravitational mass* determines its gravitational attraction for other masses. The SI unit of mass is the kilogram.

mass action, law of The principle that, at constant temperature, the rate of a chemical reaction is directly proportional to the *active mass* of the reactants, the active mass being taken as the concentration for a reaction in solution or the partial pressure for a gas reaction. For the reaction A + B → products, the law of mass action states that:

$$\text{rate} = k[A][B]$$

where [A] represents the concentration of A, [B] the concentration of B, and k is a constant dependent on the particular reaction. The interpretation of active mass as concentration or partial pressure is only valid if there is no interaction or interference between the reacting molecules. In general the concentration has to be multiplied by an ACTIVITY COEFFICIENT in order to obtain the actual active mass. *See* activity coefficient.

mass number *See* nucleon number.

mass spectrometer An instrument for producing ions and analyzing them according to their charge/mass ratio. The earliest experiments by the British physicist J. J. Thomson (1856–1940) used a stream of positive ions from a discharge tube, which were deflected by parallel electric and magnetic fields at right angles to the beam. Each type of ion formed a parabolic trace on a photographic plate (a *mass spectrograph*). The design was improved by the British chemist Francis William Aston (1877–1945). In modern instruments, the ions are usually produced by ionizing a gas with electrons. The positive ions are accelerated out of this ion source into a high-vacuum region. Here, the stream of ions is deflected and focused by a combination of electric and magnetic fields, which can be varied so that different types of ion fall on a detector. In this way, the ions can be analyzed according to their mass, giving a *mass spectrum* of the material. Mass spectrometers are used for accurate measurements of relative atomic mass and for analysis of isotope abundance. They can also be used to identify compounds and analyze mixtures. An organic compound bombarded with electrons forms a number of fragment ions. (Ethane (C_2H_6), for instance, might form CH^+, $C_2H_5^+$, CH_2^+, etc.) The relative proportions of different types of ions may be used to find the structure of new compounds. The characteristic spectrum can also identify compounds by comparison with standard spectra.

mass spectrum *See* mass spectrometer.

matrix A continuous solid phase in which particles of a different solid phase are embedded.

mechanism A step-by-step description of the events taking place in a chemical reaction. It is a theoretical framework accounting for the fate of bonding electrons and illustrates which bonds are broken and which are formed. For example, in the chlorination of methane to give chloromethane:

step 1

$$Cl{:}Cl \rightarrow 2Cl\bullet$$

step 2

$$Cl\bullet + CH_4 \rightarrow HCl + CH_3\bullet$$

step 3

$$CH_3\bullet + Cl{:}Cl \rightarrow CH_3Cl + Cl\bullet$$

See also nucleophilic substitution.

mega- Symbol: M A prefix denoting 10^6. For example, 1 megahertz (MHz) = 10^6 hertz (Hz).

melamine (triaminotriazine; $C_3N_3(NH_2)_3$) A white solid organic compound whose molecules consist of a six-membered heterocyclic ring of alternate carbon and nitrogen atoms with three amino groups attached to the carbons. Condensation polymerization with methanal or other aldehydes produces *melamine resins*, which are important thermosetting plastics.

melting (fusion) The process by which a solid is converted into a liquid by heat or pressure.

melting point The temperature at which a solid is in equilibrium with its liquid phase at standard pressure and above which the solid melts. This temperature is always the same for a particular solid. Ionically bonded solids generally have much higher melting points than those in which the forces are covalent or intermolecular.

membrane A thin pliable sheet of tissue or other material acting as a boundary. The membrane may be either natural (as in cells, skin, etc.) or synthetic modifications of natural materials (cellulose derivatives or rubbers). In many physicochemical studies membranes are supported on porous materials, such as porcelain, to provide mechanical strength. Membranes are generally permeable to some degree.

Membranes can be prepared to permit the passage of other molecules and micromolecular material. Because of permeability effects, concentration differences at a membrane give rise to a whole range of membrane-equilibrium studies, of which osmosis, dialysis, and ultrafiltration are examples. *See also* semipermeable membrane.

menaquinone *See* vitamin K.

Mendius reaction The reduction of the cyanide group to a primary amine group using sodium in alcohol:

$$RCN + 2H_2 \rightarrow RCH_2NH_2$$

It is a method of increasing the chain length of compounds in ascending a homologous series of compounds.

mer *See* polymer.

mercaptan *See* thiol.

meso-form *See* optical activity.

mesomerism *See* resonance.

meta- **1.** Designating a benzene compound with substituents in the 1,3 positions. The position on a benzene ring that is two carbon atoms away from a substituent is the meta position. This was used in the systematic naming of benzene deriv-

atives. For example, meta-dinitrobenzene (or *m*-dinitrobenzene) is 1,3-dinitrobenzene.
2. Certain inorganic acids regarded as formed from an anhydride and water are named meta acids to distinguish them from the more hydrated ortho acids. For example, H_2SiO_3 ($SiO_2 + H_2O$) is metasilicic acid; H_4SiO_4 ($SiO_2 + 2H_2O$) is orthosilicic acid.

See also ortho-; para-.

metabolic pathway *See* metabolism.

metabolism The biochemical reactions that take place in cells. The molecules taking part in these reactions, either as a reactant or a product, are termed *metabolites*. Some are synthesized within the organism itself, whereas others have to be taken in as food. Metabolic reactions characteristically occur in small steps that together make up a *metabolic pathway*. They involve the breaking down of molecules to provide energy (*catabolism*) and the building up of more complex molecules and structures from simpler molecules (*anabolism*).

metabolite *See* metabolism.

metaldehyde *See* ethanal.

metallic bond A bond formed between atoms of a metallic element in its zero oxidation state and in an array of similar atoms. The outer electrons of each atom are regarded as contributing to an 'electron gas', which occupies the whole crystal of the metal. It is the attraction of the positive atomic cores for the negative electron gas that provides the strength of the metallic bond.

metallocene A SANDWICH COMPOUND in which a metal atom or ion is coordinated to two cyclopentadienyl ions. Ferrocene ($Fe(C_5H_5)_2$) is the commonest example.

metalloid Any of a class of chemical elements that are intermediate in properties between metals and nonmetals. Examples are germanium, arsenic, and tellurium.

metastable species An excited state of an atom, ion, or molecule, that has a relatively long lifetime before reverting to the ground state. Metastable species are intermediates in some chemical reactions.

metastable state A condition of a system or body in which it appears to be in stable equilibrium but, if disturbed, can settle into a lower energy state. For example, supercooled water is liquid below 0°C (at standard pressure). When a small crystal of ice or dust (for example) is introduced, rapid freezing occurs.

meter Symbol: m The SI base unit of length, defined as the distance traveled by light in vacuum in $1/(2.99\ 792\ 458 \times 10^8)$ second. This definition was adopted in 1983 to replace the 1967 definition of a length equal to 1 650 763.73 wavelengths in vacuum corresponding to the transition between the levels $2p^{10}$ and $5d^5$ of the ^{86}Kr atom.

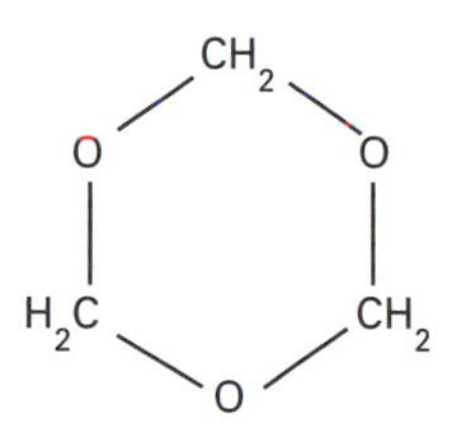

Methanal: methanal trimer

methanal (formaldehyde; HCOH) A colorless gaseous aldehyde. It is manufactured by the oxidation of methanol (500°C and a silver catalyst):

$$2CH_3OH + O_2 \rightarrow 2HCOH + 2H_2O$$

The compound is used in the manufacture of UREA–FORMALDEHYDE RESINS. A solution of methanal (40%) in water is called *formalin*. It is extensively used as a preservative for biological specimens.

If an aqueous solution of methanal is evaporated a polymer – *polymethanal* – is formed:

$$\text{–O–CH}_2\text{–O–CH}_2\text{–O–CH2–}$$

This was formerly called *paraformaldehyde*. If methanal is distilled from acidic solutions a cyclic *methanal trimer* ($C_3O_3H_6$) is produced.

methanal trimer *See* methanal.

methane (CH_4) A gaseous alkane. Natural gas is about 99% methane and this provides an important starting material for the organic-chemicals industry. Methane can be chlorinated directly to produce the more reactive chloromethanes, or it can be 'reformed' by partial oxidation or using steam to give mixtures of carbon oxides and hydrogen. Methane is the first member of the homologous series of alkanes and the simplest organic compound.

methanoate (formate) A salt or ester of methanoic acid.

methanoic acid (formic acid; HCOOH) A liquid carboxylic acid made by the action of sulfuric acid on sodium methanoate (NaOOCH). It is a strong reducing agent. Methanoic acid is the substance responsible for the stings of ants and nettles.

methanol (methyl alcohol; wood alcohol; CH_3OH) A colorless liquid alcohol, which is used as a solvent and in the manufacture of methanal (formaldehyde) for the plastics and drugs industries. Methanol was originally produced from the distillation of wood. Now it is manufactured by the catalytic oxidation of methane from natural gas.

methionine *See* amino acids.

methoxy group The group CH_3O–.

methyl acetate *See* methyl ethanoate.

methyl alcohol *See* methanol.

methylamine (CH_3NH_2) A colorless flammable gas that smells like ammonia. It is the simplest primary amine, used for making herbicides and other organic chemicals.

methylaniline *See* toluidine.

methylated spirits Ethanol to which is added methanol (about 9.5%), pyridine (about 0.5%), and a blue dye. The ethanol is denatured in this way so that it can be sold without excise duty for use as a fuel and solvent.

methylation A reaction in which a methyl group (CH_3–) is introduced into a compound.

methylbenzene (toluene; $C_6H_5CH_3$) A colorless liquid hydrocarbon, similar to benzene both in structure and properties. As methylbenzene is much less toxic than benzene it is more widely used, especially as a solvent. Large quantities are required for the manufacture of TNT (trinitrotoluene).

Methylbenzene can be obtained by the fractional distillation of coal tar or synthesized from methylcyclohexane (a constituent of some crude oils):

$$C_6H_{11}CH_3 \rightarrow C_6H_5CH_3 + 3H_2$$

A catalyst of aluminum and molybdenum oxides is employed at high temperatures and pressures.

methyl bromide *See* bromomethane.

2-methylbuta-1,3-diene (isoprene; CH_2-CH:CH_2) A colorless unsaturated liquid hydrocarbon, which occurs in terpenes and natural rubber. Is is used to make synthetic rubber.

methyl chloride *See* chloromethane.

methyl cyanide (acetonitrile; CH_3CN) A pleasant-smelling poisonous colorless liquid organic NITRILE. It is a polar solvent, widely used for dissolving both inorganic and organic compounds.

methylene *See* carbene.

methylene group The group :CH_2.

methyl ethanoate (methyl acetate; CH_3COOCH_3) A colorless liquid ester with a fragrant odor, used as a solvent.

methyl ethyl ketone *See* butanone.

methyl group The group CH_3–.

methyl 2-hydroxybenzoate *See* methyl salicylate.

methyl iodide *See* iodomethane.

methyl methacrylate (CH_2:CCH_3-$COOCH_3$) The methyl ester of methacrylic acid, used to make acrylic polymers such as Plexiglas (polymethylmethacrylate).

methyl orange An acid–base indicator that is red in solutions below a pH of 3 and yellow above a pH of 4.4. As the transition range is clearly on the acid side, methyl orange is suitable for the titration of an acid with a moderately weak base, such as sodium carbonate.

methylphenol (cresol; $HOC_6H_4CH_3$) A compound with both methyl and hydroxyl groups substituted onto the benzene ring. There are three isomers, with the methyl group in the 2–, 3–, and 4– positions, respectively. A mixture of the isomers can be obtained from coal tar. It is used as a germicide (known as Lysol).

methyl red An acid–base indicator that is red in solutions below a pH of 4.2 and yellow above a pH of 6.3. It is often used for the same types of titration as methyl orange but the transition range of methyl red is nearer neutral (pH7) than that of methyl orange. The two molecules are structurally similar.

methyl salicylate (oil of wintergreen; methyl 2-hydroxybenzoate; $C_8H_8O_3$) The methyl ester of SALICYLIC ACID, which occurs in certain plants. It is absorbed through the skin and used medicinally to relieve rheumatic symptoms. It is also used in perfumes and as a flavoring agent in various foods.

metric system A system of units based on the meter and the kilogram and using multiples and submultiples of 10. SI units, c.g.s. units, and m.k.s. units are all scientific metric systems of units.

mho *See* siemens.

micelle An aggregate of molecules in a COLLOID.

Michaelis constant Symbol: K_m For an enzyme-catalyzed reaction obeying MICHAELIS KINETICS under steady-state conditions (when the reaction intermediates have reached a steady concentration):

$$[ES] = [E][S]/K_m$$

where [ES] is the concentration of enzyme–substrate complex, [E] is the concentration of enzyme, [S] is the concentration of substrate, and K_m is the Michaelis constant.

K_m gives the concentration of substrate at which half the active sites are filled and also gives an indication of the strength of the enzyme–substrate complex if the rate of product formation is much slower than the rate of dissociation of the enzyme–substrate complex into enzyme and substrate. If this is the case then a high K_m indicates weak binding of the complex and a low K_m indicates strong binding.

Michaelis kinetics (**Michaelis–Menten kinetics**) A simple and useful model of the kinetics of enzyme-catalyzed reactions. It assumes the formation of a specific enzyme–substrate complex. Many enzymes obey Michaelis kinetics and a plot of reaction velocity (V) against substrate concentration [S] gives a characteristic curve showing that the rate increases quickly at first and then levels off to a maximum value. When substrate concentration is low, the rate of reaction is almost proportional to substrate concentration. When substrate concentration is high, the rate is at a maximum, V_{max}, and independent of substrate concentration. The Michaelis constant K_m is the concentration of substrate at half the maximum rate and can be determined experimentally by measuring reaction rate at varying substrate concentrations. Different types of inhibition can also be distinguished in this way. Allosteric enzymes do not obey Michaelis kinetics.

micro- Symbol: μ A prefix denoting 10^{-6}. For example, 1 micrometer (μm) = 10^{-6} meter (m).

micron Symbol: μ A unit of length equal to 10^{-6} meter.

microwaves A form of electromagnetic radiation, ranging in wavelength from about 1 mm (where it merges with infrared) to about 120 mm (bordering on radio waves). Microwaves are produced by various electronic devices; they are often carried over short distances in tubes of rectangular section called *waveguides*. Spectra in the microwave region can give information on the rotational energy levels of certain molecules. *See also* electromagnetic radiation.

migration 1. The movement of an atom, group, or double bond from one position to another in a molecule.
2. The movement of ions in an electric field.

milk sugar *See* lactose.

Miller, Stanley Lloyd (1930–) American chemist. Stanley Miller is famous for an experiment, the results of which he published in 1953, concerning the possible origin of life. Miller, who was then a graduate student of Harold Urey, mixed water vapour, ammonia, methane and hydrogen in a flask so as to simulate the early atmosphere of the Earth and then put a powerful electric discharge through it to simulate lightning. He discovered that after a short time organic molecules of biological interest, including some simple amino acids, were formed. Since amino acids are the 'building blocks' for proteins it has been suggested that this experiment may give a clue as to the origin of life on Earth.

milli- Symbol: m A prefix denoting 10^{-3}. For example, 1 millimeter (mm) = 10^{-3} meter (m).

millimeter of mercury *See* mmHg.

mineral acid An inorganic acid, especially an acid used commercially in large quantities. Examples are hydrochloric, nitric, and sulfuric acids.

mirror image A shape that is identical to another except that its structure is reversed as if viewed in a mirror. If an object is not symmetrical it cannot be superimposed on its mirror image. For example, the left hand is the mirror image of the right hand. *See* chirality.

miscible Denoting combinations of substances that, when mixed, give rise to only one phase; i.e. substances that dissolve in each other.

mixed indicator A mixture of two or more indicators so as to decrease the pH range or heighten the color change, etc.

mixture Two or more substances forming a system in which there is no chemical bonding between the two. In homogeneous mixtures (e.g. solutions or mixtures of gases) the molecules of the substances are mixed, and there is only one phase. In heterogeneous mixtures (e.g. gunpowder or certain alloys) different phases can be distinguished. Mixtures differ from chemical compounds in that:

1. The chemical properties of the components of a mixture are the same as those of the pure substances.
2. The mixture can be separated by physical means (e.g. distillation or crystallization) or mechanically.
3. The proportions of the components can vary. Some mixtures (e.g. certain solutions) can only vary in proportions between definite limits.

m.k.s. system A system of units based on the meter, the kilogram, and the second. It formed the basis for SI units.

mmHg (millimeter of mercury) A former unit of pressure defined as the pressure that will support a column of mercury one millimeter high under specified conditions. It is equal to 133.322 4 Pa, and is almost identical to the torr.

moiety A part of a molecule; for example, the sugar moiety in a nucleoside.

molal concentration *See* concentration.

molar **1.** Denoting a physical quantity divided by the amount of substance. In almost all cases the amount of substance will be in moles. For example, volume (*V*) divided by the number of moles (*n*) is molar volume $V_m = v/n$.
2. Denoting a solution that contains one mole of solute per cubic decimeter of solvent.

molarity A measure of the concentration of solutions based upon the number of molecules or ions present, rather than on the mass of solute, in any particular volume of solution. The molarity (M) is the number of moles of solute in one cubic decimeter (litre). Thus a 0.5M solution of hydrochloric acid contains 0.5 × (1 + 35.5)g HCl per dm^3 of solution.

mole Symbol: mol The SI base unit of amount of substance, defined as the amount of substance that contains as many elementary entities as there are atoms in 0.012 kilogram of ^{12}C. The elementary entities may be atoms, molecules, ions, electrons, photons, etc., and they must be specified. The amount of substance is proportional to the number of entities, the constant of proportionality being the Avogadro number. One mole contains $6.022\,045 \times 10^{23}$ entities. One mole of an element with relative atomic mass *A* has a mass of *A* grams (this mass was formerly called one *gram-atom* of the element).

molecular crystal A crystal in which molecules, as opposed to atoms, occupy lattice points. Because the forces holding the molecules together are weak, molecular crystals have low melting points. When the molecules are small, the crystal structure approximates to a close-packed arrangement.

molecular formula *See* formula.

molecularity The total number of reacting molecules in the individual steps of a chemical reaction. Thus, a unimolecular step has molecularity 1, a bimolecular step 2, etc. Molecularity is always an integer, whereas the order of a reaction need not necessarily be so. The molecularity of a reaction gives no information about the mechanism by which it takes place.

molecular orbital *See* orbital.

molecular sieve A substance through which molecules of a limited range of sizes can pass, enabling volatile mixtures to be separated. Zeolites and other metal aluminum silicates can be manufactured with pores of constant dimensions in their molecular structure. When a sample is passed through a column packed with granules of this material, some of the molecules enter these pores and become trapped. The remainder of the mixture passes through the interstices in the column. The trapped molecules can be recovered by heating. *Molecular-sieve chromatography* is widely used in chemistry and biochemistry laboratories. A modified form of molecular sieve is used in *gel filtration.* The sieve is a continuous gel made from a polysaccharide. In this case, molecules larger than the largest pore size are totally excluded from the column.

molecular-sieve chromatography *See* molecular sieve.

molecular spectrum The absorption or emission SPECTRUM that is characteristic of a molecule. Molecular spectra are usually band spectra.

molecular weight *See* relative molecular mass.

molecule A particle formed by the combination of atoms in a whole-number ratio. A molecule of an element (combining atoms are the same, e.g. O_2) or of a compound (different combining atoms, e.g. HCl) retains the properties of that element or compound. Thus, any quantity of a compound is a collection of many identical molecules. Molecular sizes are characteristically 10^{-10} to 10^{-9} m.

Many molecules of natural products are so large that they are regarded as giant molecules (macromolecules); they may contain thousands of atoms and have complex structural formulae that require very advanced techniques to identify. *See also* formula; relative molecular mass.

mole fraction The number of moles of a given component in a mixture divided by the total number of moles present of all the components. The mole fraction of component A is

$$n_A/(n_A + n_B + n_C + \ldots)$$

where n_A is the number of moles of A, etc.

Molisch's test *See* alpha-naphthol test.

monobasic acid An acid that has only one acidic hydrogen. Ethanoic acid, CH_3COOH, is an example of a monobasic acid. *See also* dibasic acid.

monochlorobenzene *See* chlorobenzene.

monohydric alcohol *see* alcohol.

monomer The molecule, group, (or compound) from which a dimer, trimer, or POLYMER is formed.

monosaccharide A SUGAR that cannot be hydrolyzed to simpler carbohydrates of smaller carbon content. Glucose and fructose are examples.

monosodium glutamate (MSG) A white crystalline solid compound, made from soya-bean protein. It is a sodium salt of glutamic acid (*see* amino acid) used as a flavor enhancer, particularly in Chinese food. Monosodium glutamate can cause an allergic reaction in certain people.

monoterpene *See* terpene.

monovalent (univalent) Having a valence of one.

mordant An inorganic compound used to fix dye in cloth. The mordant (e.g. aluminum hydroxide or chromium salts) is precipitated in the fibers of the cloth, and the dye then absorbs in the particles.

morphine An alkaloid present in *opium* (from the poppy *Papaver somniferum*). It is used for the relief of severe pain. The drug *heroin* is a derivative.

mother liquor The solution remaining after the formation of crystals.

MSG *See* monosodium glutamate.

mucopolysaccharide *See* glycosaminoglycan.

mucoprotein *See* proteoglycan.

multicenter bond A two-electron bond formed by the overlap of orbitals from more than two atoms (usually 3). The bridging in diborane is believed to take place by overlap of an sp^3 hybrid orbital from each boron atom with the 1s orbital on the hydrogen atom. This multicenter bond is called a two-electron three-center bond. *See also* electron-deficient compound.

multidentate ligand A LIGAND that possesses at least two sites at which it can coordinate.

multiple bond A bond between two atoms involving more than one pair of electrons; i.e. a double bond or a triple bond. This additional bonding arises from overlap of atomic orbitals that are perpendicular to the internuclear axis and gives rise to an increase in electron density above and below this axis. Such bonds are called pi bonds. (The bond along the axis is a sigma bond.)

multiple proportions, law of Proposed by Dalton in 1804, the principle that when two elements A and B combine to form more than one compound, the weights of B that combine with a fixed weight of A are in small whole-number ratios. For example, in dinitrogen oxide, N_2O, nitrogen monoxide, NO, and dinitrogen tetroxide, N_2O_4, the amounts of nitrogen combined with a fixed weight of oxygen are in the ratio 4:2:1.

multiple-range indicator *See* universal indicator.

mustard gas ($(CH_2ClCH_2)_2S$) A poisonous vesicant gas used as a war gas. The systematic name is 2,2′-dichlorodiethyl sulfide.

mutarotation A change in the optical rotation of a solution with time, caused by the conversion of one optical isomer into another. *See* optical activity.

myoglobin A globular protein formed of a heme group and a single polypeptide chain. It occurs in muscle tissue, where it acts as an oxygen store.

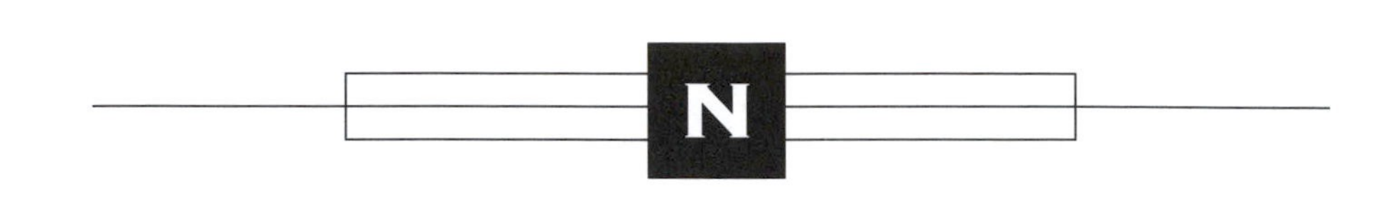

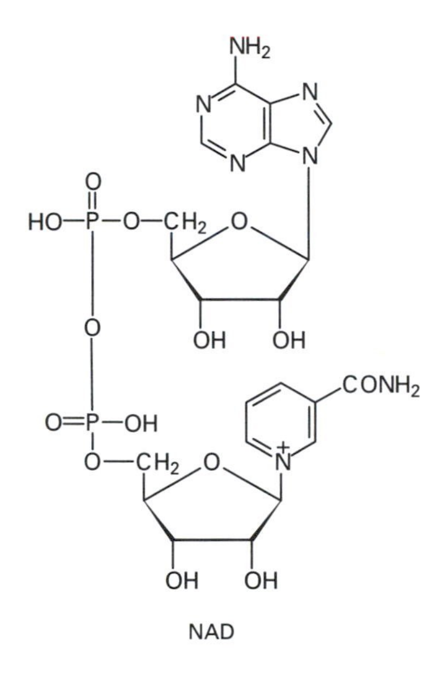

NAD

NAD (nicotinamide adenine dinucleotide) A derivative of nicotinic acid that acts as a coenzyme in electron-transfer reactions (e.g. the electron-transport chain). Its role is to carry hydrogen atoms; the reduced form is written NADH.

NADP Nicotinamide adenine dinucleotide phosphate; a coenzyme similar in its action to NAD. The reduced form is written NADPH, which acts as an electron donor in many synthetic reactions.

nano- Symbol: n A prefix denoting 10^{-9}. For example, 1 nanometer (nm) = 10^{-9} meter (m).

nanotube A tubular structure with a diameter of a few nanometers (1nm = 10^{-9} m). Examples of nanotubes are the carbon structures known as 'bucky tubes', which have a structure similar to that of buckminsterfullerene. Interest has been shown in nanotubes as possible microscopic probes in experiments, as semiconductor materials, and as a component of composite materials. Nanotubes can also be produced by joining amino acids to give tubular polypeptide structures. *See also* buckminsterfullerene.

naphtha (solvent naphtha) A mixture of hydrocarbons obtained from coal and petroleum. It has a boiling range of 70–160°C and is used as a solvent and as a raw material for making various other organic chemicals.

naphthalene ($C_{10}H_8$) A white crystalline solid with a distinctive smell of

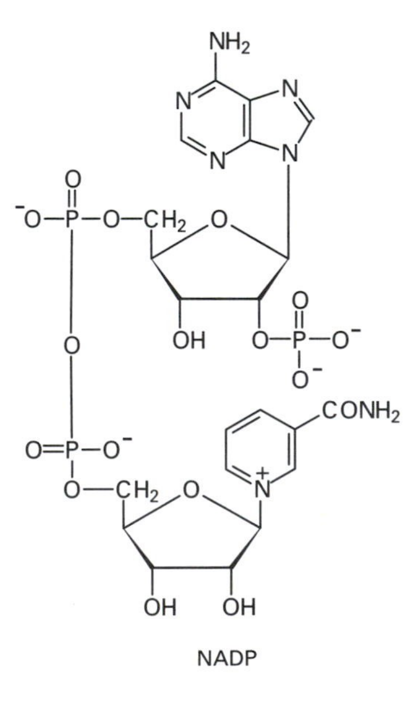

NADP

mothballs. Naphthalene is found in both the middle- and heavy-oil fractions of crude oil and is obtained by fractional crystallization. It is used in the manufacture of benzene-1,2-dicarboxylic anhydride (phthalic anhydride) and thence in the production of plastics and dyes.

The structure of naphthalene is 'benzene-like', having two six-membered rings fused together. The reactions are characteristic of AROMATIC COMPOUNDS.

nascent hydrogen A particularly reactive form of hydrogen, which is believed to exist briefly between its generation (e.g. by the action of dilute acid on magnesium) and its appearance as bubbles of normal hydrogen gas. It is thought that part of the free energy of the production reaction remains with the hydrogen molecules for a short time. Nascent hydrogen may be used to produce the hydrides of phosphorus, arsenic, and antimony, which are not readily formed from ordinary hydrogen.

Natta, Giulio (1903–79) Italian chemist. The early part of Natta's career was devoted to x-ray crystallography and catalysts. In 1938 he started research on synthetic rubbers. In 1953 he developed methods for using catalysts to produce polymers which had been initiated by Karl ZIEGLER to form polypropene. In 1954 he found that polymers produced in this way are very specific and regular in their stereochemistry. This meant that these polymers have desirable properties such as high melting points and high strength. After 1954 he continued to investigate this type of polymerization. Natta and Ziegler shared the 1963 Nobel Prize for chemistry for their work on polymers.

Natta process A method for the manufacture of isotactic polypropene using Ziegler catalysts. It is named for the Italian chemist Giulio Natta. *See* Ziegler process.

natural gas Gas obtained from underground deposits and often associated with sources of petroleum. It contains a high proportion of methane (about 85%) and other volatile hydrocarbons (ethane, propane, and butane).

neighboring-group participation An effect in an organic reaction in which groups close to the point at which reaction occurs affect the rate of reaction or stereochemistry of the products in some way.

neoprene A type of synthetic RUBBER made by polymerization of 2-chlorobuta-1,2,-diene ($H_2C{:}CHCCl{:}CH_2$). It is more resistant to oil, solvents, and temperature than natural rubbers.

neutralization The stoichiometric reaction of an acid and a base in volumetric analysis. The neutralization point or end point is detected with indicators.

neutron diffraction A method of structure determination used for solids, liquids, and gases that makes use of the quantum mechanical wave nature of neutrons. Thermal neutrons with average kinetic energies of about 0.025eV have a wavelength of about 0.1 nanometer, making them suitable for investigating the structure of matter at the atomic level.

Since a neutron has a nonzero magnetic moment, it interacts both with nuclear magnetic moments and with the magnetic moments of unpaired electrons. This property is particularly useful in identifying the positions of hydrogen atoms in a molecule. These positions are difficult to establish using x-rays because x-rays interact with electrons, and hence are scattered weakly by hydrogen atoms. Protons scatter neutrons strongly, and the positions of protons can readily be determined by neutron diffraction.

neutron number Symbol: N The number of neutrons in the nucleus of an atom; i.e. the nucleon number (A) minus the proton number (Z).

Newman projection A type of projection FORMULA in which the molecule is viewed along a bond between two of its atoms. *See illustration at* conformation.

newton Symbol: N The SI unit of force, equal to the force needed to accelerate one kilogram by one meter second^{-2}. 1 N = 1 kg m s^{-2}. It is named for Sir Isaac Newton (1642–1727).

niacin *See* nicotinic acid.

nicotinamide adenine dinucleotide *See* NAD.

nicotinic acid (niacin) One of the water-soluble B-group of vitamins. Its deficiency in man causes pellagra. Nicotinic acid functions as a constituent of two coenzymes, NAD and NADP, which operate as hydrogen and electron transfer agents and play a vital role in metabolism. *See also* vitamin B complex.

ninhydrin A colorless organic compound that gives a blue coloration with amino acids. It is used as a test for amino acids, in particular to show the positions of spots of amino acids in paper chromatography.

nitrate A salt or ester of nitric acid.

nitration A reaction introducing the nitro ($-NO_2$) group into an organic compound. Nitration of aromatic compounds is usually carried out using a mixture of concentrated nitric and sulfuric acids, although the precise conditions differ from compound to compound. The attacking species is NO_2^+ (the *nitryl ion*), and the reaction is an example of electrophilic substitution.

nitric acid (HNO_3) A colorless fuming corrosive liquid that is a strong acid. Nitric acid can be made in a laboratory by the distillation of a mixture of an alkali metal nitrate and concentrated sulfuric acid. Commercially it is prepared by the catalytic oxidation of ammonia and is supplied as concentrated nitric acid, which contains 68% of the acid and is often colored yellow by dissolved oxides of nitrogen.

Nitric acid is a strong oxidizing agent. Most metals are converted to their nitrates with the evolution of oxides of nitrogen (the composition of the mixture of the oxides depends on the temperature and on the concentration of the nitric acid used). Some nonmetals (e.g. sulfur and phosphorus) react to produce oxyacids. Organic substances (e.g. sawdust and ethanol) react violently, but the more stable aromatic compounds, such as benzene and toluene, can be converted to NITRO COMPOUNDS in controllable reactions.

nitrile (cyanide) A type of organic compound containing the –CN group. Nitriles are colorless liquids with pleasant smells. They can be prepared by refluxing an organic halogen compound with an alcoholic solution of potassium cyanide:

$$R{+}Cl + KCN \rightarrow RCN + KCl$$

Alternatively it is possible to dehydrate an amide using a dehydrating agent such as phosphorus(V) oxide:

$$RCONH_2 - H_2O \rightarrow RCN$$

Nitriles can be hydrolyzed to give the amide. Another reaction is hydrogenation to give amines:

$$RCN + 2H_2 \rightarrow RCH_2NH_2$$

nitrile rubber A copolymer of butadiene and propenonitrile (acrylonitrile; $CH_2{=}CHCN$). It is a useful type of rubber because of its resistance to oil and solvents.

nitrobenzene ($C_6H_5NO_2$) A yellow organic oil obtained by refluxing benzene with a mixture of concentrated nitric and sulfuric acids. The reaction is a typical electrophilic substitution on the benzene ring by the nitryl ion (NO_2^+).

nitrocellulose *See* cellulose trinitrate.

nitro compound A type of organic compound containing the nitro ($-NO_2$) group attached to an aromatic ring. Nitro compounds can be prepared by nitration using a mixture of concentrated nitric and sulfuric acids. They can be reduced to aromatic amines:

$$RNO_2 + 3H_2 \rightarrow RNH_2 + 2H_2O$$

They can also undergo further substitution on the benzene ring. The nitro group directs substituents into the 3 position.

nitrogen The first element of group 15 (formerly group VA) of the periodic table; a very electronegative element existing in the uncombined state as gaseous diatomic N_2 molecules. The nitrogen atom has the electronic configuration $[He]2s^22p^3$. It is typically nonmetallic and its bonding is primarily by polarized covalent bonds. With electropositive elements the nitride ion N^{3-} may be formed. It is present in many organic compounds including amines, amides, nitriles, and nitro compounds.

Nitrogen has two isotopes; ^{14}N, the common isotope, and ^{15}N (natural abundance 0.366%), which is used as a label in mass spectrometric studies.

Symbol: N; m.p. –209.86°C; b.p. –195.8°C; d. 1.2506 kg m^{-3} (0°C); p.n. 7; r.a.m. 14.

nitrogenous base A basic compound in which a nitrogen atom can accept a proton. The term is used especially for the cyclic ring compounds adenine, guanine, cytosine, thymine, and uracil, which occur in nucleic acids. *See also* quaternary ammonium compound.

nitroglycerine (glyceryl trinitrate) A highly explosive substance used in dynamite. It is obtained by treating glycerol (1,2,3-trihydroxypropane) with a mixture of concentrated nitric and sulfuric acids. It is not a nitro compound, but a nitrate ester

$$CH_2(NO_3)CH(NO_3)CH_2(NO_3)$$

nitro group The group $-NO_2$; the functional group of nitro compounds.

nitronium ion *See* nitryl ion.

nitrophenols ($C_6H_4(OH)NO_2$) Organic compounds formed directly or indirectly by the nitration of phenol. Three isomeric forms are possible. The 2 and 4 isomers are produced by the direct nitration of phenol and can be separated by steam distillation, the 2 isomer being steam volatile. The 3 isomer is produced from nitrobenzene by formation of 1,3-dinitrobenzene, conversion to 3-nitrophenylamine, and thence by diazotization to 3-nitrophenol.

nitryl ion (nitronium ion) The ion NO_2^+, occurring in NITRATION reactions.

NMR *See* nuclear magnetic resonance.

nonbenzenoid aromatic *See* aromatic compound.

nonessential amino acid *See* amino acid.

nonlocalized bond *See* delocalized bond.

nonmetal Any of a class of chemical elements. Non-metals lie in the top right-hand region of the periodic table. They are electronegative elements with a tendency to form covalent compounds or negative ions. They have acidic oxides and hydroxides.

nonpolar compound A compound that has molecules with no permanent dipole moment. Examples of nonpolar compounds are hydrogen, tetrachloromethane, and carbon dioxide.

nonpolar solvent *See* solvent.

noradrenaline *See* norepinephrine.

norepinephrine (noradrenaline) A catecholamine, secreted as a hormone by the adrenal medulla, that regulates heart muscle, smooth muscle, and glands. It causes narrowing of arterioles and hence raises blood pressure. It is also secreted by nerve endings of the sympathetic nervous system in which it acts as a neurotransmitter. In the brain, levels of norepinephrine are related to mental function; lowered levels lead to mental depression.

normality The number of gram equivalents per cubic decimeter of a given solution.

normal solution A solution that contains one gram equivalent weight per liter of solution. Values are designated by the symbol N, e.g. 0.2N, N/10, etc. Because there is not a clear definition of equivalent weight suitable for all reactions, a solution may have one value of normality for one reaction and another value in a different reaction. Because of this it is now more usual to use the molar solution notation.

NTP *See* STP.

nuclear magnetic resonance (NMR) A method of investigating nuclear spin. In an external magnetic field the nucleus can have certain quantized energy states, corresponding to certain orientations of the spin magnetic moment. Hydrogen nuclei, for instance, can have two energy states, and transitions between the two occur by absorption of radiofrequency radiation. In chemistry, this is the basis of a spectroscopic technique for investigating the structure of molecules. In the basic technique, radiofrequency radiation is fed to a sample and the magnetic field is changed slowly. Absorption of the radiation is detected when the difference between the nuclear levels corresponds to absorption of a quantum of radiation. This difference depends slightly on the electrons around the nucleus – i.e. the position of the atom in the molecule. The difference in frequency of absorption caused by the distribution of electrons is known as a *chemical shift*. Thus a different absorption frequency is seen for each type of hydrogen atom. In ethanol, for example, there are three frequencies, corresponding to hydrogen atoms on the CH_3, the CH_2, and the OH. The intensity of absorption also depends on the number of hydrogen atoms (3:2:1). NMR spectroscopy is a powerful method of finding the structures of organic compounds. It is most often used to detect hydrogen atoms but certain other nuclei can be investigated (e.g. ^{13}C).

nucleic acids Organic acids whose molecules consist of chains of alternating sugar and phosphate units, with nitrogenous bases attached to the sugar units. They occur in the cells of all organisms. In DNA the sugar is deoxyribose; in RNA it is ribose. *See* DNA; RNA.

nucleon number (mass number) Symbol: *A* The number of nucleons (protons plus neutrons) in an atomic nucleus.

nucleophile An electron-rich ion or molecule that takes part in an organic reaction. The nucleophile can be a negative ion (Br^-, CN^-) or a molecule with a lone pair of electrons (NH_3, H_2O). The nucleophile attacks positively charged parts of molecules, which usually arise from the presence of an electronegative atom elsewhere in the molecule. *Compare* electrophile.

nucleophilic addition A class of reaction involving the addition of a small molecule to the double bond in an unsaturated organic compound. The initial part of the reaction is attack by a nucleophile and the unsaturated bond must contain an electronegative atom, which creates an electron-deficient area in the molecule. Nucleophilic addition is a characteristic reaction of aldehydes and ketones where polarization of the C=O carbonyl causes a positive charge on the carbon. This is the site at which the nucleophile attacks. Addition is often followed by the subsequent elimination of a different small molecule, particularly water. *See* condensation reaction.

nucleophilic substitution A reaction involving the substitution of an atom or group of atoms in an organic compound with a nucleophile as the attacking substituent. Since nucleophiles are electron-rich species, nucleophilic substitution occurs in compounds in which a strongly electronegative atom or group leads to a dipolar bond. The electron-deficient center can then be attacked by the electron-rich nucleophile causing the electronegative atom or group to be displaced. In general terms:

$$R\text{–}Le + Nu^- \rightarrow R\text{–}Nu + Le^-$$

where Nu^- represents the incoming nucleophile and Le^- represents the LEAVING GROUP.

There are two possible mechanisms for nucleophilic substitution. In the S_N1 (substitution, nucleophilic, monomolecular) re-

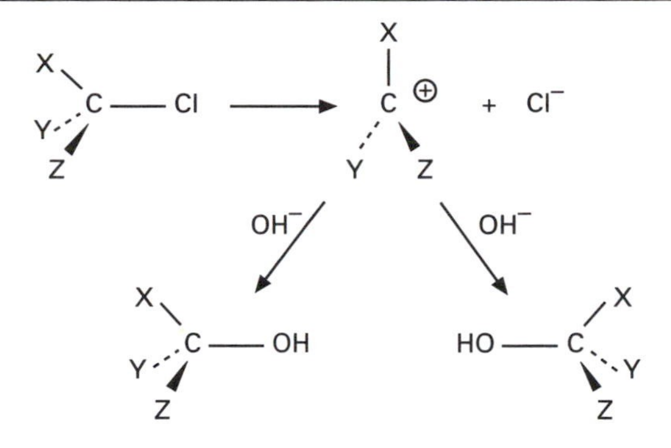

Nucleophilic substitution: the S_N1 reaction

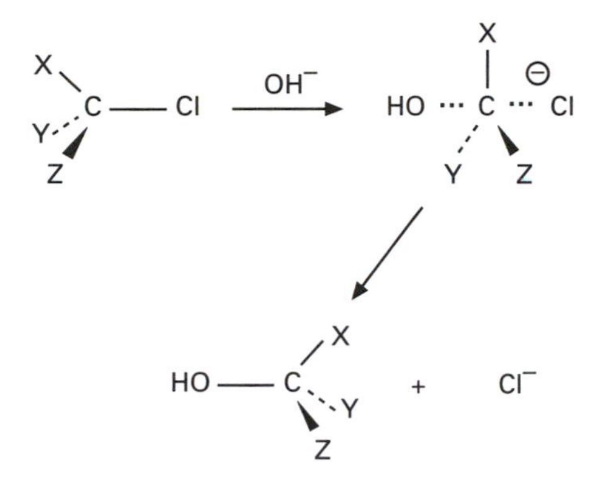

Nucleophilic substitution: the S_N2 reaction

action the molecule first forms a carbonium ion; for example:

$$RCH_2Cl \rightarrow RCH_2^+ + Cl^-$$

The nucleophile then attaches itself to this carbonium ion:

$$RCH_2^+ + OH^- \rightarrow RCH_2OH$$

In the S_N2 (substitution, nucleophilic, bimolecular) reaction the nucleophile approaches as the other group leaves, forming a transition state in which the carbon has five attached groups.

The preferred mechanism depends on several factors:

1. The stability of the intermediate in the S_N1 mechanism.
2. Steric factors affecting the formation of the transition state in the S_N2 mechanism.
3. The solvent in which the reaction occurs: polar solvents will stabilize polar intermediates and so favor the S_N1 mechanism.

There is a difference in the two mechanisms in that, for an optically active reactant, the S_N1 mechanism gives a racemic mixture of products, whereas an S_N2 mechanism gives an optically active product (*see* illustration).

See also substitution reaction.

nucleoside A molecule consisting of a purine or pyrimidine base linked to a sugar, either ribose or deoxyribose. ADENOSINE, CYTIDINE, GUANOSINE, THYMIDINE, and URIDINE are common nucleosides.

nucleotide The compound formed by condensation of a nitrogenous base (a purine, pyrimidine, or pyridine) with a sugar (ribose or deoxyribose) and phos-

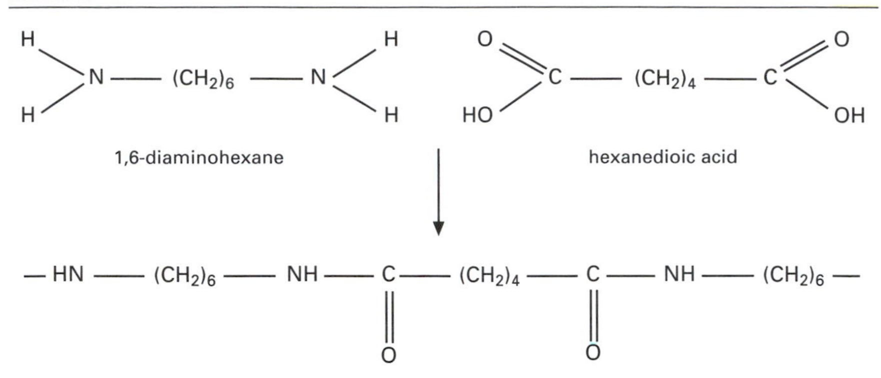

Nylon: formation of nylon by a condensation reaction

phoric acid. The coenzymes NAD and FAD are *dinucleotides* (consisting of two linked nucleotides) while the nucleic acids are *polynucleotides* (consisting of chains of many linked nucleotides).

nucleus The compact positively charged center of an atom made up of one or more nucleons (protons and neutrons) around which is a cloud of electrons. The density of nuclei is about 10^{15} kg m^{-3}. The number of protons in the nucleus defines the element, being its proton number (or atomic number). The nucleon number, or atomic mass number, is the sum of the protons and neutrons. The simplest nucleus is that of a hydrogen atom, ^{1}H, being simply one proton (mass 1.67×10^{-27} kg). The most massive naturally occurring nucleus is ^{238}U of 92 protons and 146 protons (mass 4×10^{-25} kg, radius 9.54×10^{-15} m). Only certain combinations of protons and neutrons form stable nuclei. Others undergo spontaneous decay.

A nucleus is depicted by a symbol indicating nucleon number (mass number), proton number (atomic number), and element name. For example, $^{23}_{11}$Na represents a nucleus of sodium having 11 protons and mass 23, hence there are (23 – 11) = 12 neutrons.

nuclide A nuclear species with a given number of protons and neutrons; for example, ^{23}Na, ^{24}Na, and ^{24}Mg are all different nuclides. Thus:

$^{23}_{11}$Na has 11 protons and 12 neutrons
$^{24}_{11}$Na has 11 protons and 13 neutrons
$^{24}_{12}$Mg has 12 protons and 12 neutrons

The term is applied to the nucleus and often also to the atom.

nylon A type of synthetic polymer linked by amide groups –NH.CO–. Nylon polymers can be made by copolymerization of a molecule containing two amine groups with one containing two carboxylic acid groups.

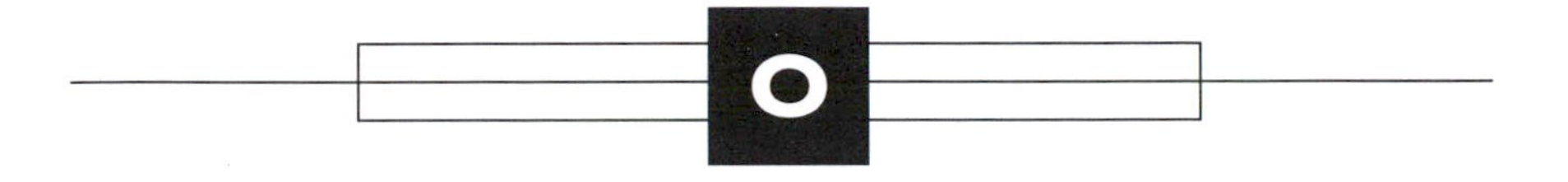

OAA *See* oxaloacetic acid.

occlusion 1. The process in which small amounts of one substance are trapped in the crystals of another; for example, pockets of liquid occluded during crystallization from a solution.
2. Absorption of a gas by a solid; for example, the occlusion of hydrogen by palladium.

octadecanoic acid (stearic acid; $CH_3(CH_2)_{16}COOH$) A solid carboxylic acid present in fats and oils as the glyceride.

octadecenoic acid (oleic acid) A naturally occurring unsaturated carboxylic acid present (as glycerides) in fats and oils:

$$CH_3(CH_2)_7CH{:}CH(CH_2)_7COOH$$

The naturally occurring form is *cis*-9-octadecenoic acid.

octane (C_8H_{18}) A liquid alkane obtained from the light fraction of crude oil. Octane and its isomers are the principal constituents of gasoline, which is obtained as the refined light fraction from crude oil. *See also* octane rating.

octane number *See* octane rating.

octane rating (octane number) A rating for the performance of gasoline in internal-combustion engines. The octane rating measures the freedom from 'knocking' – i.e. preignition of the fuel in the engine. This depends on the relative proportions of branched-chain and straight-chain hydrocarbons present. High proportions of branched-chain alkanes are better in high-performance engines. In rating fuels, 2,2,4-trimethylpentane (isooctane) is given a value of 100 and heptane is given a value 0. The performance of a fuel is compared with a mixture of these hydrocarbons.

octet A stable shell of eight electrons in an atom. The completion of the octet gives rise to particular stability and this is the basis of the *Lewis octet theory*, thus:

1. The rare gases have complete octets and are chemically inert.
2. The bonding in small covalent molecules is frequently achieved by the central atom completing its octet by sharing electrons with surrounding atoms, e.g. CH_4, H_2O.
3. The ions formed by electropositive and electronegative elements are generally those with a complete octet, e.g. Na^+, Ca^{2+}, O^{2-}, Cl^-.

ohm Symbol: Ω The SI unit of electrical resistance, equal to a resistance that passes a current of one ampere when there is an electric potential difference of one volt across it. $1\ \Omega = 1\ V\ A^{-1}$. Formerly, it was defined in terms of the resistance of a column of mercury under specified conditions. The unit is named for the German physicist Georg Ohm (1787–1854).

oil Any of various viscous liquids. *Mineral oils* are mainly composed of mixtures of hydrocarbons (*see* petroleum). Natural oils secreted by plants and animals are either mixtures of esters and terpenes (*see* essential oil) or are GLYCERIDES of fatty acids.

oil of wintergreen *See* methyl salicylate.

oil shale A sedimentary rock that includes in its structure 30–60% of organic matter, mainly in the form of bitumen.

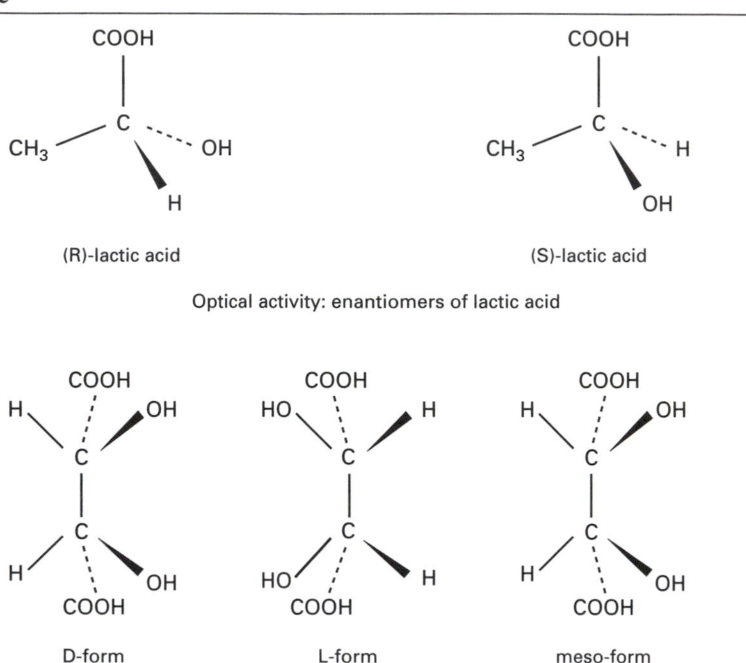

Optical activity: enantiomers of lactic acid

Optical activity: tartaric acid

Heated in the absence of air it produces an oily substance resembling petroleum, which is rich in nitrogen and sulfur compounds.

oleate A salt or ester of oleic acid; i.e. an octadecenoate.

olefin *See* alkene.

oleic acid *See* octadecenoic acid.

oligomer A polymer formed from a relatively few monomer molecules. *See* polymer; polymerization.

oligopeptide *See* peptide.

oligosaccharide A carbohydrate formed of a small number of monosaccharide units (up to around 20). *See* sugar.

one-pot synthesis A synthesis of a chemical compound in which the reactants form the required product in a single reaction mixture.

onium ion An ion formed by addition of a proton (H^+) to a molecule. The hydronium (or hydroxonium) ion (H_3O^+) and ammonium ion (NH_4^+) are examples.

Oparin, Alexandr Ivanovich (1894–1980) Russian biochemist. Oparin was one of the first people to put forward a theory of the origin of life on Earth. In 1922 he postulated that life originated in the seas, with there being many suitable organic molecules being present there, that there is a large supply of external energy and that life is characterized by a high degree of order. He expounded his views in more detail in his book *The Origin of Life on Earth* (1936). His ideas stimulated a great deal of further work such as the experiment of Stanley MILLER.

optical activity The ability of certain compounds to rotate the plane of POLARIZATION of plane-polarized light when the light is passed through them. Optical activity can be observed in crystals, gases, liquids, and solutions. The amount of rota-

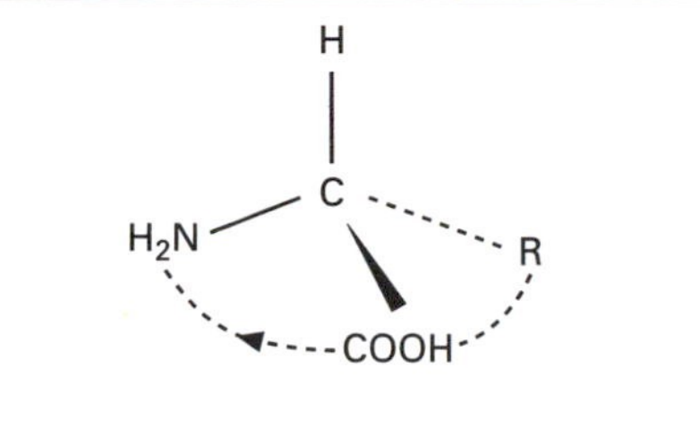

The corn rule for absolute configuration of alpha amino acids

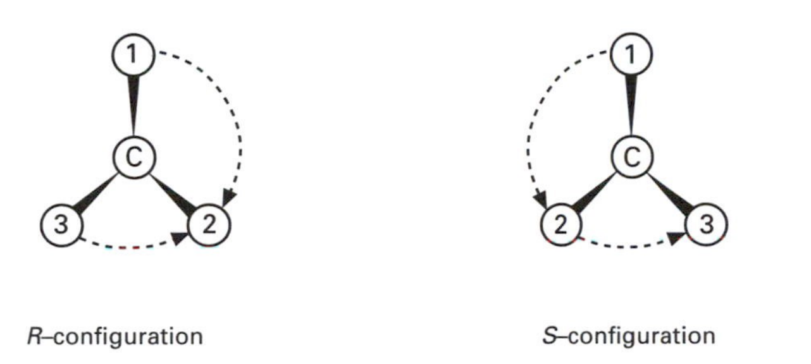

R–configuration

S–configuration

Corn rule

tion depends on the concentration of the active compound.

Optical activity is caused by the interaction of the varying electric field of the light with the electrons in the molecule. It occurs when the molecules are asymmetric – i.e. they have no plane of symmetry. Such molecules have a mirror image that cannot be superimposed on the original molecule. The two forms of the molecule are *optical isomers* or *enantiomers*. (Stereoisomers that are not mirror images of each other are called *diastereoisomers*.) In organic compounds this usually means that the molecule contains a carbon atom attached to four different groups, forming a chiral center. One isomer will rotate the polarized light in one sense and the other by the same amount in the opposite sense. Such isomers are described as *dextrorotatory* or *levorotatory*, according to whether they rotate the plane to the 'right' or 'left' respectively (rotation to the left is clockwise to an observer viewing the light coming toward the observer). Dextrorotatory compounds are given the symbol *d* or (+) and levorotatory compounds *l* or (–). A mixture of the two isomers in equal amounts does not show optical activity. Such a mixture is sometimes called the (±) or *dl*-form, a *racemate*, or a *racemic mixture*

Optical isomers have identical physical properties (apart from optical activity) and cannot be separated by fractional crystallization or distillation. Their general chemical behavior is also the same, although they do differ in reactions involving other optical isomers. Many naturally occurring substances are optically active (only one optical isomer exists naturally) and biochemical reactions occur only with the natural isomer. For instance, the natural form of glucose is *d*-glucose and living organisms cannot metabolize the *l*-form.

The terms 'dextrorotatory' and 'levorotatory' refer to the effect on polarized light. A more common method of distinguishing two optical isomers is by their *D-form* (*dextro-form*) or *L-form* (*levo-form*). This convention refers to the absolute structure of the isomer according to specific rules. Sugars are related to a particular configuration of glyceraldehyde (2,3-dihydroxypropanal). For alpha amino acids the *corn rule* is used: the structure of the acid $RC(NH_2)(COOH)$ H is drawn with H at the top; viewed from the top the groups spell CORN in a clockwise direction for all D-amino acids (i.e. the clockwise order is –COOH,*R*,NH_2). The opposite is true for L-amino acids. Note that this convention refers to absolute configuration, not to optical activity: D-alanine is dextrorotatory but D-cystine is levorotatory.

An alternative is the R-S convention for showing configuration. There is an order of priority of attached groups based on the proton number of the attached atom:
I, Br, Cl, SO_3H, $OCOCH_3$, OCH_3, OH, NO_2, NH_2, $COOCH_3$, $CONH_2$, $COCH_3$, CHO, CH_2OH, C_6H_5, C_2H_5, CH_3, H

Hydrogen has the lowest priority (*see* CIP system). The chiral carbon is viewed such that the group of lowest priority is hidden behind it. If the other three groups are in descending priority in a clockwise direction, the compound is R-. If descending priority is anticlockwise it is S-.

The existence of a carbon atom bound to four different groups is not the strict condition for optical activity. The essential point is that the molecule should be asymmetric. Inorganic octahedral complexes, for example, can show optical isomerism. It is also possible for a molecule to contain asymmetric carbon atoms and still have a plane of symmetry. One structure of tartaric acid has two parts of the molecule that are mirror images, thus having a plane of symmetry. This (called the *meso-form*) is not optically active. *See also* resolution.

optical isomer *See* optical activity.

optical rotary dispersion (ORD) The phenomenon in which the amount of rotation of plane-polarized light by an optically active substance depends on the wavelength of the light. Plots of rotation against wavelength can be used to give information about the molecular structure of optically active compounds.

optical rotation Rotation of the plane of polarization of plane-polarized light by an optically active substance.

orbit The path of an electron as it moves around the nucleus in an atom.

orbital A region around an atomic nucleus in which there is a high probability of finding an electron. The modern picture of the atom according to quantum mechanics does not have electrons moving in fixed elliptical orbits. Instead, there is a finite probability that the electron will be found in any small region at all possible distances from the nucleus. In the hydrogen atom the probability is low near the nucleus, increases to a maximum, and falls off to infinity. It is useful to think of a region in space around the nucleus – in the case of hydrogen the region within a sphere – within which there is a high chance of finding the electron. Each of these, called an *atomic orbital*, corresponds to a subshell and can 'contain' a single electron or two electrons with opposite spins. Another way of visualizing an orbital is as a cloud of electron charge (the average distribution with time).

Similarly, in molecules the electrons move in the combined field of the nuclei and can be assigned to *molecular orbitals*. In considering bonding between atoms it is useful to treat molecular orbitals as formed by overlap of atomic orbitals.

It is possible to calculate the shapes and energies of atomic and molecular orbitals by quantum theory. The shapes of atomic orbitals depend on the orbital angular momentum (the subshell). For each shell there is one s orbital, three p orbitals, five d orbitals, etc. The s orbitals are spherical, the p orbitals each have two lobes; d orbitals have more complex shapes, typically with four lobes.

Molecular orbitals are formed by overlap of atomic orbitals, and again there are different types. If the orbital is completely symmetrical about an axis between the nuclei, it is a *sigma orbital*. This can occur, for instance, by overlap of two s orbitals, as in the hydrogen atom, or two p orbitals with their lobes along the axis. However,

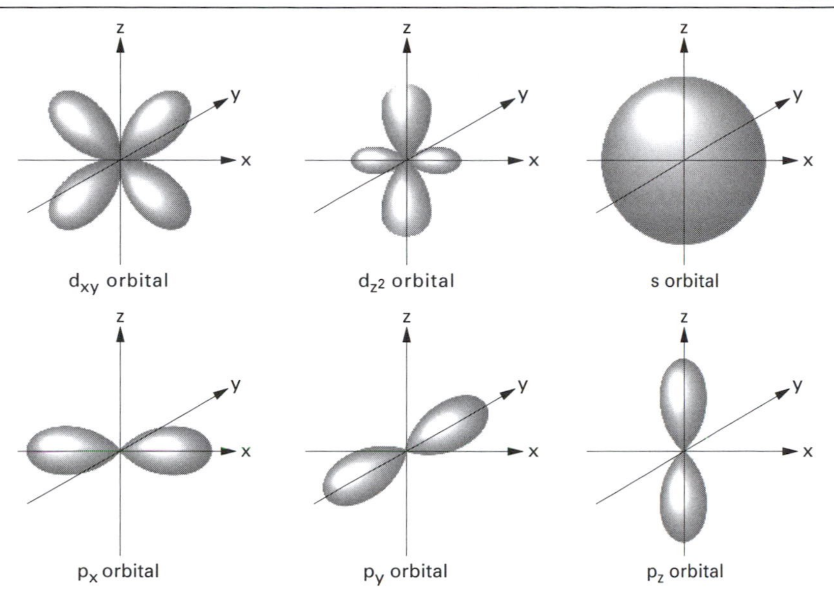

Orbital: atomic orbitals

two p orbitals overlapping at right angles to the axis form a different type of molecular orbital – a *pi orbital* – with regions above and below the axis. Pi orbitals are also formed by overlap of d orbitals. Each molecular orbital can contain a pair of electrons, forming a sigma bond or pi bond. A double bond, for example the bond in ethene, is a combination of a sigma bond and a pi bond. The triple bond in ethyne is one sigma bond and two pi bonds.

Hybrid orbitals are atomic orbitals formed by combinations of s, p, and d atomic orbitals, and are useful in describing the bonding in compounds. There are various types. In carbon, for instance, the electron configuration is $1s^2 2s^2 2p^2$. Carbon, in its outer (valence) shell, has one filled s orbital, two filled p orbitals, and one 'empty' p orbital. These four orbitals may *hybridize* (sp^3 hybridization) to act as four equal orbitals arranged tetrahedrally, each with one electron. In methane, each hybrid orbital overlaps with a hydrogen s orbital to form a sigma bond. Alternatively, the s and two of the p orbitals may hybridize (sp^2 hybridization) and act as three orbitals in a plane at 120°. The remaining p orbital is at right angles to the plane, and can form pi bonds. Finally, sp hybridization may occur, giving two orbitals in a line. More complex types of hybridization, involving d orbitals, explain the geometries of inorganic complexes.

The combination of two atomic orbitals in fact produces two molecular orbitals. One – the *bonding orbital* – has a concentration of electron density between the nuclei, and thus tends to hold the atoms together. The other – the *antibonding orbital* – has slightly higher energy and tends to repel the atoms. If both atomic orbitals are filled, the two molecular orbitals are also filled and cancel each other out – there is no net bonding effect. If each atomic orbital has one electron, the pair occupies the lower energy bonding orbital, producing a net attraction.

order The sum of the indices of the concentration terms in the expression that determines the rate of a chemical reaction. For example, in the expression:

$$\text{rate} = k[A]^x[B]^y$$

x is called the order with respect to A, y the order with respect to B, and $(x + y)$ the order overall. The values of x and y are not

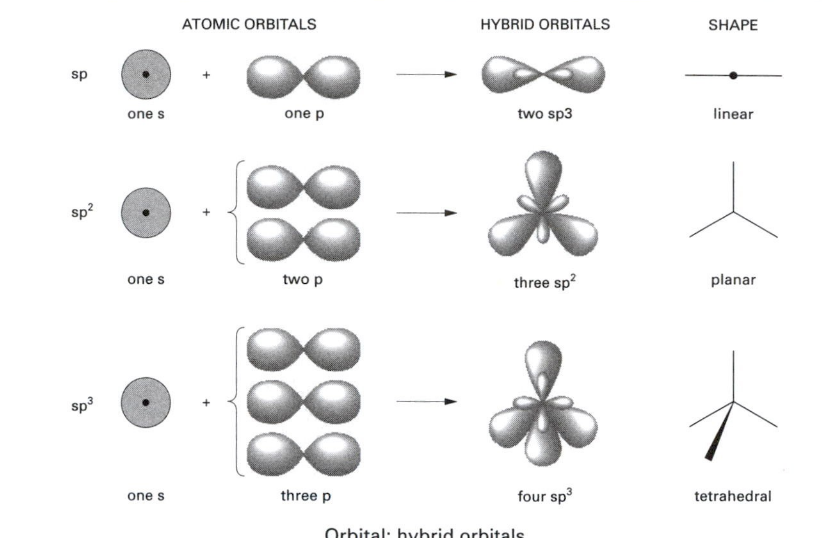

Orbital: hybrid orbitals

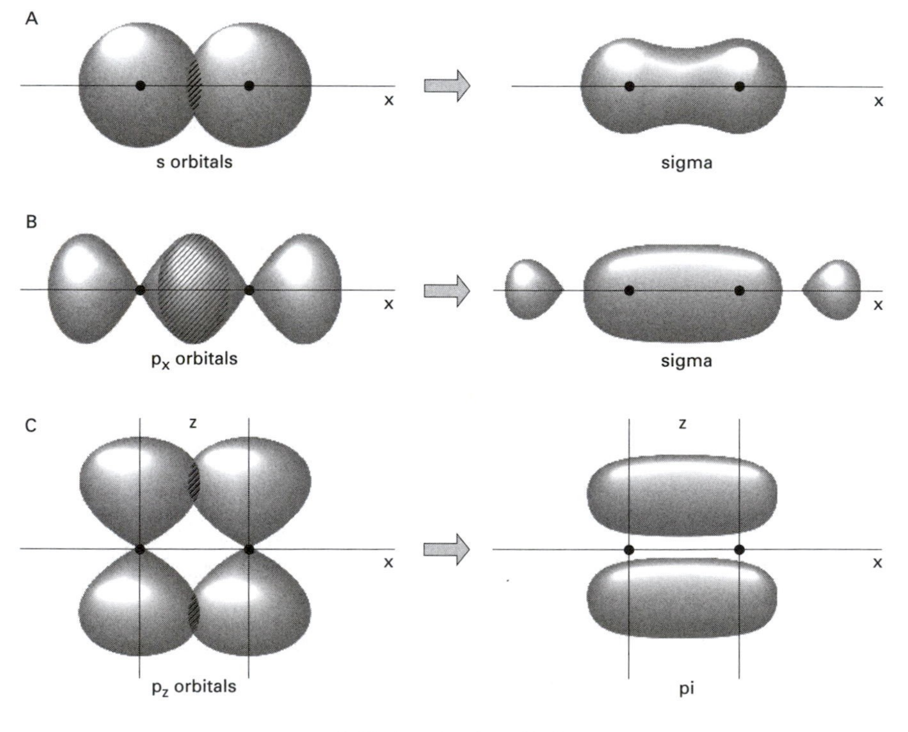

Orbital: molecular orbitals

necessarily equal to the coefficients of A and B in the molecular equation. Order is an experimentally determined quantity derived without reference to any equation or mechanism. Fractional orders do occur. For example, in the reaction:

$$CH_3CHO \rightarrow CH_4 + CO$$

the rate is proportional to $[CH_3CHO]^{1.5}$ i.e. it is of order 1.5.

organic acid An organic compound that can release hydrogen ions (H^+) to a base, such as a carboxylic acid or a phenol. *See* acid; carboxylic acid; phenol.

organic base An organic compound that can function as a base. Organic bases are typically amines that gain H^+ ions. *See* amine.

organic chemistry The chemistry of compounds of carbon. Originally the term *organic chemical* referred to chemical compounds present in living matter, but now it covers any carbon compound with the exception of certain simple ones, such as the carbon oxides, carbonates, cyanides, and cyanates. These are generally studied in inorganic chemistry. The vast numbers of synthetic and natural organic compounds exist because of the ability of carbon to form chains of atoms (catenation). Other elements are involved in organic compounds: principally hydrogen and oxygen but also nitrogen, halogens, sulfur, and phosphorus.

organometallic compound An organic compound containing a carbon–metal bond. Tetraethyl lead, $(C_2H_5)_4Pb$, is an example of an organometallic compound, formerly used as an additive in petrol.

ornithine cycle (urea cycle) The sequence of enzyme-controlled reactions by which urea is formed as a breakdown product of amino acids. It occurs in cells of the liver. The amino acid ornithine is combined with ammonia (from amino acids) and carbon dioxide, forming another amino acid, arginine, which is then split into urea (which is excreted) and ornithine.

ortho- **1.** Designating a benzene compound with substituents in the 1,2 positions. The position next to a substituent is the ortho position on the benzene ring. This was used in the systematic naming of benzene derivatives. For example, orthodinitrobenzene (or *o*-dinitrobenzene) is 1,2-dinitrobenzene.
2. Certain acids, regarded as formed from an anhydride and water, were named ortho acids to distinguish them from the less hydrated meta acids. For example, H_4SiO_4 (from SiO_2 + $2H_2O$) is orthosilicic acid; H_2SiO_3 (SiO_2 + H_2O) is metasilicic acid.
3. Designating the form of a diatomic molecule in which both nuclei have the same spin direction; e.g. orthohydrogen, orthodeuterium.

See also meta-; para-.

osmium(IV) oxide (osmium tetroxide; OsO_4) A volatile yellow crystalline solid with a penetrating odor, used as an oxidizing agent and, in aqueous solution, as a catalyst for organic reactions.

osmium tetroxide *See* osmium(IV) oxide.

osmosis Systems in which a solvent is separated from a solution by a SEMIPERMEABLE MEMBRANE approach equilibrium by solvent molecules on the solvent side of the membrane migrating through it to the solution side; this process is called osmosis and always leads to dilution of the solution. The phenomenon is quantified by measurement of the osmotic pressure. The process of osmosis is of fundamental importance in transport and control mechanisms in biological systems; for example, plant growth and general cell function. *See* osmotic pressure.

osmotic pressure Symbol: π The pressure that must be exerted on a solution to prevent the passage of solvent molecules into it when the solvent and solution are separated by a semipermeable membrane. The osmotic pressure is therefore the pressure required to maintain equilibrium between the passage of solvent molecules through the membrane in either direction and thus prevent the process of osmosis proceeding. The osmotic pressure can be measured by placing the solution, contained in a small perforated thimble covered by a semipermeable membrane and fitted with a length of glass tubing, in a beaker of the pure solvent. Solvent molecules pass through the membrane, diluting the solution and thereby increasing the volume on the solution side and forcing the solution to rise up the glass tubing. The process continues until the pressure exerted by the solvent molecules on the membrane is balanced by the hydrostatic pressure of the solution in the tubing. A sample of the solution is then removed and its concentration determined. Osmosis is a colligative property; therefore the method can be applied to the determination of relative molecular masses, particularly for large molecules, such as proteins, but it is restricted by the difficulty of preparing good semipermeable membranes.

As the osmotic pressure is a colligative property it is directly proportional to the molar concentration of the solute if the temperature remains constant; thus π is proportional to the concentration n/V, where n is the number of moles of solute, and V the solvent volume. The osmotic pressure is also proportional to the absolute temperature. Combining these two proportionalities gives $\pi V = nCT$, which has the same form as the gas equation, $PV = nRT$, and experimental values of C are similar to those for R, the universal gas constant. This gives considerable support to the kinetic theory of colligative properties.

Ostwald's dilution law *See* dissociation constant.

oxalate A salt or ester of ethanedioic acid (oxalic acid). *See* ethanedioic acid.

oxalic acid *See* ethanedioic acid.

oxaloacetic acid (OAA) A water-soluble carboxylic acid, structurally related to fumaric acid and maleic acid. Oxaloacetic acid forms part of the Krebs cycle, it is produced from L-malate in an NAD-requiring reaction and itself is a step towards the formation of citric acid in a reaction involving pyruvate ion and coenzyme A.

oxidant An oxidizing agent. In rocket fuels, the oxidant is the substance that provides the oxygen for combustion (e.g. liquid oxygen or hydrogen peroxide).

oxidative metabolism *See* aerobic respiration.

oxidative phosphorylation The production of ATP from phosphate and ADP as electrons are transferred along the electron-transport chain from NADH or $FADH_2$ to oxygen. Most of the NADH and $FADH_2$ is formed in the mitochondrial matrix by the Krebs cycle and fatty acid oxidation. Oxidative phosphorylation occurs in mitochondria and is the main source of ATP in aerobes. *See* electron-transport chain.

oxidation An atom, an ion, or a molecule is said to undergo oxidation or to be oxidized when it loses electrons. The process may be effected chemically, i.e. by reaction with an *oxidizing agent*, or electrically, in which case oxidation occurs at the anode. For example,

$$2Na + Cl_2 \rightarrow 2Na^+ + 2Cl^-$$

where chlorine is the oxidizing agent and sodium is oxidized, and

$$4CN^- + 2Cu^{2+} \rightarrow C_2N_2 + 2CuCN$$

where Cu^{2+} is the oxidizing agent and CN^- is oxidized.

The *oxidation state* of an atom is indicated by the number of electrons lost or effectively lost by the neutral atom, i.e. the oxidation number. The oxidation number of a negative ion is negative. The process of oxidation is the converse of reduction. *See also* redox.

oxidizing agent *See* oxidation.

oxime A type of organic compound containing the C:NOH group, formed by reaction of an ALDEHYDE or KETONE with hydroxylamine (NH_2OH). The reaction is a condensation reaction, in which a molecule of water is lost.

oxo process A method of manufacturing aldehydes by passing a mixture of carbon monoxide, hydrogen, and alkanes over a cobalt catalyst at high pressure (100 atmospheres and 150°C). The aldehydes can subsequently be reduced to alcohols, making the process a useful source of alcohols of high molecular weight.

2-oxopropanoic acid *See* pyruvic acid.

oxyacid An acid in which the replaceable hydrogen atom is part of a hydroxyl group, including carboxylic acids, phenols and inorganic acids such as phosphoric(V) acid and sulfuric(VI) acid. *See* acid.

oxygen A colorless odorless diatomic gas; the first member of group 16 (formerly VIA) of the periodic table. It has the electronic configuration $[He]2s^22p^4$ and its chemistry involves the acquisition of electrons to form either the di-negative ion, O^{2-}, or two covalent bonds. In each case the oxygen atom attains the configuration of the rare gas neon. Oxygen is the most plentiful element in the Earth's crust accounting for over 40% by weight. It is present in the atmosphere (20%) and is a constituent of the majority of minerals and rocks (e.g. sandstones, SiO_2, carbonates, $CaCO_3$, aluminosilicates) as well as the major constituent of the sea. Oxygen is an essential element for almost all living things. Elemental oxygen has two forms: the diatomic molecule O_2 and the less stable molecule trioxygen (OZONE), O_3, which is formed by passing an electric discharge through oxygen gas.

Oxygen occurs in three natural isotopic forms, ^{16}O (99.76%), ^{17}O (0.0374%), ^{18}O (0.2039%); the rarer isotopes are used in detailed studies of the behavior of oxygen-containing groups during reactions (tracer studies).

Symbol: O; m.p. –218.4°C; b.p. –182.962°C; d. 1.429 kg m^{-3} (0°C); p.n. 8; r.a.m. 15.9994.

ozone (trioxygen; O_3) A poisonous, blue-colored allotrope of oxygen made by passing oxygen through a silent electric discharge. Ozone is unstable and decomposes to oxygen on warming. It is a strong oxidizing agent. It is present in the upper layers of the atmosphere, where it screens the Earth from harmful short-wave ultraviolet radiation. There is concern that the ozone layer is possibly being depleted by fluorocarbons and other compounds produced by industry.

ozonide *See* ozonolysis.

ozonolysis The addition of ozone (O_3) to alkenes and the subsequent hydrolysis of the ozonide into hydrogen peroxide and a mixture of carbonyl compounds. The carbonyl compounds can be separated and identified, which in turn, identifies the groups and locates the position of the double bond in the original alkene. Ozonolysis was formerly an important analytical technique.

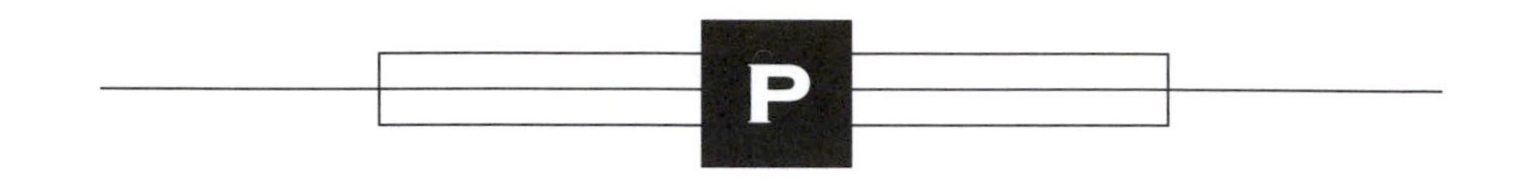

palmitate A salt or ester of palmitic acid.

palmitic acid *See* hexadecanoic acid.

pantothenic acid (vitamin B_5) One of the water-soluble B-group of vitamins. Sources of the vitamin include egg yolk, kidney, liver, and yeast. As a constituent of coenzyme A, pantothenic acid is essential for several fundamental reactions in metabolism. A deficiency results in symptoms affecting a wide range of tissues; the overall effects include fatigue, poor motor coordination, and muscle cramps.

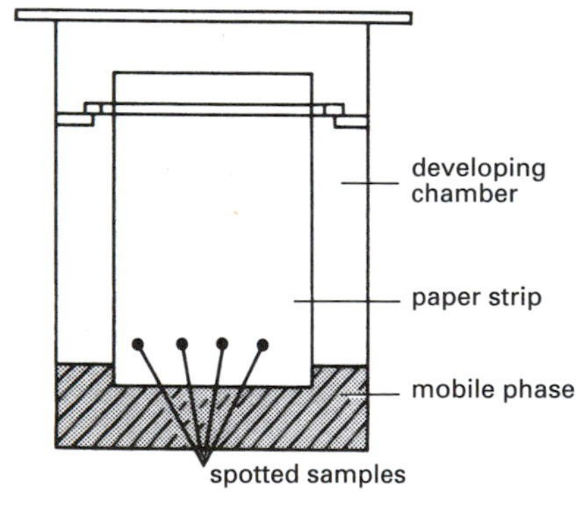

Paper chromatography

paper chromatography A form of CHROMATOGRAPHY widely used for the analysis of mixtures. Paper chromatography usually employs a specially produced paper as the stationary phase. A base line is marked in pencil near the bottom of the paper and a small sample of the mixture is spotted onto it using a capillary tube. The paper is then placed vertically in a suitable solvent, which rises up to the base line and beyond by capillary action. The components within the sample mixture dissolve in this mobile phase and are carried up the paper. However, the paper holds a quantity of moisture and some components will have a greater tendency than others to dissolve in this moisture rather than in the mobile phase. In addition, some components will preferentially cling to the surface of the paper. Therefore, as the solvent moves through the paper, certain components will be left behind and components in the mixture will become separated from each other.

When the solvent has almost reached the top of the paper, the paper is removed and quickly dried. The paper is developed to locate the positions of colorless fractions by spraying with a suitable chemical, e.g. ninhydrin, or by exposure to ultraviolet radiation. The components are identified by comparing the distance they have traveled up the paper with standard solutions that have been run simultaneously, or by computing an R_F VALUE. A simplified version of paper chromatography uses a piece of filter paper.

para- **1.** Designating a benzene compound with substituents in the 1,4 positions. The position on a benzene ring directly opposite a substituent is the para position. This was used in the systematic naming of benzene compounds. For example, para-dinitrobenzene (or *p*-dinitrobenzene) is 1,4-dinitrobenzene.
2. Designating the form of a diatomic molecule in which both nuclei have opposite spin directions; e.g. parahydrogen, paradeuterium.

See also meta-, ortho-.

paraffin 1. *See* petroleum.
2. *See* alkane.

paraffin oil *See* petroleum.

paraffin wax A solid mixture of hydrocarbons obtained from petroleum. *See also* wax.

paraformaldehyde *See* methanal.

paraldehyde *See* ethanal.

partial ionic character The electrons of a covalent bond between atoms or groups with different electronegativities will be polarized towards the more electronegative constituent; the magnitude of this effect can be measured by the ionic character of the bond. When the effect is small the bond is referred to simply as a polar bond and is adequately treated using DIPOLE MOMENTS; as the effect grows the theoretical treatment requires other contributions to ionic character.

partial pressure In a mixture of gases, the contribution that one component makes to the total pressure. It is the pressure that the gas would have if it alone were present in the same volume. *See* Dalton's law.

particulate matter (PM) An airborne pollutant consisting of small particles of silicate, carbon, or large polyaromatic hydrocarbons (*PAHs*). These pollutants are often referred to as *particulates*. They are classified according to size; e.g. PM10 is particulate matter formed of particles less than 10 μm diameter.

particulates *See* particulate matter.

partition coefficient If a solute dissolves in two nonmiscible liquids, the partition coefficient is the equilibrium ratio of the concentration in one liquid to the concentration in the other liquid.

pascal Symbol: Pa The SI unit of pressure, equal to a pressure of one newton per square meter (1 Pa = 1 N m^{-2}). The unit is named for the French mathematician Blaise Pascal (1623–62).

Pasteur, Louis (1822–95) French chemist and biologist. Pasteur is famous for his investigations on stereochemistry, fermentation, and vaccines. In his early work he discovered the phenomenon of optical isomers and invented methods for separating such isomers. In 1860 he postulated that optical isomers existed because of the arrangements of atoms in the molecules. This idea was a major stimulus to the development of structural chemistry. His work on fermentation led to the process known as *pasteurization*, which uses elevated temperatures to kill organisms that spoil milk or wine. He also established the role of micro-organisms such as yeast in fermentation.

Pauli exclusion principle *See* exclusion principle.

Pauling, Linus Carl (1901–94) American chemist. Pauling was one of the greatest scientists of the twentieth century. His early work was on determining the structure of complex minerals such as molybdenite by x-ray diffraction. In 1928–29 this led to *Pauling's rules* governing the structure of complex minerals. Pauling was a major pioneer in the application of quantum mechanics to chemical bonding. In 1931 he published a classic paper entitled *The Nature of the Chemical Bond* that explained how a chemical bond is formed from a pair of electrons. Pauling also introduced the concept of hybridization to explain the chemical bonding of the carbon atom. Pauling also considered partially ionic bonds. Pauling put together his ideas about chemical bonding in his book *The Nature of the Chemical Bond*, the first edition of which was published in 1939. In the mid-1930s Pauling turned his attention to molecules of biological interest and was one of the founders of molecular biology. Together with Robert Corey, he showed that many proteins have helical shapes. He also worked on sickle-cell anemia. He coauthored the book *Introduction to Quantum Mechanics* (1935) and wrote the influential books *General Chemistry* (1948) and *college Chemistry* (1950). Pauling was awarded the 1954 Nobel Prize for chemistry. In the 1950s he became concerned with nuclear weapons, particularly their testing in the atmosphere. This led to his winning the 1962 Nobel Prize for peace.

PCB *See* polychlorinated biphenyl.

pectic substances Polysaccharides that, together with hemicelluloses, form the matrix of plant cell walls. They serve to cement the cellulose fibers together. Fruits are a rich source.

They are principally made from the group of sugar acids known as uronic acids. *Pectic acids*, the basis of the other pectic substances, are soluble unbranched chains of α-1,4 linked galacturonic acid units (derived from the sugar galactose). The acid is precipitated as insoluble calcium or magnesium pectate in the middle lamella of plant cells. *Pectinic acids* are slightly modified pectic acids. Under suitable conditions pectinic acids and *pectins* form gels with sugar and acid. Pectins are used commercially as gelling agents, e.g. in jams. Insoluble pectic substances are termed *protopectin* and this is the most important group in normal cell walls. Protopectin is hydrolyzed to soluble pectin by pectinase in ripening fruits, changing the fruit consistency.

pectin *See* pectic substances.

pentane (C_5H_{12}) A straight-chain alkane obtained by distillation of crude oil.

pentanoic acid (valeric acid; $CH_3(CH_2)_3$-COOH) A colorless liquid carboxylic acid, used in making perfumes.

pentose A SUGAR that has five carbon atoms in its molecules.

pentose phosphate pathway (hexose monophosphate shunt) A pathway of glucose breakdown in which pentoses are produced, in addition to reducing power (NADPH) for many synthetic reactions. It is an alternative to glycolysis and is much more active in adipose tissue (where large amounts of NADPH are consumed) than in skeletal muscle.

pentyl group (amyl group) The group $CH_3CH_2CH_2CH_2CH_2$–

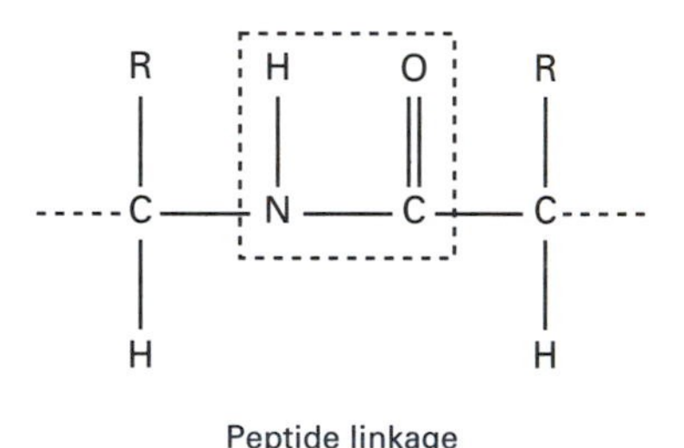

Peptide linkage

peptide A compound formed by linkage of two or more amino-acid groups. Peptides can be formed by reaction of the carboxylic acid group on one amino acid with the amino group on another amino acid, with elimination of water. The amino acids are joined together by a bond of the type –CO–NH–, known as a *peptide linkage.* Simple peptides consisting of a small number of amino-acid units (less than about 10) are known as *oligopeptides* and are designated as *dipeptides*, *tripeptides*, etc., according to the number of amino-acid units present. Peptides with large numbers of amino-acid units are called POLYPEPTIDES.

peptide linkage *See* peptide.

percentage composition A way of expressing the composition of a chemical compound in terms of the percentage (by mass) of each of the elements that make it up. It is calculated by dividing the mass of each element (taking into account the number of atoms present) by the relative molecular mass of the whole molecule. For example, methane (CH_4) has a relative molecular mass of 16 and its percentage composition is 12/16 = 75% carbon and (4 × 1)/16 = 25% hydrogen.

perfect gas (ideal gas) *See* gas laws.

pericyclic reaction A type of CONCERTED REACTION in which the TRANSITION STATE is cyclic and can be regarded as formed by movement of electrons in a circle, from one bond to an adjacent bond. It is usual to distinguish three types of pericyclic reaction:
Cycloadditions. In these reactions a conjugated diene adds to a double bond to form a ring. This involves the formation of two new sigma bonds. The DIELS–ALDER REACTION is a cycloaddition. The reverse reaction, in which a ring containing a double bond breaks to a diene and a compound containing a double bond, is also a pericyclic reaction.
Sigmatropic rearrangements. In these reactions a sigma bond breaks and another is formed.
Electrocyclic reactions. In these reactions a ring is formed across the ends of a conjugated system of double bonds. The reverse reaction, in which a ring containing two double bonds breaks by movement of electrons in a cycle, is also an electrocyclic reaction.
Pericyclic reactions are often treated using FRONTIER ORBITAL theory or the WOODWARD–HOFFMANN RULES.

period One of the horizontal rows in the conventional periodic table. Each period represents the elements arising from progressive filling of the outer shell (i.e. the addition of one extra electron for each new element), the elements being arranged in order of ascending proton number. In a strict sense hydrogen and helium represent one period but convention refers to the elements lithium to neon (8 elements) as the first *short period* ($n = 2$), and the elements sodium to argon (8 elements) as the second *short period* ($n = 3$). With entry to the $n = 4$ level there is filling of the 4s, then back filling of the 3d, before the 4p are filled. Thus this set contains a total of 18 electrons (potassium to krypton) and is called a *long period.* The next set, rubidium to xenon, is similarly a long period.

periodic law The law upon which the modern periodic table is based. Enunciated in 1869 by the Russian chemist Dmitri Mendeléev (1834–1907), this law stated that the properties of the elements are a periodic function of their atomic weights: if arranged in order of increasing atomic weight then elements having similar properties occur at fixed intervals. Certain exceptions or gaps in the table lead to the view that the nuclear charge is a more characteristic function, thus the modern statement of the periodic law is that the physical and chemical properties of elements are a periodic function of their proton number.

periodic table A table of the elements arranged in order of increasing proton number to show similarities in chemical behavior between elements. Horizontal rows of elements are called *periods.* Across a period there is a general trend from metallic to nonmetallic behavior. Vertical columns of related elements are called *groups.* Down a group there is an increase in atomic size and in electropositive (metallic) behavior.

Originally the periodic table was arranged in eight groups with the alkali metals as group I, the halogens as group VII, and the rare gases as group 0. The transition elements were placed in a block in the middle of the table. Groups were split into two sub-groups. For example, group I contained the main-group elements, Li, Na, K, Rb, Cs, in subgroup IA and the subgroup IB elements Cu, Ag, Au. The system was confusing because there was a difference in usage for subgroups and a current form of the table has 18 groups.

See also group; period.

Perkin, Sir William Henry (1838–1907) British organic chemist. Perkin became famous for his discovery of mauve, the first synthetic dye. This discovery originated in 1856 when Perkin attempted to synthesize quinine. He did not succeed in doing so. However, when he used chromic acid to oxidize toluidine he obtained a dark precipitate. When he repeated the experiment using aniline he again obtained a dark precipitate. He found that adding alcohol to this precipitate produced a bright purple solution. He also found that this bright purple substance was a dye which did not fade readily in light. This synthetic dye was called mauve. Perkin patented the process for the manufacture of mauve and set up a factory for its production. This venture was so successful that he was able to retire in 1874 and devote his life to scientific research. His son William Henry Perkin (1860–1929) also became an eminent organic chemist.

peroxide 1. A compound containing the –O–O– group. Organic peroxides tend to be unstable and form free radicals. They are used to initiate free-radical polymerization reactions.
2. An oxide containing the $^{-}O–O^{-}$ ion.

peta- Symbol: P A prefix denoting 10^{15}.

petrochemicals Chemicals that are obtained from crude oil or from natural gas.

petrol *See* petroleum.

petroleum A mixture of hydrocarbons formed originally from marine animals and plants, found beneath the ground trapped between layers of rock. It is obtained by drilling (also called *crude oil*). Different oilfields produce petroleum with differing compositions. The mixture is separated into fractions by fractional distillation in a vertical column. The main fractions are:
Diesel oil (*gas oil*) in the range 220–350°C, consisting mainly of C_{13}–C_{25} hydrocarbons. It is used as fuel in diesel engines.
Kerosene (*paraffin*) in the range 160–250°C, consisting mainly of C_{11} and C_{12} hydrocarbons. It is a fuel both for domestic heating and jet engines.
Gasoline (*petrol*) in the range 40–180°C, consisting mainly of C_5–C_{10} hydrocarbons. It is used as motor fuel and as a raw material for making other chemicals.
Refinery gas, consisting of C_1–C_4 gaseous hydrocarbons.

In addition lubricating oils and paraffin wax are obtained from the residue. The black material left is bitumen tar.

petroleum ether A flammable mixture of hydrocarbons, mainly pentane and hexane, used as a solvent. Note that it is not an ether.

pH The logarithm to base 10 of the reciprocal of the hydrogen-ion concentration of a solution. In pure water at 25°C the concentration of hydrogen ions is 1.00×10^{-7} mol l^{-1}, thus the pH equals 7 at neutrality. An increase in acidity increases the value of $[H^+]$, decreasing the value of the pH below 7. An increase in the concentration of hydroxide ion $[OH^-]$ proportionately decreases $[H^+]$, therefore increasing the value of the pH above 7 in basic solutions. pH values can be obtained approximately by using indicators. More precise measurements use electrode systems. The term 'pH' is short for 'potential of hydrogen'.

phase One of the physically separable parts of a chemical system. For example, a mixture of ice (solid phase) and water (liquid phase) consists of two phases. A system consisting of only one phase is said to be homogeneous. A system consisting of more than one phase is said to be heterogeneous.

phase diagram A graphical representation of the state in which a substance will occur at a given pressure and temperature. The lines show the conditions under which more than one phase can coexist at equilibrium. For one-component systems (e.g. water) the point at which all three phases can coexist at equilibrium is called the triple point and is the point on the graph at which the pressure–temperature curves intersect.

phenol 1. (carbolic acid; hydroxybenzene; C_6H_5OH) A white crystalline solid used to make a variety of other organic compounds. It is usually made using the CUMENE PROCESS or the RASHIG PROCESS.
2. A type of organic compound in which at least one hydroxyl group is bound directly to one of the carbon atoms of an aromatic ring. Phenols do not show the behavior typical of alcohols. In particular they are

more acidic because of the electron-withdrawing effect of the aromatic ring.
Phenol ionizes in water:

$$C_6H_5OH \rightarrow C_6H_5O^- + H^+$$

The preparation of phenol itself is by fusing the sodium salt of the sulfonic acid with sodium hydroxide:

$$C_6H_5SO_2.ONa + 2NaOH \rightarrow C_6H_5ONa + Na_2SO_3 + H_2O$$

The phenol is then liberated by sulfuric acid:

$$2C_6H_5ONa + H_2SO_4 \rightarrow 2C_6H_5OH + Na_2SO_4$$

Reactions of phenol include:

1. Replacement of the hydroxyl group with a chlorine atom using phosphorus(V) chloride:
$$ROH \rightarrow RCl$$
2. Reaction with acyl halides to form esters of carboxylic acids:
$$R^1OH + R^2COCl \rightarrow R^2COOR^1$$
3. Reaction with haloalkanes under alkaline conditions to give mixed alkyl–aryl ethers:
$$R^1OH + R^2Cl \rightarrow R^1OR^2$$

In addition phenol can undergo further substitution on the benzene ring. The hydroxyl group directs other substituents into the 2- and 4-positions.

phenolphthalein An acid–base indicator that is colorless in acid solutions and becomes red if the pH rises above the transition range of 8–9.6. It is used as the indicator in titrations for which the end point lies clearly on the basic side ($pH > 7$), e.g. oxalic acid or potassium hydrogentartrate against caustic soda.

phenoxy resin A type of thermoplastic resin made by condensation of phenols.

phenylalanine *See* amino acid.

phenylamine *See* aniline.

phenylethene (styrene; $C_6H_5CHCH_2$) A liquid hydrocarbon used as the starting material for the production of polystyrene and some synthetic rubbers. The manufacture of phenylethene is by dehydrogenation of ethyl benzene using various metal oxide catalysts:

$$C_6H_5C_2H_5 \rightarrow C_6H_5CH{:}CH_2 + H_2$$

phenyl group The group C_6H_5–, derived from benzene.

phenylhydrazone *See* hydrazone.

phenylmethanol (benzyl alcohol; $C_6H_5CH_2OH$) An aromatic primary alcohol used as a solvent. Phenylmethanol is synthesized by the CANNIZZARO REACTION, which involves the simultaneous oxidation and reduction of benzenecarbaldehyde (benzaldehyde) by refluxing in an aqueous solution of sodium hydroxide:

$$2C_6H_5CHO \rightarrow C_6H_5CH_2OH + C_6H_5COOH$$

Benzoic acid is the other product.

Phenylmethanol undergoes the reactions characteristic of alcohols, especially those in which the formation of a stable carbonium ion as an intermediate ($C_6H_5CH_2^+$) enhances the reaction. Substitution onto the benzene ring is also possible; the $-CH_2OH$ group directs into the 2- or 4-position by the donation of electrons to the ring.

phenyl methyl ketone (acetophenone; $C_6H_5COCH_3$) A colorless sweet-smelling organic liquid, which solidifies below 20°C. It is used as a solvent for methyl and ethyl cellulose plastics.

3-phenylpropenoic acid (**cinnamic acid**) A white pleasant-smelling crystalline carboxylic acid, $C_6H_5CH{:}CHCOOH$. It occurs in amber but can be synthesized and is used in perfumes and flavorings.

pheromone A substance that is excreted by an animal and causes a response in other animals of the same species (e.g. sexual attraction, development). *Compare* kairomone.

Phillips process A method for the manufacture of high-density polyethene using a catalyst of chromium(III) oxide on a promoter of silica and alumina. The reaction conditions are 150°C and 30 atm pressure. *See also* Ziegler process.

phosgene *See* carbonyl chloride.

phosphate A salt or ester of a phosphorus(V) oxoacid, especially one of phosphoric(V) acid, H_3PO_4. Polymeric phosphates occur containing P–O–P bridges.

phosphide A compound of phosphorus with a more electropositive element.

phosphine (phosphorus(III) hydride; PH_3) A colorless gas that is slightly soluble in water. It has a characteristic fishy smell. It can be made by reacting water and calcium phosphide or by the action of yellow phosphorus on a concentrated alkali. Phosphine usually ignites spontaneously in air because of contamination with diphosphine.

phospholipid A lipid with a phosphate group attached by an ester linkage. They are the major class of lipid in all biological membranes and, together with glycolipids and cholesterol, are the main structural components. All membrane phosphlipids, except sphingosine, are derived from glycerol and are called *glycerophospholipids*. They consist of a glycerol backbone with two fatty acid chains esterified to carbons 1 and 2 and a phosphorylated alcohol esterified at carbon 3. The simplest glycerophospholipid is diacylglycerol 3-phosphate or *phosphatidate*, which has no alcohol part. Only small amounts exist naturally, but it is a key intermediate in the biosynthesis of other glycerophospholipids with the phosphate being esterified to one of several alcohols (serine, ethanolamine, choline, or glycerol) to form the major membrane glycerophospholipids: phosphatidylserine, phosphatidylethanolamine, phosphatidylcholine, and phosphatidylglycerol. Sphingomylein (like glycolipids) is derived from sphingosine and has a phosphorylcholine esterified to the primary hydroxyl group of sphingosine.

Another important glycerophospholipid is phosphatidylinositol, which is phosphorylated by specific kinases to phosphatidylinositol 4,5-bisphosphate. This is a key molecule in signal transduction that mediates the action of several hormones and other effectors that control important cellular functions, e.g. glycogenolysis, insulin secretion, the aggregation of platelets, and smooth muscle contraction.

phosphonic acid (phosphorous acid; H_3PO_3) A colorless deliquescent solid that can be prepared by the action of water on phosphorus(III) oxide or phosphorus(III) chloride. It is a dibasic acid producing the anions $H_2PO_3^-$ and HPO_3^{2-} in water. The acid and its salts are slow reducing agents.

phosphonium ion The ion PH_4^+ derived from phosphine.

phosphoprotein A conjugated protein formed by the combination of protein with phosphate groups. Casein is an example.

phosphoric(V) acid (orthophosphoric acid; H_3PO_4) A white solid that can be made by reacting phosphorus(V) oxide with water or by heating yellow phosphorus with nitric acid. The naturally occurring *phosphates* (orthophosphates, M_3PO_4) are salts of phosphoric(V) acid.

phosphorous acid *See* phosphonic acid.

phosphorescence **1.** The absorption of energy by atoms followed by emission of electromagnetic radiation. Phosphorescence is a type of luminescence, and is distinguished from fluorescence by the fact that the emitted radiation continues for some time after the source of excitation has been removed. In phosphorescence the excited atoms have relatively long lifetimes before they make transitions to lower energy states. However, there is no defined time distinguishing phosphorescence from fluorescence.
2. In general usage the term is applied to the emission of 'cold light' – light produced without a high temperature. The name comes from the fact that white phosphorus glows slightly in the dark as a result of a chemical reaction with oxygen. The light comes from excited atoms produced directly in the reaction – not from the heat

produced. It is thus an example of *chemiluminescence*. There are also a number of biochemical examples termed bioluminescence; for example, phosphorescence is sometimes seen in the sea from marine organisms, or on rotting wood from certain fungi (known as 'fox fire').

phosphorus A reactive solid nonmetallic element; the second element in group 15 (formerly VA) of the periodic table. It has the electronic configuration $[Ne]3s^2 3p^3$ and is therefore formally similar to nitrogen. It is however very much more reactive than nitrogen and is never found in nature in the uncombined state. Phosphorus is widespread throughout the world; economic sources are phosphate rock ($Ca_3(PO_4)_2$) and the apatites, variously occurring as both fluoroapatite ($3Ca_3(PO_4)_2.CaF_2$) and as chloroapatite ($3Ca_3(PO_4)_2CaCl_2$). Guano formed from the skeletal phosphate of fish in sea-bird droppings is also an important source of phosphorus. The largest amounts of phosphorus compounds produced are used as fertilizers, with the detergents industry producing increasingly large tonnages of phosphates. Phosphorus is an essential constituent of living tissue and bones, and it plays a very important part in metabolic processes and muscle action.

Symbol: P; m.p. 44.1°C (white) 410°C (red under pressure); b.p. 280.5°C; r.d. 1.82 (white) 2.2 (red) 2.69 (black) (all at 20°C); p.n. 15; r.a.m. 30.973762.

phosphorus(III) bromide (phosphorus tribromide; PBr_3) A colorless liquid made by reacting phosphorus with bromine. It is readily hydrolyzed by water to phosphonic acid and hydrogen bromide. Phosphorus(III) bromide is important in organic chemistry, being used to replace a hydroxyl group with a bromine atom.

phosphorus(V) bromide (phosphorus pentabromide; PBr_5) A yellow crystalline solid that sublimes easily. It can be made by the reaction of bromine and phosphorus(III) bromide. Phosphorus(V) bromide is readily hydrolyzed by water to phosphoric(V) acid and hydrogen bromide. Its main use is in organic chemistry to replace a hydroxyl group with a bromine atom.

phosphorus(III) chloride (phosphorus trichloride; PCl_3) A colorless liquid formed from the reaction of phosphorus with chlorine. It is rapidly hydrolyzed by water to phosphonic acid and hydrogen chloride. Phosphorus(III) chloride is used in organic chemistry to replace a hydroxyl group with a chlorine atom.

phosphorus(V) chloride (phosphorus pentachloride; PCl_5) A white easily sublimed solid formed by the action of chlorine on phosphorus(III) chloride. It is hydrolyzed by water to phosphoric(V) acid and hydrogen chloride. Its main use is as a chlorinating agent in organic chemistry to replace a hydroxyl group with a chlorine atom.

phosphorus(III) chloride oxide (phosphorus trichloride oxide; phosphorus oxychloride, phosphoryl chloride, $POCl_3$) A colorless liquid that can be obtained by reacting phosphorus(III) chloride with oxygen or by distilling phosphorus(III) chloride with potassium chlorate. The reactions of phosphorus(III) chloride oxide are similar to those of phosphorus(III) chloride. The chlorine atoms can be replaced by alkyl groups using Grignard reagents or by alkoxo groups using alcohols. Water hydrolysis yields phosphoric(V) acid.

phosphorus(III) hydride *See* phosphine.

phosphorus(V) oxide (phosphorus pentoxide, P_2O_5) A white powder that is soluble in organic solvents. It usually exists as P_4O_{10} molecules. Phosphorus(V) oxide can be prepared by burning phosphorus in a plentiful supply of oxygen. It readily combines with water to form phosphoric(V) acid and is therefore used as a drying agent for gases. It is a useful dehydrating agent because it is able to remove the elements of water from compounds containing oxygen and hydrogen

phosphorus oxychloride *See* phosphorus(III) chloride oxide.

phosphorus pentabromide *See* phosphorus(V) bromide.

phosphorus pentachloride *See* phosphorus(V) chloride.

phosphorus pentoxide *See* phosphorus(V) oxide.

phosphorus tribromide *See* phosphorus(III) bromide.

phosphorus trichloride *See* phosphorus(III) chloride.

phosphorus trichloride oxide *See* phosphorus(III) chloride oxide.

phosphorylation The transfer of a phosphoryl group $-PO(OH)_2$ from ATP to a protein by a protein kinase. Many metabolic enzymes are regulated by phosphorylation as are several signal pathways involved in cell growth. The removal of the phosphoryl group (dephosphorylation) is brought about by enzymes known as phosphatases.

phosphoryl chloride *See* phosphorus(III) chloride oxide.

photochemical reaction A reaction brought about by light or ultraviolet radiation; examples include the bleaching of colored material, the reduction of silver halides (in photography), and the photosynthesis of carbohydrates. Chemical changes occur only when the reacting atoms or molecules absorb photons of the appropriate energy to produce excited species or when the photons have sufficient energy to produce free radicals or ions. The amount of substance that reacts is proportional to the quantity of energy absorbed. For example, in the reaction between hydrogen and chlorine, it is not the concentrations of hydrogen or chlorine that dictate the rate of reaction but the intensity of the radiation.

photochemistry The branch of chemistry dealing with reactions induced by light or ultraviolet radiation.

photoelectron spectroscopy *See* photoionization.

photoemission The emission of photoelectrons by the photoelectric effect or by photoionization.

photoionization The ionization of atoms or molecules by electromagnetic radiation. Photons absorbed by an atom may have sufficient photon energy to free an electron from its attraction by the nucleus. The process is

$$M + h\nu \rightarrow M^+ + e^-$$

As in the photoelectric effect, the radiation must have a certain minimum threshold frequency. The energy of the photoelectrons ejected is given by $W = h\nu - I$, where I is the ionization potential of the atom or molecule. Analysis of the energies of the emitted electrons gives information on the ionization potentials of the substance – a technique known as *photoelectron spectroscopy*.

photolysis A chemical reaction that is produced by electromagnetic radiation (light or ultraviolet radiation). Many photolytic reactions involve the formation of free radicals. *See also* flash photolysis.

photosynthesis The synthesis of organic compounds using light energy absorbed by chlorophyll. With the exception of a small group of bacteria, organisms photosynthesize from inorganic materials. All green plants photosynthesize as well as certain prokaryotes (some bacteria). In green plants, photosynthesis takes place in chloroplasts, mainly in leaves. Directly or indirectly, photosynthesis is the source of carbon and energy for all except chemoautotrophic organisms. The mechanism is complex and involves two sets of stages: *light reactions* followed by *dark reactions*. The overall reaction in green plants can be summarized by the equation:

$$CO_2 + 4H_2O \rightarrow [CH_2O] + 3H_2O + O_2$$

In the light reactions, light energy is absorbed by chlorophyll (and other pigments), setting off a chain of chemical reactions in which water is split and gaseous oxygen evolved. The hydrogen from the water is attached to other molecules, and used to reduce carbon dioxide to carbohydrates in the later dark reactions. These involve a complex cycle of reactions (the Calvin cycle) in which sugar phosphates are formed.

See electron-transport chain; photosynthetic pigments.

photosynthetic pigments Pigments that absorb the light energy required in photosynthesis. They are located in the chloroplasts of plants and algae, whereas in most photosynthetic bacteria they are located in thylakoid membranes, typically distributed around the cell periphery. All photosynthetic organisms contain chlorophylls and carotenoids; some also contain phycobilins. Chlorophyll *a* is the *primary pigment* since energy absorbed by this is used directly to drive the light reactions of photosynthesis. The chlorophyll *a* that forms the reaction center of photosystem II has an absorption peak at 680 nm and that of photosystem I at 700 nm. The other pigments (chlorophylls *b*, *c*, and *d*, and the carotenoids and phycobilins) are *accessory pigments* that pass the energy they absorb on to chlorophyll *a*. They broaden the spectrum of light used in photosynthesis. *See* absorption spectrum.

phthalic acid *See* benzene-1,2-dicarboxylic acid.

phthalic anhydride *See* benzene-1,2-dicarboxylic acid.

phycobilins A group of accessory photosynthetic pigments found in Cyanobacteria and red algae. Chemically they are linear tetrapyrroles in contrast to chlorophyll, which is a cyclic tetrapyrrole. They absorb light in the middle of the spectrum not absorbed by chlorophyll, an important function in algae living under water where blue and red light are absorbed in the surface layers. They comprise the blue *phycocyanins*, which absorb extra orange and red light, and the red *phycoerythrins*, which absorb green light, enabling red algae to grow at depth in the sea. *See also* absorption spectrum; photosynthetic pigments.

phycocyanin A photosynthetic pigment. *See* phycobilins.

phycoerythrin A photosynthetic pigment. *See* phycobilins.

phylloquinone *See* vitamin K.

physical change A change to a substance that does not alter its chemical properties. Physical changes (e.g. melting, boiling, and dissolving) are comparatively easy to reverse.

physical chemistry The branch of chemistry concerned with the physical properties of compounds and how these depend on the chemical bonding. It includes such topics as chemical thermodynamics and electrochemistry.

physisorption *See* adsorption.

phytohormone *See* plant hormone.

pi bond *See* orbital.

pico- Symbol: p A prefix denoting 10^{-12}. For example, 1 picofarad (pF) = 10^{-12} farad (F).

picrate A salt of picric acid; metal picrates are explosive. *See also* picric acid.

picric acid (2,4,6-trinitrophenol; $C_6H_2(NO_3)_3OH$) A yellow crystalline solid made by nitrating phenolsulfonic acid. It is used as a dye and as an explosive. With aromatic hydrocarbons picric acid forms characteristic charge-transfer complexes (misleadingly called *picrates*), used in analysis for identifying the hydrocarbon.

pine-cone oil *See* turpentine.

pi orbital *See* orbital.

pipette A device used to transfer a known volume of solution from one container to another; in general, several samples of equal volume are transferred for individual analysis from one stock solution. Pipettes are of two types, bulb pipettes, which transfer a known and fixed volume, and graduated pipettes, which can transfer variable volumes. Pipettes were at one time universally mouth-operated but safety pipettes using a plunger or rubber bulb are now preferred.

pK The logarithm to the base 10 of the reciprocal of an acid's dissociation constant:

$$\log_{10}(1/K_a)$$

Planck constant Symbol: h A fundamental constant; the ratio of the energy (W) carried by a photon to its frequency (ν). A basic relationship in the quantum theory of radiation is $W = h\nu$. The value of h is $6.626\ 196 \times 10^{-34}$ J s. The Planck constant appears in many relationships in which some observable measurement is quantized (i.e. can take only specific discrete values rather than any of a range of values).

plane polarization A type of POLARIZATION of electromagnetic radiation in which the vibrations take place entirely in one plane.

plant hormone (phytohormone) One of a group of essential organic substances produced in plants. They are effective in very low concentrations and control growth and development by their interactions. Examples are auxins, gibberellins, cytokinins, abscisic acid, and ethylene.

plastic A substance that can be shaped by heat and pressure. Most plastics are made from synthetic POLYMERS, although some are based on natural materials such as cellulose. In a plastic, the synthetic resin is usually mixed with other substances such as plasticizers, fillers, stabilizers, and colorants. *See also* resin.

plasticizer A substance added to a synthetic resin to make it more flexible.

platinum A silvery-white malleable ductile transition metal. It occurs naturally in Australia and Canada, either free or in association with other platinum metals. It is resistant to oxidation and is not attacked by acids (except aqua regia) or alkalis. Platinum is used as a catalyst for ammonia oxidation (to make nitric acid), hydrocarbon cracking, and in catalytic converters. It is also used in jewelry.

Symbol: Pt; m.p. 1772°C; b.p. 3830 ± 100°C; r.d. 21.45 (20°C); p.n. 78; r.a.m. 195.08.

platinum black A finely divided black form of platinum produced, as a coating, by evaporating platinum onto a surface in an inert atmosphere. Platinum-black coatings are used as absorbents and as catalysts.

Plexiglas (*Trademark*) A widely-used acrylic resin, polymethylmethacrylate.

PM *See* particulate matter.

poison **1.** A substance that destroys catalyst activity.
2. Any substance that endangers biological activity, whether by physical or chemical means.

polar Describing a compound with molecules that have a permanent dipole moment. Hydrogen chloride and water are examples of polar compounds.

polar bond A covalent bond in which the bonding electrons are not shared equally between the two atoms. A bond between two atoms of different electronegativity is said to be polarized in the direction of the more electronegative atom, i.e. the electrons are drawn preferentially towards the atom. This leads to a small separation of charge and the development of a bond dipole moment as in, for example, hydrogen fluoride, represented as H→F or as $H^{\delta+}$–$F^{\delta-}$ (F is more electronegative).

The charge separation is much smaller than in ionic compounds; molecules in which bonds are strongly polar are said to display partial ionic character. The effect of the electronegative element can be transmitted beyond adjacent atoms, thus the C–C bonds in, for example, CCl_3CH_3 and CH_3CHO are slightly polar. *See also* dipole moment; intermolecular force.

polarimeter (polariscope) An instrument for measuring optical activity. *See* optical activity.

polariscope *See* polarimeter.

polarizability The ease with which an electron cloud is deformed (polarized). In ions, an increase in size or negative charge leads to an increase in polarizability.

polarization **1.** The restriction of the vibrations in a transverse wave so that the vibration occurs in a single plane. Electromagnetic radiation, for instance, is a transverse wave motion. It can be thought of as an oscillating electric field and an oscillating magnetic field, both at right angles to the direction of propagation and at right angles to each other. Usually, the electric vector is considered since it is the electric field that interacts with charged particles of matter and causes the effects. In 'normal' unpolarized radiation, the electric field oscillates in all possible directions perpendicular to the wave direction. On reflection or on transmission through certain substances (e.g. Polaroid) the field is confined to a single plane. The radiation is then said to be *plane-polarized*. If the tip of the electric vector describes a circular helix as the wave propagates, the light is said to be *circularly polarized*.
2. *See* polarizability.

polar molecule A molecule in which the individual polar bonds are not perfectly symmetrically arranged and are therefore not 'in balance'. Thus the charge separation in the bonds gives rise to an overall charge separation in the molecule as, for example, in water. Such molecules possess a DIPOLE MOMENT.

polarography An analytical method in which current is measured as a function of potential. A special type of cell is used in which there is a small easily polarizable cathode (the dropping mercury electrode) and a large non-polarizable anode (reference cell). The analytical reaction takes place at the cathode and is essentially a reduction of the cations, which are discharged according to the order of their electrode potential values. The data is expressed in the form of a *polarogram*, which is a plot of current against applied voltage. As the applied potential is increased a point is reached at which the ion is discharged. There is a step-wise increase in current, which levels off because of polarization effects. The potential at half the step height (called the *half-wave potential*) is used to identify the ion. Most elements can be determined by polarography. The optimum concentrations are in the range 10^{-2}–10^{-4}M; modified techniques allow determinations in the parts per million range.

polar solvent *See* solvent.

pollution Any damaging or unpleasant change in the environment that results from the physical, chemical, or biological side-effects of human industrial or social activities. Pollution can affect the atmosphere, rivers, seas, and the soil.

Air pollution is caused by the domestic and industrial burning of carbonaceous fuels, by industrial processes, and by car exhausts. Among recent problems are industrial emissions of sulfur(IV) oxide causing *acid rain*, and the release into the atmosphere of chlorofluorocarbons, used in refrigeration, aerosols, etc., has been linked to the depletion of ozone in the stratosphere. Carbon dioxide, produced by burning fuel and by car exhausts, is slowly building up in the atmosphere, which could result in an overall increase in the temperature of the atmosphere (greenhouse effect). Car exhausts also contain carbon monoxide and lead. The former has not yet reached dangerous levels, but vegetation near main roads contains a high proportion of lead and levels are sufficiently high in urban areas to cause concern about the effects on children. Lead-free gasoline is widely available. Photochemical smog, caused by the action of sunlight on hydrocarbons and nitrogen oxides from car exhausts, is a problem in several countries. Catalytic converters reduce harmful emissions from car exhausts.

Water pollutants include those that are *biodegradable*, such as sewage effluent, which cause no permanent harm if adequately treated and dispersed, as well as those that are nonbiodegradable, such as certain chlorinated hydrocarbon pesticides (e.g. DDT) and heavy metals, such as lead, copper, and zinc in some industrial effluents (causing *heavy-metal pollution*). When these accumulate in the environment they can become very concentrated in food chains. The pesticides DDT, aldrin, and dieldrin are now banned in many countries. Water supplies can become polluted by leaching of nitrates from agricultural land. The discharge of waste heat can cause thermal pollution of the environment, but this is reduced by the use of cooling towers. In the sea, oil spillage from tankers and the inadequate discharge of sewage effluent are the main problems.

Other forms of pollution are noise from aircraft, traffic, and industry and the disposal of radioactive waste.

polyamide A synthetic polymer in which the monomers are linked by the group –NH–CO–. Nylon is an example of a polyamide.

polyamine An aliphatic compound which has two or more amino and/or imino groups. Polyamines are often found associated with DNA and RNA in bacteria and viruses. This may stabilize the nucleic acid molecule in a way analogous to the action of histones on DNA in eukaryote cells. Examples of polyamines include spermine, spermidine, cadaverine, and putrescine.

polyatomic molecule A molecule that consists of several atoms (three or more). Examples are benzene (C_6H_6) and methane (CH_4).

polybasic acid An acid that has two or more replaceable hydrogen atoms. For example, phosphorus(V) acid, H_3PO_4, is tribasic.

polycarbonate A thermoplastic polymer consisting of polyesters of carbonic acid and dihydroxy compounds. Polycarbonates are tough and transparent, used for making soft-drink bottles and electrical connectors.

polychlorinated biphenyl (PCB) A type of compound based on biphenyl ($C_6H_5C_6H_5$), in which some of the hydrogen atoms have been replaced by chlorine atoms. They are used in certain polymers used for electrical insulators. PCBs are highly toxic and concern has been caused by the fact that they can accumulate in the food chain.

polychloroethene (polyvinyl chloride; PVC) A synthetic polymer made from chloroethene. It is a strong material with a wide variety of uses.

polycyclic Describing a compound that has two or more rings in its molecules.

polyene An alkene with more than two double bonds in its molecules.

polyester A synthetic polymer made by reacting alcohols with acids, so that the monomers are linked by the group –O–CO–. Synthetic fibers such as Dacron are polyesters.

polyethene (polyethylene; polythene) A synthetic polymer made from ethene. It is produced in two forms – a soft material of low density and a harder, higher density form, which is more rigid. It can be made by the ZIEGLER PROCESS and the PHILLIPS PROCESS.

polyethylene *See* polyethene.

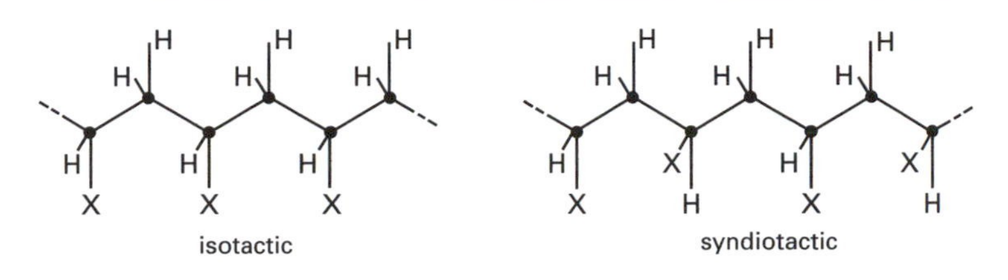

Polymerization: examples of stereospecific polymers

polyhydric alcohol An alcohol that has several –OH groups in its molecules.

polymer A compound in which there are very large molecules made up of repeating molecular units (*monomers*). A polymer has a repeated structural unit, known as a *mer*. Polymers do not usually have a definite relative molecular mass, because there are variations in the lengths of different chains. They may be natural substances (e.g. polysaccharides and proteins) or synthetic materials (e.g. nylon and polyethene). The two major classes of synthetic polymers are thermosetting (e.g. Bakelite) and thermoplastic (e.g. polyethene). The former are infusible, and heat may only make them harder, whereas the latter soften on heating. *See also* polymerization.

polymerization The process in which one or more compounds react to form a POLYMER. *Homopolymers* are formed by polymerization of one monomer (e.g. the formation of polyethene). *Heteropolymers* or *copolymers* come from two or more monomers (as for nylon). Heteropolymers may be of different types depending on the arrangement of units. An *alternating copolymer* of two units A and B has an arrangement:

–A–B–A–B–A–B–

A *block copolymer* has an arrangement in which blocks of one monomer alternate with blocks of the other; for example:

–A–A–A–B–B–B–A–A–A–

In a *graft copolymer* there is a main choice of one monomer (–A–A–A–A–), with short side chains of the other monomer attached at regular intervals (–B–B–).

Stereospecific polymers have the subunit repeated along the chain in a regular way. These are *tactic polymers*. If one particular group is always on the same side of the chain, the polymer is said to be *isotactic*. If the group alternates in position along the chain the polymer is *syndiotactic*. If there is no regular pattern, the polymer is *atactic*.

Polymerization reactions are also classified according to the type of reaction. *Addition polymerization* occurs when the monomers undergo addition reactions, with no other substance formed. *Condensation polymerization* involves the elimination of small molecules in the formation of the polymer. *See also* cross linkage.

polymethanal *See* methanal.

polymethylmethacrylate A transparent ACRYLIC RESIN made by polymerizing methyl methacrylate, trade name Plexiglass.

polypeptide A PEPTIDE composed of a large number of amino-acid units. PROTEINS are polypeptides containing a few hundred amino-acid units.

polypropene (polypropylene) A synthetic polymer made from propene. It is similar in properties to polyethene, but stronger and lighter. The propene is polymerized by the Ziegler process.

polypropylene *See* polypropene.

polysaccharide A high-molecular-weight polymer of a monosaccharide (*see* sugar). The polysaccharides contain many repeated units in their molecular structures. They can be broken down to smaller polysaccharides, disaccharides, and monosaccharides by hydrolysis or by the appropriate enzyme. Important polysaccharides are heparin, inulin, starch, glycogen (sometimes known as animal starch), and cellulose. *See also* carbohydrates; sugar.

polystyrene A synthetic polymer made from styrene (phenylethene). Expanded polystyrene is a rigid foam used in packing and insulation.

polytetrafluoroethene (PTFE) A synthetic polymer made from tetrafluoroethene (i.e. CF_2:CF_2). It is able to withstand high temperatures without decomposing and also has a very low coefficient of friction, hence its use in non-stick pans, bearings, etc.

polythene *See* polyethene.

polyunsaturated Describing a compound that has a number of C=C bonds in its compounds.

polyurethane A synthetic polymer containing the group –NH–CO–O– linking the monomers. Polyurethanes are made by condensation of isocyanates (–NCO) with alcohols.

polyvinyl acetate (PVA; polyvinyl ethanoate) A thermoplastic polymer made by the polymerization of vinyl ethanoate, CH_2:$CHOOCH_3$. It is used as a coating for paper and cloth and as an adhesive.

polyvinyl chloride *See* polychloroethene.

polyvinyl ethanoate *See* polyvinyl acetate.

p orbital *See* orbital.

porphyrin Any of several cyclic organic structures that have the important characteristic property of forming complexes with metal ions. Examples of such *metalloporphyrins* are the iron porphyrins (e.g. heme in hemoglobin) and the magnesium porphyrin, chlorophyll, the photosynthetic pigment in plants. In nature, the majority of metalloporphyrins are conjugated to proteins to form a number of very important molecules, e.g. hemoglobin, myoglobin, and the cytochromes.

potentiometric titration A titration in which an electrode is used in the reaction mixture. The end point can be found by monitoring the electrode potential of this during the titration.

precipitate A suspension of small particles of a solid in a liquid formed by a chemical reaction.

precursor A substance from which another substance is formed in a chemical reaction.

Prelog, Vladimir (1906–98) Yugoslav-born Swiss organic chemist. The early part of Prelog's career was devoted to the structure of the alkaloids. This resulted in the determination of the structure of antimalarial alkaloids. He also corrected the formulae for strychnine alkaloids found by Sir Robert ROBINSON. Prelog subsequently became interested in the relation between stereochemistry and chemical reactivity of ring molecules with about 10 members of the ring. He showed that specific stereochemistry has an important effect on the course of chemical reactions. Together with Sir Christopher INGOLD he found a system for defining chirality of molecules. Prelog shared the 1975 Nobel Prize for chemistry with Sir John CORNFORTH for his work on the 'stereochemistry of organic molecules and reactions'.

pressure Symbol: p The pressure on a surface due to forces from another surface or from a fluid is the force acting at 90° to unit area of the surface:

$$\text{pressure} = \text{force}/\text{area}$$

The unit is the pascal (Pa).

primary alcohol *See* alcohol.

primary amine *See* amine.

primary standard A substance that can be used directly for the preparation of standard solutions without reference to some other concentration standard. Primary standards should be easy to purify, dry, capable of preservation in a pure state, unaffected by air or CO_2, of a high molecular weight (to reduce the significance of weighing errors), stoichiometric, and readily soluble. Any likely impurities should be easily identifiable.

producer gas (air gas) A mixture of carbon monoxide (25–30%), nitrogen (50–55%), and hydrogen (10–15%), prepared by passing air with a little steam through a thick layer of white-hot coke in a furnace or 'producer'. The air gas is used while still hot to prevent heat loss and finds uses in industrial heating, for example the firing of retorts and in glass furnaces. *Compare* water gas.

projection formula *See* formula.

proline *See* amino acid.

promoter (activator) A substance that improves the efficiency of a catalyst. It does not itself catalyze the reaction but assists the catalytic activity.

proof A measure of the ethanol content of intoxicating drinks. In the United States, proof spirit contains 40% ethanol by volume. In the UK, it contains 49.28% of ethanol by weight (57.1% by volume at 16°C). The degree of proof gives the number of parts of proof spirit per 100 parts of the total. 100° of proof is 50% by volume, 80° of proof is 0.8 × 50%, etc.

propanal (propionaldehyde, C_2H_5CHO) A colorless liquid aldehyde.

propane (C_3H_8) A gaseous alkane obtained either from the gaseous fraction of crude oil or by the cracking of heavier fractions. The principal use of propane is as a fuel for heating and cooking, since it can be liquefied under pressure, stored in cylinders, and transported easily. Propane is the third member of the homologous series of alkanes.

propanedioic acid (malonic acid; $CH_2(COOH)_2$) A white crystalline dibasic carboxylic acid.

propane-1,2,3-triol (glycerol; glycerine; $CH_2(OH)CH(OH)CH_2(OH)$) A colorless viscous liquid obtained as a by-product from the manufacture of soap by the reaction of animal fats with sodium hydroxide. It is used as a solvent and plasticizer. *See also* glyceride.

propanoate A salt or ester of propanoic acid.

propanoic acid (propionic acid, C_2H_5COOH) A colorless liquid carboxylic acid.

propanol Either of two alcohols: propan-1-ol ($CH_3CH_2CH_2OH$) and propan-2-ol ($CH_3CH_2(OH)CH_3$). Both are colorless volatile flammable liquids.

propanone (acetone; CH_3COCH_3) A colorless liquid ketone, used as a solvent and in the manufacture of methyl 2-methyl-propanoate (from which polymethylmethacrylate is produced). Propanone is manufactured from propene, either by the air-oxidation of propan-2-ol or as a by-product from the CUMENE PROCESS.

propenal (acrolein; $CH_2{:}CHCHO$) A colorless liquid unsaturated aldehyde with a pungent odor. It can be polymerized to make acrylic resins.

propene (propylene; C_3H_6) A gaseous alkene. Propene is not normally present in the gaseous crude-oil fraction but can be obtained from heavier fractions by catalytic cracking. This is the principal industrial source. Propene is the organic starting material for the production of propan-2-ol, required for the manufacture of propanone (acetone), and the starting material for the production of polypropene (polypropylene).

propenoate A salt or ester of propenoic acid.

propenoic acid (acrylic acid, CH_2:CH-COOH) An unsaturated liquid carboxylic acid with a pungent odor. The acid and its esters are used to make acrylic resins. *See* acrylic resin.

propenonitrile (acrylonitrile, H_2C:-CH(CN)) An organic compound from which acrylic-type polymers are produced.

propenyl group (allyl group) The organic group, CH_2=CH–CH_2–, derived by removing one hydrogen atom from propene.

propionaldehyde *See* propanal.

propionic acid *See* propanoic acid.

propylene *See* propene.

propyl group The group $CH_3CH_2CH_2$–.

prosthetic group The nonprotein component of a conjugated protein. Thus the heme group in hemoglobin is an example of a prosthetic group, as are the coenzyme components of a wide range of enzymes.

protamine One of a group of polypeptides formed from a few amino acids. They are soluble in water, dilute acids, and bases. On heating they do not coagulate. When protamines are hydrolyzed they yield a large proportion of basic amino acids, particularly arginine, alanine, and serine. They occur in the sperm of vertebrates, packing the DNA into a condensed form.

protein One of a large number of substances that are important in the structure and function of all living organisms. Proteins are polypeptides; i.e. they are made up of AMINO ACID molecules joined together by peptide links. Their molecular weight may vary from a few thousand to several million. About 20 amino acids are present in proteins. Simple proteins contain only amino acids. In conjugated proteins, the amino acids are joined to other groups. Proteins may consist of one or several polypeptide chains.

The *primary structure* of a protein is the particular sequence of amino acids present. The *secondary structure* is the way in which this chain is arranged; for example, coiled in an ALPHA HELIX or held in beta pleated sheets. The secondary structure is held by hydrogen bonds. The TERTIARY STRUCTURE of the protein is the way in which the polypeptide chain is folded into a three-dimensional structure. This may be held by cystine bonds and by attractive forces between atoms. A protein must have the correct tertiary structure in order to function properly and abnormally folded proteins can cause disease, e.g. prion diseases, Alzheimer's. QUATERNARY STRUCTURE is the interaction between polypeptide chains (or subunits).

proteoglycan (mucoprotein) A type of glycoprotein consisting of long branched heterogeneous chains of glycosaminoglycan molecules linked to a protein core of amino acids. Unlike more typical glycoproteins, they have a greater carbohydrate content, the protein core is rich in serine, and they have a higher molecular weight.

protic acid *See* acid.

proton An elementary particle with a positive charge ($+1.602\ 192 \times 10^{-19}$ C) and rest mass $1.672\ 614 \times 10^{-27}$ kg. Protons are nucleons, found in all nuclides.

proton number (atomic number) Symbol: Z The number of protons in the nucleus of an atom. The proton number determines the chemical properties of the element because the electron structure, which determines chemical bonding, depends on the electrostatic attraction to the positively charged nucleus.

prussic acid *See* hydrocyanic acid.

pseudoaromatic *See* aromatic compound.

pseudo-first order Describing a reaction that appears to exhibit first-order kinetics under special conditions, even though the 'true' order is greater than one. For example, in the hydrolysis of an ester in the presence of a *large* volume of water, the concentration of water remains approximately constant. The rate of reaction is thus found experimentally to be proportional to the concentration of the ester only (even though it also depends on the amount of water present). Such a reaction is described as 'bimolecular of the first order'.

pteroylglutamic acid *See* folic acid.

PTFE *See* polytetrafluoroethene.

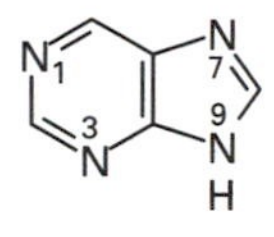

Purine

purine A simple nitrogenous organic molecule with a double ring structure. Members of the purine group include adenine and guanine, which are constituents of the nucleic acids, and certain plant alkaloids, e.g. caffeine and theobromine.

putrescine An amine, $H_2N[CH_2]_4NH_2$, produced from ornithine in decaying meat or fish.

PVA *See* polyvinyl acetate.

PVC *See* polychloroethene.

pyranose A sugar that has a six-membered ring form (five carbon atoms and one oxygen atom). *See also* sugar.

pyrazine (1,4-diazine; $C_4H_4N_2$) A heterocyclic aromatic compound with a six-membered ring containing four carbon atoms and two nitrogen atoms.

pyrazole (1,2-diazole; $C_3H_4N_2$) A heterocyclic crystalline aromatic compound with a five-membered ring containing three carbon atoms and two nitrogen atoms.

pyrene ($C_{16}H_{10}$) A solid aromatic compound whose molecules consist of four benzene rings joined together. It is carcinogenic.

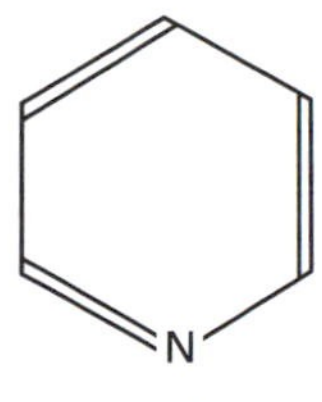

Pyridine

pyridine (C_5H_5N) An organic liquid of formula C_5H_5N. The molecules have a hexagonal planar ring and are isoelectronic with benzene. Pyridine is an example of an aromatic heterocyclic compound, with the electrons in the carbon–carbon pi bonds and the lone pair of the nitrogen delocalized over the ring of atoms. The compound is extracted from coal tar and used as a solvent and as a raw material for organic synthesis.

pyridoxine (vitamin B_6) One of the water-soluble B-group of vitamins. Good sources include yeast and certain seeds (e.g. wheat and corn), liver, and to a limited extent, milk, eggs, and leafy green vegetables. There is also some bacterial synthesis of the vitamin in the intestine. Pyridoxine gives rise to a coenzyme involved in various aspects of amino acid metabolism. *See also* vitamin B complex.

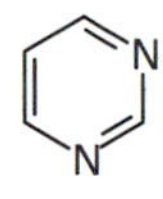

Pyrimidine

pyrimidine A simple nitrogenous organic molecule whose ring structure is contained in the pyrimidine bases cytosine, thymine, and uracil, which are constituents of the nucleic acids, and in thiamine (vitamin B_1).

pyrolysis The decomposition of chemical compounds by subjecting them to very high temperature.

pyrometer An instrument used in the chemical industry to measure high temperature, e.g. in reactor vessels.

pyrone ($C_5H_4O_2$) A compound having a six-membered ring of five carbon atoms and one oxygen, with one of the carbon atoms attached to a second oxygen in a carbonyl group. The pyrone ring system has two forms depending on the position of the carbonyl relative to the oxygen hetero atom. It occurs in many natural products.

pyrrhole ($(CH)_4NH$) A heterocyclic liquid aromatic compound with a five-membered ring containing four carbon atoms and one nitrogen atom. It has important biochemical derivatives, including chlorophyll and heme.

pyruvate An intermediate in several metaoblic pathways, including glycolysis and gluconeogenesis, with the formula CH_3COCOO^-. Pyruvate is also the precursor for the synthesis of the amino acids alanine, valine, and leucine.

pyruvic acid A carboxylic acid, CH_3-COCOOH. The systematic name is *2-oxopropanoic acid.*

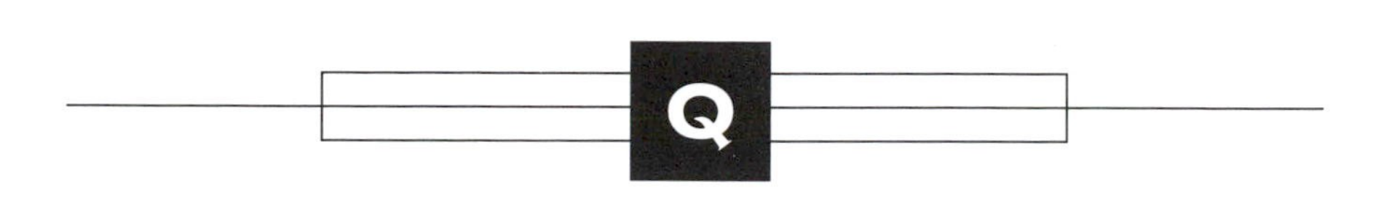

Q

qualitative analysis Analysis carried out with the purpose of identifying the components of a sample. Classical methods involved simple preliminary tests followed by a carefully devised scheme of systematic tests and procedures. Modern methods include the use of a range of spectroscopic techniques. *Compare* quantitative analysis.

quanta *See* quantum.

quantitative analysis Analysis carried out with the purpose of determining the concentration of one or more components of a sample. Classical wet methods include volumetric and gravimetric analysis. A wide range of instrumental techniques are also used, including polarography and various types of chromatography and spectroscopy. *Compare* qualitative analysis.

quantized Describing a physical quantity that can take only certain discrete values, and not a continuous range of values. Thus, in an atom or molecule the electrons around the nucleus can have certain energies, E_1, E_2, etc., and cannot have intermediate values. Similarly, in atoms and molecules, the electrons have quantized values of spin angular momentum and orbital angular momentum.

quantum (plural *quanta*) A definite amount of energy released or absorbed in a process. Energy often behaves as if it were 'quantized' in this way. The quantum of electromagnetic radiation is the photon.

quantum mechanics *See* quantum theory.

quantum number An integer or half integer that specifies the value of a quantized physical quantity (energy, angular momentum, etc.). *See* atom.

quantum states States of an atom, electron, particle, etc., specified by a unique set of quantum numbers. For example, the hydrogen atom in its ground state has an electron in the K shell specified by the four quantum numbers: $n = 1, l = 0, m = 0, m_s = ½$. In the helium atom there are two electrons:

$$n = 1, l = 0, m = 0, m_s = ½$$
$$n = 1, l = 0, m = 0, m_s = -½$$

quantum theory A mathematical theory originally introduced by the German physicist Max Planck (1848–1947) to explain the radiation emitted from hot bodies. Quantum theory is based on the idea that energy (or certain other physical quantities) can be changed only in certain discrete amounts for a given system. Other early applications were the explanations of the photoelectric effect and the Bohr theory of the atom.

Quantum mechanics is a system of mechanics that developed from quantum theory and is used to explain the behavior of atoms, molecules, etc. In one form it is based on de Broglie's idea that particles can have wavelike properties – this branch of quantum mechanics is called *wave mechanics*. *See* orbital.

quantum yield The number of reactive events per absorbed photon in a photochemical reaction.

quaternary ammonium compound A compound formed from an amine by addition of a proton to produce a positive ion. Quaternary compounds are salts, the simplest example being ammonium compounds formed from ammonia and an acid, for example:

$$NH_3 + HCl \rightarrow NH_4^+Cl^-$$

Other amines can also add protons to give analogous compounds. For instance, methylamine (CH_3NH_2) forms the compound

$$[CH_3NH_3]^+X^-$$

where X^- is an acid radical.

The formation of quarternary compounds occurs because the lone pair on the nitrogen atom can form a coordinate bond with a proton. This can also occur with heterogeneous nitrogen compounds, such as adenine, cytosine, thymine, and guanine. Such compounds are known as *nitrogenous bases*. *See also* amine salt.

quaternary structure *See* protein.

quinhydrone *See* quinone.

quinine A poisonous ALKALOID found in the bark of the cinchona tree of South America. It was used in treating malaria.

quinol *See* benzene-1,4-diol.

quinoline (C_9H_7N) A colorless two-ring heterocyclic compound with an unpleasant odor, which acts as a base and forms salts with acids. First made from the alkaloid quinine, it is found in bone oil and coal tar and used for making drugs and dyestuffs.

quinone (cyclohexadiene-1,4-dione; benzoquinone; $C_6H_4O_2$) A yellow crystalline organic compound with a pungent odor. Its molecules contain a nonaromatic six-carbon ring and it behaves as an unsaturated diketone with conjugated double bonds. It is used in making dyestuffs. The systematic name is *cyclohexadiene-1,4-dione*. A platinum electrode in an equimolar solution of quinone and hydroquinone (benzene-1,4-diol; $C_6H_4(OH)_2$) is used as a standard electrode in electrochemistry. The reaction is:

$$C_6H_4(OH)_2 \rightleftharpoons C_6H_4O_2 + 2H^+ + 2e$$

This type of electrode is called a *quinhydrone electrode*.

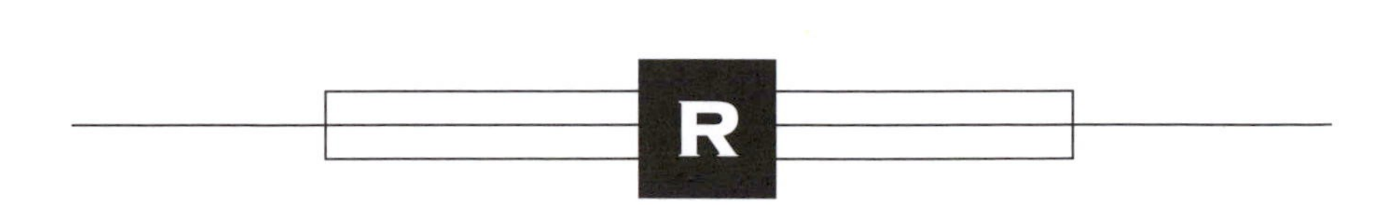

racemate *See* optical activity.

racemic mixture *See* optical activity.

racemization The conversion of an optical isomer into an equal mixture of enantiomers, which is not optically active. *See* optical activity.

rad A unit of absorbed dose of ionizing radiation, defined as being equivalent to an absorption of 10^{-2} joule of energy in one kilogram of material.

radian Symbol: rad The SI unit of plane angle; 2π radian is one complete revolution (360°).

radiation In general the emission of energy from a source, either as waves (light, sound, etc.) or as moving particles (beta rays or alpha rays).

radical 1. *See* free radical.
2. A group of atoms in a molecule. *See also* functional group.

radioactive Describing an element or nuclide that exhibits natural radioactivity.

radioactive dating (radiometric dating) A technique for dating archaeological specimens, rocks, etc., by measuring the extent to which some radionuclide has decayed to give a product.

radioactive decay *See* radioactivity.

radioactivity The disintegration of certain unstable nuclides with emission of radiation. The emission of alpha particles, beta particles, and gamma rays are the three most important forms of radiation that occur. The process by which one nuclide changes to another is *radioactive decay*.

radiocarbon dating *See* carbon dating.

radiochemistry The chemistry of radioactive isotopes of elements. Radiochemistry involves such topics as the preparation of radioactive compounds, the separation of isotopes by chemical reactions, the use of radioactive LABELS in studies of mechanisms, and experiments on the chemical reactions and compounds of transuranic elements.

radiogenic Caused by radioactive decay.

radioisotope A radioactive isotope of an element. Tritium, for instance, is a radioisotope of hydrogen. Radioisotopes are extensively used in research as souces of radiation and as tracers in studies of chemical reactions. Thus, if an atom in a compound is replaced by a radioactive nuclide of the element (a *label*) it is possible to follow the course of the chemical reaction. Radioisotopes are also used in medicine for diagnosis and treatment.

radiolysis A chemical reaction produced by high-energy radiation (x-rays, gamma rays, or particles).

radiometric dating *See* radioactive dating.

radio waves A form of electromagnetic radiation with wavelengths greater than a few millimeters. *See also* electromagnetic radiation.

raffinate The liquid remaining after the solvent extraction of a dissolved substance. *See* solvent extraction.

raffinose A SUGAR occurring in sugar beet. It is a trisaccharide consisting of fructose, galactose, and glucose units.

r.a.m. *See* relative atomic mass.

Raman effect A change in the frequency of electromagnetic radiation, such as light, which occurs when a photon of radiation undergoes an inelastic collision with a molecule. This type of scattering is called *Raman scattering*, in contrast to normal (Rayleigh) scattering. The intensity of Raman scattering is much smaller (about 1/1000) than the intensity of Rayleigh scattering. The Raman effect was first observed by the Indian physicists Sir Chandrasekhara Venkata Raman (1888–1970) and his colleague Sir Kariamanikkam Srinivasa Krishnan in 1928, having been predicted theoretically on the basis of quantum mechanics by Hendrik Anton Kramers and Werner Heisenberg in 1925. The Raman effect has been used extensively in *Raman spectroscopy* for the determination of molecular structure, particularly since then advent of the laser. In some cases it is possible to draw conclusions about molecular structure by observing whether certain lines are present or absent.

Raney nickel A catalytic form of nickel produced by treating a nickel–aluminum alloy with caustic soda. The aluminum dissolves (as aluminate) and Raney nickel is left as a spongy mass, which is pyrophoric when dry. It is used especially for catalyzing hydrogenation reactions. It was discovered by the US chemist M. Raney in 1927.

Raoult's law A relationship between the pressure exerted by the vapor of a solution and the presence of a solute. It states that the partial vapor pressure of a solvent above a solution (p) is proportional to the mole fraction of the solvent in the solution (X) and that the proportionality constant is the vapor pressure of pure solvent, (p_0), at the given temperature: i.e. $p = p_0X$. Solutions that obey Raoult's law are said to be *ideal*. There are some binary solutions for which Raoult's law holds over all values of X for either component. Such solutions are said to be *perfect* and this behavior occurs when the intermolecular attraction between molecules within one component is almost identical to the attraction of molecules of one component for molecules of the other (e.g. chlorobenzene and bromobenzene). Because of solvation forces this behavior is rare and in general Raoult's law holds only for dilute solutions.

For solutions that are ideal but not perfect the solute behavior is similar in that the partial pressure of the solute, p_s, is proportional to the mole fraction of the solute, X_s, but in this case the proportionality constant, p', is not the vapor pressure of the pure solute but must be determined experimentally for each system. This solute equivalent of Raoult's law has the form $p_s = p'X_s$, and is called HENRY'S LAW. Because of intermolecular attractions p' is usually less than p_0. It is named for the French chemist François-Marie Raoult (1830–1901), who formulated it in 1882.

Rashig process A method for the manufacture of chlorobenzene, and thence phenol, from benzene. Benzene vapor, hydrogen chloride, and air are passed over a copper(II) chloride catalyst (230°C):

$$2C_6H_6 + 2HCl + O_2 \rightarrow 2C_6H_5Cl + 2H_2O$$

The chlorobenzene is converted to phenol by reaction with water over a silicon catalyst (425°C):

$$C_6H_5Cl + H_2O \rightarrow HCl + C_6H_5OH$$

It is named for the German chemist Fritz Rashig (1863–1928).

rate constant (velocity constant; specific reaction rate) Symbol: k The constant of proportionality in the rate expression for a chemical reaction. For example, in a reaction A + B → C, the rate may be proportional to the concentration of A multiplied by that of B; i.e.

$$rate = k[A][B]$$

where k is the rate constant for this particular reaction. The constant is independent of the concentrations of the reactants but depends on temperature; consequently the temperature at which k is recorded must be stated. The units of k vary depending on the number of terms in the rate expression, but are easily determined remembering that rate has the units s^{-1}.

rate-determining step (limiting step) The slowest step in a multistep reaction. Many chemical reactions are made up of a number of steps in which the one with the lowest rate is the one that determines the rate of the overall process.

The overall rate of a reaction cannot exceed the rate of the slowest step. For example, the first step in the reaction between acidified potassium iodide solution and hydrogen peroxide is the rate-determining step:

$$H_2O_2 + I^- \rightarrow H_2O + OI^- \text{ (slow)}$$
$$H^+ + OI^- \rightarrow HOI \text{ (fast)}$$
$$HOI + H^+ + I^- \rightarrow I_2 + H_2O \text{ (fast)}$$

rate of reaction A measure of the amount of reactant consumed in a chemical reaction in unit time. It is thus a measure of the number of effective collisions between reactant molecules. The rate at which a reaction proceeds can be measured by the rate the reactants disappear or by the rate at which the products are formed. The principal factors affecting the rate of reaction are temperature, pressure, concentration of reactants, light, and the action of a catalyst. The units usually used to measure the rate of a reaction are mol dm^{-3} s^{-1}. *See also* mass action, law of.

rationalized units A system of units in which the equations have a logical form related to the shape of the system. SI units form a rationalized system of units. In it formulae concerned with circular symmetry contain a factor of 2π; those concerned with radial symmetry contain a factor of 4π.

raw material A substance from which other substances are made. In the chemical industry it may be simple (such as nitrogen from air used to make ammonia) or complex (such as coal and petroleum, used to make a wide range of products).

rayon An artificial fiber formed from wood pulp (cellulose). There are two types. *Viscose rayon* is made by dissolving the cellulose in carbon disulfide and sodium hydroxide. The solution is forced through a fine nozzle into an acid bath, which regenerates the fibers. *Acetate rayon* is made by dissolving cellulose acetate in an organic solvent, and forcing the solution through a nozzle. The solvent is evaporated, and the cellulose acetate thus obtained as fibers.

reactant A compound taking part in a CHEMICAL REACTION.

reaction *See* chemical reaction.

reactive dye A dye that sticks to the fibers of a fabric by forming covalent chemical bonds with the substance of the fabric. The dyes used to color the cellulose fibers in rayon are examples of reactive dyes.

reagent A compound that reacts with another (the substrate). The term is often also used for common laboratory chemicals – sodium hydroxide, hydrochloric acid, etc. – used for experiment and analysis.

rearrangement A reaction in which the groups of a compound rearrange themselves to form a different compound. An example is the BECKMANN REARRANGEMENT.

reciprocal proportions, law of *See* equivalent proportions, law of.

recrystallization The repeated crystallization of a compound to ensure purity of the sample.

rectified spirit A constant-boiling mixture of ethanol and water that contains about 6% water; no more water can be removed by further distillation. It is used as an industrial solvent.

redox Relating to the process of oxidation and reduction, which are intimately connected in that during oxidation by chemical agents the oxidizing agent itself becomes reduced, and vice versa. Thus an oxidation process is always accompanied by a reduction process. In electrochemical processes this is equally true, oxidation taking place at the anode and reduction at the cathode. These systems are often called *redox systems*, particularly when the interest centers on both compounds.

Oxidizing and reducing power is indicated quantitatively by the *redox potential* or standard electrode potential, $E^{\ominus}$. Redox potentials are normally expressed as reduction potentials. They are obtained by electrochemical measurements and the values are referred to the H^+/H_2 couple for which $E^{\ominus}$ is set equal to zero. Thus increasingly negative potentials indicate increasing ease of oxidation or difficulty of reduction. Thus in a redox reaction the half reaction with the most positive value of $E^{\ominus}$ is the reduction half and the half reaction with the least value of $E^{\ominus}$ (or most highly negative) becomes the oxidation half.

reducing agent *See* reduction.

reduction The gain of electrons by atoms, molecules, ions, etc. It often involves the loss of oxygen from a compound, or addition of hydrogen. Reduction can be effected chemically, i.e. by the use of *reducing agents* (electron donors), or electrically, in which case the reduction process occurs at the cathode. *See also* redox.

refining The process of removing impurities from a substance or of extracting a substance from a mixture. The refining of PETROLEUM is a particularly important industrial process.

refluxing The process of boiling a liquid in a vessel connected to a condenser, so that the condensed liquid runs back into the vessel. By using a reflux condenser, the liquid can be maintained at its boiling point for long periods of time, without loss. The technique is a standard method of carrying out reactions in organic chemistry.

reforming The cyclization of straight-chain hydrocarbons from crude oil by heating under pressure with a catalyst, usually platinum on alumina. For example, the manufacture of methylbenzene from heptane:

$$C_7H_{16} \rightarrow C_6H_{11}CH_3$$

This first step is the production of methyl cyclohexane, which then loses six hydrogen atoms to give methylbenzene:

$$C_6H_{11}CH_3 \rightarrow C_6H_5CH_3 + 3H_2$$

See also steam reforming.

Regnault's method A method used for the determination of the density of gases. A bulb of known volume is evacuated and weighed then the gas is admitted at a known pressure (from a vacuum line) and the bulb weighed again. The temperature is also noted and the data corrected to STP. The method is readily applicable to the determination of approximate relative molecular masses of gaseous samples. It is named for the French chemist Henri Victor Regnault (1810–78).

relative atomic mass (r.a.m.) Symbol: A_r The ratio of the average mass per atom of the naturally occurring element to 1/12 of the mass of an atom of nuclide ^{12}C. It was formerly called *atomic weight*

relative density Symbol: d The ratio of the density of a given substance to the density of some reference substance. The relative densities of liquids are usually measured with reference to the density of water at 4°C. Relative densities are also specified for gases; usually with respect to air at STP. The temperature of the substance is stated or is understood to be 20°C. Relative density was formerly called *specific gravity*.

relative molecular mass Symbol: M_r The ratio of the average mass per molecule of the naturally occurring form of an element or compound to 1/12 of the mass of an atom of nuclide ^{12}C. This was formerly called *molecular weight*. It does not have to be used only for compounds that have discrete molecules; for ionic compounds (e.g. NaCl) and giant-molecular structures (e.g. BN) the formula unit is used.

relaxation The process by which an excited species loses energy and falls to a lower energy level (such as the ground state).

rem (*r*adiation *e*quivalent *m*an) A unit for measuring the effects of radiation dose on the human body. One rem is equivalent to an average adult male absorbing one rad of radiation. The biological effects depend on the type of radiation as well as the energy deposited per kilogram.

resin A yellowish insoluble organic compound exuded by trees as a viscous liquid that gradually hardens on exposure to air to form a brittle amorphous solid. Synthetic resins are artificial polymers used in making adhesives, insulators, and paints. *See also* rosin.

resolution The separation of a racemate into the two optical isomers. This cannot be done by normal methods, such as crystallization or distillation, because the isomers have identical physical properties. The main methods are:

1. Mechanical separation. Certain optically active compounds form crystals with distinct left- and right-handed shapes. The crystals can be sorted by hand.
2. Chemical separation. The mixture is reacted with an optical isomer. The products are then not optical isomers of each other, and can be separated by physical means. For instance, a mixture of D- and L-forms of an acid, acting with a pure L-base, produces two salts that can be separated by fractional crystallization and then reconverted into the acids.
3. Biochemical separation. Certain organic compounds can be separated by using bacteria that feed on one form only, leaving the other.

resonance (mesomerism) The behavior of many compounds cannot be adequately explained by a single structure using simple single and double bonds. The bonding electrons of the compound have a different distribution in the molecules. The actual bonding in the molecule can be regarded as a hybrid of two or more conventional forms of the molecule, called *resonance forms* or *canonical forms*. The result is a resonance hybrid. For example, the carbonyl group in a ketone has negative charge on the oxygen atom. It can be described as a resonance hybrid, somewhere between =C=O, in which a pair of electrons is shared between the C and the O, and $=C^{+}–O^{-}$, in which the electrons are localized on the O atom. Note that the two canonical forms do not contribute equally in the hybrid. The bonding of benzene can be represented by a resonance hybrid of two Kekulé structures and, to a lesser extent, three Dewar structures. It is conventional to represent a resonance hybrid by two or more conventional structures joined by double-headed arrows, as in:

$$R_2C=O \leftrightarrow R_2C^{+}O^{-}$$

The double-headed arrow should not be confused with the equilibrium symbol ($\rightleftharpoons$). The two forms are not in equilibrium but represent different classical structures that contribute to the actual structure.

resonance ionization spectroscopy (RIS) A type of spectroscopy that detects specific types of atoms using lasers. The laser ionizes the atoms of interest. The frequency of the laser is chosen so that only the atoms of interest in a sample are excited by the laser. This method is very selective because ionization only occurs for those atoms those whose energy levels fit in with the frequency of the laser light. This selectivity has led to many practical uses for this technique.

resorcinol *See* benzene-1,3-diol.

respiration The oxidation of organic molecules to provide energy in plants and animals. In animals, food molecules are respired, but autotrophic plants respire molecules that they have themselves synthesized by photosynthesis. The energy from respiration is used to attach a high-energy phosphate group to ADP to form the short-term energy carrier ATP, which can then be used to power energy-requiring processes within the cell. The actual chemical reactions of respiration are known as *internal* (*cell* or *tissue*) *respiration* and they normally require oxygen from the environment (*aerobic respiration*). Some organisms are able to respire, at least for a short period, without the use of oxygen (*anaerobic respiration*), although this process produces far less energy than aerobic respiration, e.g. 38 molecules of ATP are generated for each molecule of glucose oxidized in aerobic respiration, compared with only 2 in anerobic respiration. Respiration usually involves an exchange of gases with the environment; this is known as *external respiration*. In small animals and all plants exchange by diffusion is adequate, but larger animals generally have special respiratory organs with large moist and ventilated surfaces (e.g. lungs, gills) and there is often a circulatory system to transport gases internally to and from the respiratory organs. *See also* electron-transport chain; glycolysis; Krebs cycle.

respiratory chain The electron-transport chain in aerobic respiration.

respiratory pigments Colored compounds that can combine reversibly with oxygen. HEMOGLOBIN is the blood pigment in all vertebrates and a wide range of invertebrates. Other blood pigments, such as haemoerythrin (containing iron) and hemocyanin (containing copper), are found in lower animals, and in many cases are dissolved in the plasma rather than present in cells. Their affinity for oxygen is comparable with hemoglobin, though oxygen capacity is generally lower.

retinal (retinene) An aldehyde derivative of retinol (vitamin A). Retinal is a constituent of the light-sensitive conjugated protein, rhodopsin, which occurs in the rod cells of the retina. *See* rhodopsin.

retinene *See* retinal.

retinol *See* vitamin A.

retort A piece of laboratory apparatus consisting of a glass bulb with a long narrow neck. In industrial chemistry, various metallic vessels in which distillations or reactions take place are called retorts.

retrosynthetic analysis A technique for planning the synthesis of an organic molecule. The structure of the molecule is considered and it is divided into imaginary parts, which could combine by known reactions. These *disconnections* of the molecule suggest charged fragments, known as *synthons*, which give a guide to possible reagents for the synthesis.

reversible change In thermodynamics, a change in the pressure, volume, or other properties of a system, in which the system remains at equilibrium throughout the change. Such processes could be reversed; i.e. returned to the original starting position through the same series of stages. They are never realized in practice. An isothermal reversible compression of a gas, for example, would have to be carried out infinitely slowly and involve no friction, etc. Ideal energy transfer would have to take place between the gas and the surroundings to maintain a constant temperature.

In practice, all real processes are *irreversible changes* in which there is not an equilibrium throughout the change. In an irreversible change, the system can still be returned to its original state, but not through the same series of stages. For a closed system, there is always an entropy increase involved in an irreversible change.

reversible reaction A chemical reaction that can proceed in either direction. An example is the reaction of an acid with an alcohol to form an ester and water:

$$R^1COOH + R^2OH \rightleftharpoons R^1COOR^2 + H_2O$$

In general, there will be an equilibrium mixture of reactants and products.

R_F value The distance traveled by a given component divided by the distance travelled by the solvent front in chromatography.

rheology The study of the ways in which matter can flow. This topic is of particular interest in the study of polymers.

riboflavin (vitamin B_2) One of the water-soluble B-group of vitamins. It is found in cereal grains, peas, beans, liver, kidney, and milk. Riboflavin is a constituent of several enzyme systems (flavoproteins), acting as a coenzyme for hydrogen transfer in the reactions catalyzed by these enzymes. Two forms of phosphorylated riboflavin are known to exist in various enzyme systems: FMN (flavin mononucleotide) and FAD (flavin adenine dinucleotide). *See also* vitamin B complex.

ribonucleic acid *See* RNA.

ribose ($C_5H_{10}O_5$) A monosaccharide SUGAR. It rarely occurs naturally in the free state but ribose is an important component of RNA. The derivative compound, *deoxyribose*, is a component of DNA.

ring A closed loop of atoms in a molecule, as in benzene or cyclohexane. A *fused ring* is one joined to another ring in such a way that they share two atoms. Naphthalene is an example of a fused-ring compound.

ring closure A reaction in which one part of an open chain in a molecule reacts with another part, so that a ring of atoms is formed. For example, an amino group on one end of a long molecule can react with a carboxyl group on the other end to form a cyclic amide containing the –NH.CO– group (*see* lactam). Another example of ring closure is the conversion of the straight-chain form of a SUGAR into the cyclic form. There are many examples of ring-closure reactions in organic chemistry.

RNA (ribonucleic acid) A nucleic acid found mainly in the cytoplasm and involved in protein synthesis. It is a single polynucleotide chain similar in composition to a single strand of DNA except that the sugar ribose replaces deoxyribose and the pyrimidine base uracil replaces thymine. RNA is synthesized on DNA in the nucleus and exists in three forms. In certain viruses, RNA is the genetic material.

Robinson, Sir Robert (1886–1975) British organic chemist. Sir Robert Robinson made many important contributions to experimental and theoretical organic chemistry. His experimental work concerned plant products, particularly the alkaloids. Robinson won the 1947 Nobel Prize for chemistry for this work. His theoretic work concerned the role of electrons in organic reactions. He originated the representation of electronic transfer by curly arrows. In particular, he applied these ideas for benzene and its chemical reactions using concepts such as electrophilic reagents and nucleophilic reagents.

roentgen Symbol: R A unit of radiation, used for x-rays and gamma rays, defined in terms of the ionizing effect on air. One roentgen induces 2.58×10^{-4} coulomb of charge in one kilogram of dry air. The unit is named for the German physicist W. K. Roentgen (1845–1923).

rosin A brittle yellow or brown resin that remains after the distillation of turpentine. It is used as a flux in soldering and in making paints and varnishes. Powdered rosin gives a 'grip' to violin bows and boxers' shoes. *See* resin.

rotamer *See* conformation.

rotary dryers Devices commonly used in the chemical industry for the drying, mixing, and sintering of solids. They consist essentially of a rotating inclined cylinder, which is longer in length than in diameter. Gases flow through the cylinder in either a countercurrent or cocurrent direction to regulate the flow of solids, which are fed into the end of the cylinder. Rotary dryers can be applied to both batch and continuous processes.

R-S convention *See* optical activity.

rubber A natural or synthetic polymeric elastic material. Natural rubber is a polymer of methylbuta-1,3-diene (isoprene). Various synthetic rubbers are made by polymerization; for example CHLOROPRENE rubber (from 2-chlorobuta-1,3-diene) and silicone rubbers. *See also* vulcanization.

S

Sabatier, Paul (1854–1941) French organic chemist. Starting in 1897, Sabatier performed experiments in which he studied the hydrogenation of ethene when it is passed over nickel. When nickel is finely divided it catalyses this reaction. It also catalyses other hydrogenation reactions such as the conversion of benzene into cyclohexane. Sabatier explained his results in terms of chemisorption of unstable compounds onto the surface of a catalyst and produced evidence for this idea. He discussed his findings in the book *Catalysis in Organic Chemistry* (1912). Sabatier shared the 1912 Nobel Prize with François GRIGNARD for his work on catalysis.

Sabatier–Senderens process A method of hydrogenating unsaturated vegetable oils to make margarine, using hydrogen and a nickel catalyst. It is named for the French chemists Paul Sabatier and Jean-Baptiste Senderens (1856–1937). *See also* hardening.

saccharide *See* sugar.

saccharin ($C_7H_5NO_3S$) A white crystalline organic compound used as an artificial sweetener; it is about 550 times as sweet as sugar (sucrose). It is nearly insoluble in water and so generally used in the form of its sodium salt. Possible links with cancer in animals has restricted its use in some countries.

saccharose *See* sucrose.

Sachse reaction A process for the manufacture of ethyne (acetylene) from natural gas (methane). Part of the methane is burned in two stages to raise the furnace temperature to about 1500°C. Under these conditions, the rest of the methane is converted to ethyne and hydrogen:

$$2CH_4 \rightarrow C_2H_2 + 3H_2$$

The process is important because it provides a source of ethyne from readily available natural gas, thus avoiding the expensive carbide process.

salicylate (hydroxybenzoate) A salt or ester of salicylic acid.

salicylic acid (2-hydroxybenzoic acid; $C_6H_4(OH)COOH$) A crystalline aromatic carboxylic acid. It is used in medicines, as an antiseptic, and in the manufacture of azo dyes. Its ethanoyl (acetyl) ester is aspirin. *See also* aspirin; methyl salicylate.

Sandmeyer reaction A method of producing chloro- and bromo-substituted derivatives of aromatic compounds by using the DIAZONIUM SALT with a copper halide. The reaction starts with an amine, which is diazotized at low temperature by using hydrochloric acid and sodium nitrite (to produce nitrous acid, HNO_2). For example:

$$C_6H_5NH_2 + NaNO_2 + HCl \rightarrow$$
$$C_6H_5N_2^+ + Cl^- + OH^- + Na^+ + H_2O$$

The copper halide (e.g. CuCl) acts as a catalyst to give the substituted benzene derivative:

$$C_6H_5N_2^+ + Cl^- \rightarrow C_6H_5Cl + N_2$$

This reaction was discovered by the German chemist Traugott Sandmeyer (1854–1922) in 1884. A variation of the reaction, in which the catalyst is freshly precipitated copper powder, was reported in 1890 by the German chemist Ludwig Gatterman (1860–1920). This is known as the *Gatter-*

man reaction (or *Gatterman–Sandmeyer reaction*).

sandwich compound A type of complex formed between transition-metal ions and aromatic compounds, in which the metal ion is 'sandwiched' between the rings. The bonding is between the d orbitals of the metal and the pi electrons on the ring. Four-, five-, six-, seven-, and eight-membered rings are known to complex with a number of elements including V, Cr, Mn, Co, Ni, and Fe. Ferrocene ($Fe(C_5H_5)_2$) is the best-known example. Compounds of this type are also known as *metallocenes*.

Sanger, Frederick (1918–) British biochemist. Sanger is one of the very few people to win two Nobel Prizes. He was awarded the 1958 Nobel Prize for chemistry for determining the sequence of amino acids in the insulin of cows. This work started in 1943. By 1945 he had discovered a compound now known as *Sanger's reagent* which attaches itself to the amino acids, and breaks up the protein chain. After the discovery of the structure of DNA in 1953 Sanger and his colleagues started to investigate the sequence of the nucleotides in DNA. Sanger and his colleagues developed techniques for splitting up the DNA. In 1977 they announced the complete sequence of more than 5000 nucleotides of the DNA of a bacterial virus. This resulted in Sanger sharing the 1980 Nobel Prize for chemistry with Paul Berg and Walter Gilbert.

saponification The process of hydrolyzing an ester with a hydroxide. The carboxylic acid forms a salt. For instance, with sodium hydroxide:

$$R^1COOR^2 + NaOH \rightarrow NaOOCR^1 + R^2OH$$

The saponification of fats, which are esters of propane-1,2,3-triol (glycerol) with long-chain carboxylic acids, forms soaps.

saturated compound An organic compound that does not contain any double or triple bonds in its structure. A saturated compound will undergo substitution reactions but not addition reactions since each atom in the structure will already have its maximum possible number of single bonds. *Compare* unsaturated compound.

saturated solution A solution that contains the maximum equilibrium amount of solute at a given temperature. A solution is saturated if it is in equilibrium with its solute. If a saturated solution of a solid is cooled slowly, the solid may stay temporarily in solution; i.e. the solution may contain *more* than the equilibrium amount of solute. Such solutions are said to be *supersaturated.*

saturated vapor A vapor that is in equilibrium with the solid or liquid. A saturated vapor is at the maximum pressure (the saturated vapor pressure) at a given temperature. If the temperature of a saturated vapor is lowered, the vapor condenses. Under certain circumstances, the substance may stay temporarily in the vapor phase; i.e. the vapor contains *more* than the equilibrium concentration of the substance. The vapor is then said to be *supersaturated.*

scavenger A compound or chemical species that removes a trace component from a system or that removes a reactive intermediate from a reaction.

Schiff's base A type of compound formed by reacting an aldehyde or ketone (e.g. RCOR) with an aryl amine (e.g. $ArNH_2$). The product, an *N*-arylimide, which is usually crystalline, has the formula $R_2C{:}NAr$. Schiff's bases were formerly used to identify aldehydes and ketones (by forming the crystalline base and measuring its melting point). They are named after the German organic chemist Hugo Schiff (1834–1915).

Schiff's reagent An aqueous solution of magenta dye decolorized by reduction with sulfur(IV) oxide (sulfur dioxide; SO_2). It is a test for aldehydes and ketones. Aliphatic aldehydes restore the color quickly; aliphatic ketones and aromatic aldehydes slowly; aromatic ketones give no reaction.

Schultze's solution A solution of zinc chloride, potassium iodide, and iodine used mainly for testing for cellulose and hemicellulose. Both materials stain a blue color with the reagent, that of hemicellulose being weaker.

scleroprotein One of a group of proteins obtained from the exoskeletal structures of animals. They are insoluble in water, salt solutions, dilute acids, and alkalis. This group exhibits a wide range of both physical and chemical properties. Typical examples of scleroproteins are keratin (hair), elastin (elastic tissue), and collagen (connective tissue).

SCP *See* single-cell protein.

scrubber A part of an industrial chemical plant that removes impurities from a gas by passing it through a liquid.

secondary alcohol *See* alcohol.

secondary amine *See* amine.

secondary structure *See* protein.

second-order reaction A reaction in which the rate of reaction is proportional to the product of the concentrations of two of the reactants or to the square of the concentration of one of the reactants:

$$\text{rate} = k[\text{A}][\text{B}]$$

or

$$\text{rate} = k[\text{A}]^2$$

For example, the hydrolysis by dilute alkali of an ester is a second-order reaction:

$$\text{rate} = k[\text{ester}][\text{alkali}]$$

The rate constant for a second-order reaction has the units $\text{mol}^{-1}\ \text{dm}^3\ \text{s}^{-1}$. Unlike a first-order reaction, the time for a definite fraction of the reactants to be consumed is dependent on the original concentrations.

sedimentation The settling of a suspension, either under gravity or in a centrifuge. The speed of sedimentation can be used to estimate the average size of the particles. This technique is used with an ULTRACENTRIFUGE to find the relative molecular masses of macromolecules.

seed A small crystal added to a gas or liquid to assist solidification or precipitation from a solution. The seed, usually a crystal of the substance to be formed, enables particles to pack into predetermined positions so that a larger crystal can form.

self-assembly *See* supramolecular chemistry.

Seliwanoff's test A test for ketose SUGARS in solution. The reagent used consists of benzene-1,3-diol (resorcinol) dissolved in hydrochloric acid. A few drops are added to the solution and a red precipitate indicates a ketonic sugar. The test is named for the Russian chemist F. F. Seliwanoff.

semicarbazide *See* semicarbazone.

semicarbazone A type of organic compound containing the $C{:}N.NH.CO.NH_2$ grouping, formed by reaction of an ALDEHYDE or KETONE with *semicarbazide* ($H_2N.NH.CO.NH_2$). The compounds are crystalline solids with sharp melting points, which were formerly used to characterize the original aldehyde or ketone.

semipermeable membrane A membrane that, when separating a solution from a pure solvent, permits the solvent molecules to pass through but does not allow the transfer of solute molecules. Synthetic semipermeable membranes are generally supported on a porous material, such as unglazed porcelain or fine wire screens, and are commonly formed of cellulose or related materials. They are used in osmotic studies, gas separations, and medical applications.

Equilibrium is reached at a semipermeable membrane if the chemical potentials on both sides become identical; migration of solvent molecules towards the solution is an attempt by the system to reach equilibrium. The pressure required to halt this migration is the OSMOTIC PRESSURE.

semipolar bond A coordinate bond.

sequence rule *See* CIP system.

serine *See* amino acid.

serotonin (hydroxytryptamine) A substance that serves as a neurohormone that acts on muscles and nerves, and a neurotransmitter found in both the central and peripheral nervous systems. It controls dilation and constriction of blood vessels and affects peristalsis and gastrointestinal tract motility. Within the brain it plays a role in mood behavior. Many hallucinogenic compounds (e.g. LSD) antagonize the effects of serotonin in the brain.

sesquiterpene *See* terpene.

shell A group of electrons that share the same principal quantum number. Early work on x-ray emission studies used the terms K, L, M, and these are still sometimes used for the first three shells: $n = 1$, K-shell; $n = 2$, L-shell; $n = 3$, M-shell.

shikimic acid pathway The main metabolic pathway in the biosynthesis of the aromatic amino acids tyrosine, phenylalanine, and tryptophan. The first step is erythrose-4-phosphate, which comes from the pentose phosphate pathway, condensing with phosphenolpyruvate, from glycolysis. This product cyclizes to shikimate, which is then converted by various pathways into the amino acids.

short period *See* period.

SI *See* SI units.

side chain In an organic compound, an aliphatic group or radical attached to a longer straight chain of atoms in an acyclic compound or to one of the atoms in the ring of a cyclic compound. *See also* chain.

side reaction A chemical reaction that takes place to a limited extent at the same time as a main reaction. Thus the main product of a reaction may contain small amounts of other compounds.

siemens (mho) Symbol: S The SI unit of electrical conductance, equal to a conductance of one ohm^{-1}. The unit is named for the German physicist Ernst Werner von Siemens (1816–92).

sievert Symbol: Sv The SI unit of dose equivalent. It is the dose equivalent when the absorbed dose produced by ionizing radiation multiplied by certain dimensionless factors is 1 joule per kilogram ($1\ J\ kg^{-1}$). The dimensionless factors are used to modify the absorbed dose to take account of the fact that different types of radiation cause different biological effects. The unit is named for the Swedish physicist Rolf Sievert (1896–1966).

sigma bond *See* orbital.

sigmatropic rearrangement *See* pericyclic reaction.

silica gel A gel made by coagulating sodium silicate sol. The gel is dried by heating and used as a catalyst support and as a drying agent. The silica gel used in desiccators and in packaging to remove moisture is often colored with a cobalt salt to indicate whether it is still active (blue = dry; pink = moist).

silicones Polymeric synthetic silicon compounds containing chains of alternating silicon and oxygen atoms, with organic groups bound to the silicon atoms. Silicones are used as lubricants and water repellants and in waxes and varnishes. Silicone rubbers are superior to natural rubbers in their resistance to both high and low temperatures and chemicals.

silver-mirror test A test for the aldehyde group. A few drops of the sample are warmed with TOLLEN'S REAGENT. An aldehyde reduces Ag^+ to silver metal, causing a brilliant silver mirror to coat the inside wall of the tube.

single bond A covalent bond between two elements that involves one pair of electrons only. It is represented by a single line, for example H–Br, and is usually a sigma bond, although it can be a pi bond. *Compare* multiple bond. *See also* orbital.

single-cell protein (SCP) Protein produced from microorganisms, such as bacteria, yeasts, mycelial fungi, and unicellular algae, used as food for man and other animals.

singlet state *See* carbene.

SI units (*S*ystème *I*nternational d'Unités) The internationally adopted system of units used for scientific purposes. It has seven base units and two dimensionless units, formerly called supplementary units. Derived units are formed by multiplication and/or division of base units. Standard prefixes are used for multiples and submultiples of SI units. The SI system is a coherent rationalized system of units.

S_N1 reaction *See* nucleophilic substitution.

S_N2 reaction *See* nucleophilic substitution.

SNG Substitute (or synthetic) natural gas. A mixture of hydrocarbons manufactured from coal or the naphtha fraction of petroleum for use as a fuel.

soap A substance consisting of sodium or potassium compounds of fatty acids used to improve the cleansing properties of water. Soap is a SURFACTANT and was the earliest known DETERGENT. The basic method of making soap involved treating animal fat (mainly beef tallow), which is a triglyceride of octadecanoic acid (stearic acid), with caustic soda (sodium hydroxide; NaOH) to produce sodium octadecanoate (sodium stearate). More refined soaps are made from vegetable oils, such as palm oil, which contains hexadecanoic acid (palmitic acid). Liquid soaps (*soft soap*) are made using potassium hydroxide rather than sodium hydroxide. Other metal salts of long-chain carboxylic acids are also known as 'soaps'.

soft soap *See* soap.

sol A COLLOID consisting of solid particles distributed in a liquid medium. A wide variety of sols are known; the colors often depend markedly on the particle size. The term *aerosol* is used for solid or liquid phases dispersed in a gaseous medium.

solid The state of matter in which the particles occupy fixed positions, giving the substance a definite shape. The particles are held in these positions by bonds. Three kinds of attraction fix the positions of the particles: ionic, covalent, and intermolecular. Since these bonds act over short distances the particles in solids are packed closely together. The strengths of these three types of bonds are different and so, therefore, are the mechanical properties of different solids.

solid solution A solid composed of two or more substances mixed together at the molecular level. Atoms, ions, or molecules of one component in the crystal are at lattice positions normally occupied by the other component.

solubility The amount of one substance that could dissolve in another to form a saturated solution under specified conditions of temperature and pressure. Solubilities are stated as moles of solute per 100 grams of solvent or as mass of solute per unit volume of solvent.

solubility product Symbol: K_s If an ionic solid is in contact with its saturated solution, there is a dynamic equilibrium between solid and solution:

$$AB(s) \rightleftharpoons A^+(aq) + B^-(aq)$$

The equilibrium constant for this is given by

$$[A^+][B^-]/[AB]$$

The concentration of undissolved solid [AB] is also constant, so

$$K_s = [A^+][B^-]$$

K_s is the solubility product of the salt (at a given temperature). For a salt A_2B_3, for instance:

$$K_s = [A^+]^2[B^-]^3, \text{etc.}$$

Solubility products are meaningful only for sparingly soluble salts. If the product of ions exceeds the solubility product, precipitation occurs.

BASE AND DIMENSIONLESS SI UNITS

Physical quantity	*Name of SI unit*	*Symbol for SI unit*
length	meter	m
mass	kilogram(me)	kg
time	second	s
electric current	ampere	A
thermodynamic temperature	kelvin	K
luminous intensity	candela	cd
amount of substance	mole	mol
*plane angle	radian	rad
*solid angle	steradian	sr

*supplementary units

DERIVED SI UNITS WITH SPECIAL NAMES

Physical quantity	*Name of SI unit*	*Symbol for SI unit*
frequency	hertz	Hz
energy	joule	J
force	newton	N
power	watt	W
pressure	pascal	Pa
electric charge	coulomb	C
electric potential difference	volt	V
electric resistance	ohm	Ω
electric conductance	siemens	S
electric capacitance	farad	F
magnetic flux	weber	Wb
inductance	henry	H
magnetic flux density	tesla	T
luminous flux	lumen	lm
illuminance (illumination)	lux	lx
absorbed dose	gray	Gy
activity	becquerel	Bq
dose equivalent	sievert	Sv

DECIMAL MULTIPLES AND SUBMULTIPLES USED WITH SI UNITS

Submultiple	*Prefix*	*Symbol*	*Multiple*	*Prefix*	*Symbol*
10^{-1}	deci-	d	10^{1}	deca-	da
10^{-2}	centi-	c	10^{2}	hecto-	h
10^{-3}	milli-	m	10^{3}	kilo-	k
10^{-6}	micro-	μ	10^{6}	mega-	M
10^{-9}	nano-	n	10^{9}	giga-	G
10^{-12}	pico-	p	10^{12}	tera-	T
10^{-15}	femto-	f	10^{15}	peta-	P
10^{-18}	atto-	a	10^{18}	exa-	E
10^{-21}	zepto-	z	10^{21}	zetta-	Z
10^{-24}	yocto-	y	10^{24}	yotta-	Y

solute A material that is dissolved in a solvent to form a solution.

solution A liquid system of two or more species that are intimately dispersed within each other at a molecular level. The system is therefore totally homogeneous. The major component is called the solvent (generally liquid in the pure state) and the minor component is called the solute (gas, liquid, or solid).

The process occurs because of a direct intermolecular interaction of the solvent with the ions or molecules of the solute. This interaction is called SOLVATION. Part of the energy of this interaction appears as a change in temperature on dissolution. *See also* heat of solution; solid solution; solubility.

solvation The attraction of a solute species (e.g. an ion) for molecules of solvent. In water, for example, a positive ion will be surrounded by water molecules, which tend to associate around the ion because of attraction between the positive charge of the ion, and the negative part of the polar water molecule. The energy of this solvation (hydration in the case of water) is the 'force' needed to overcome the attraction between positive and negative ions when an ionic solid dissolves. The attraction of the dissolved ion for solvent molecules may extend for several layers.

solvent A liquid capable of dissolving other materials (solids, liquids, or gases) to form a solution. The solvent is generally the major component of the solution. Solvents can be divided into classes, the most important being:

Polar. A solvent in which the molecules possess a moderate to high dipole moment and in which polar and ionic compounds are easily soluble. Polar solvents are usually poor solvents for nonpolar compounds. For example, water is a good solvent for many ionic species, such as sodium chloride or potassium nitrate, and polar molecules, such as the sugars, but does not dissolve paraffin wax.

Nonpolar. A solvent in which the molecules do not possess a permanent dipole moment and consequently will solvate nonpolar species in preference to polar species. For example, benzene and tetrachloromethane are good solvents for iodine and paraffin wax, but do not dissolve sodium chloride.

Amphiprotic. A solvent that undergoes self-ionization and can act both as a proton donator and as an acceptor. Water is a good example and ionizes according to:

$$2H_2O = H_3O^+ + OH^-$$

Aprotic. A solvent that can neither accept nor yield protons. An aprotic solvent is therefore the opposite to an amphiprotic solvent.

solvent extraction (liquid–liquid extraction) A method of removing a substance from solution by shaking it with and dissolving it in another (better) solvent that is not miscible with the original solvent.

solvent naphtha *See* naphtha.

solvolysis A reaction between a compound and the solvent in which it is dissolved. *See also* hydrolysis.

s orbital *See* orbital.

sorbitol ($HOCH_2(CHOH)_4CH_2OH$) A hexahydric alcohol that occurs in rose hips and rowan berries. It can be synthesized by the reduction of glucose. Sorbitol is used to make vitamin C (ascorbic acid) and surfactants. It is also used in medicines and as a sweetener (particularly in foods for diabetics). It is an isomer of mannitol.

sorption Absorption of gases by solids.

specific Denoting a physical quantity per unit mass. For example, volume (V) per unit mass (m) is called specific volume:

$$V = V_m$$

In certain physical quantities the term does not have this meaning: for example, *specific gravity* is more properly called relative density.

specific gravity *See* relative density.

specific reaction rate *See* rate constant.

specific rotatory power Symbol: α_m The rotation of plane-polarized light in degrees produced by a 10 cm length of solution containing 1 g of a given substance per milliliter of stated solvent. The specific rotatory power is a measure of the optical activity of substances in solution. It is measured at 20°C using the D-line of sodium.

spectra *See* spectrum.

spectral line A particular wavelength of light emitted or absorbed by an atom, ion, or molecule. *See* line spectrum.

spectral series A group of related lines in the absorption or emission spectrum of a substance. The lines in a spectral series occur when the transitions all occur between one particular energy level and a set of different levels.

spectrograph An instrument for producing a photographic record of a spectrum.

spectrographic analysis A method of analysis in which the sample is excited electrically (by an arc or spark) and emits radiation characteristic of its component atoms. This radiation is passed through a slit, dispersed by a prism or a grating, and recorded as a spectrum, either photographically or photoelectrically. The photographic method was widely used for qualitative and semiquantitative work but photoelectric detection also allows wide quantitative application.

spectrometer **1.** An instrument for examining the different wavelengths present in electromagnetic radiation. Typically, spectrometers have a source of radiation, which is collimated by a system of lenses and/or slits. The radiation is dispersed by a prism or grating, and recorded photographically or by a photocell. There are many types for producing and investigating spectra over the whole range of the electromagnetic spectrum. Often spectrometers are called *spectroscopes*.
2. Any of various other instruments for analyzing the energies, masses, etc., of particles. *See* mass spectrometer.

spectrophotometer A form of spectrometer able to measure the intensity of radiation at different wavelengths in a spectrum, usually in the visible, infrared, or ultraviolet regions.

spectroscope An instrument for examining the different wavelengths present in electromagnetic radiation. *See also* spectrometer.

spectroscopy **1.** The production and analysis of spectra. There are many spectroscopic techniques designed for investigating the electromagnetic radiation emitted or absorbed by substances. Spectroscopy, in various forms, is used for analysis of mixtures, for identifying and determining the structures of chemical compounds, and for investigating energy levels in atoms, ions, and molecules. In the visible and longer wavelength ultraviolet, transitions correspond to electronic energy levels in atoms and molecules. The shorter wavelength ultraviolet corresponds to transitions in ions. In the x-ray region, transitions in the inner shells of atoms or ions are involved. The infrared region corresponds to vibrational changes in molecules, with rotational changes at longer wavelengths.
2. Any of various techniques for analysing the energy spectra of beams of particles or for determining mass spectra.

spectrum (plural *spectra*) **1.** A range of electromagnetic radiation emitted or absorbed by a substance under particular circumstances. In an *emission spectrum*, light or other radiation emitted by the body is analyzed to determine the particular wavelengths produced. The emission of radiation may be induced by a variety of methods; for example, by high temperature, bombardment by electrons, absorption of higher-frequency radiation, etc. In an *absorption spectrum* a continuous flow of radiation is passed through the sample. The radiation is then analyzed to deter-

mine which wavelengths are absorbed. *See also* band spectrum; continuous spectrum; line spectrum.
2. In general, any distribution of a property. For instance, a beam of particles may have a spectrum of energies. A beam of ions may have a mass spectrum (the distribution of masses of ions). *See* mass spectrometer.

spin A property of certain elementary particles whereby the particle acts as if it were spinning on an axis; i.e. it has an intrinsic angular momentum. Such particles also have a magnetic moment. In a magnetic field the spins line up at an angle to the field direction and precess around this direction. Certain definite orientations to the field direction occur such that $m_s h/2\pi$ is the component of angular momentum along this direction. m_s is the spin quantum number, and for an electron it has values +1/2 and –1/2. h is the Planck constant.

spontaneous combustion The self-ignition of a substance that has a low ignition temperature. It occurs when slow oxidation of the substance (such as a heap of damp straw or oily rags) builds up sufficient heat for ignition to take place.

stabilization energy The difference in energy between the delocalized structure and the conventional structure for a compound. For example, the stabilization energy of benzene is 150 kJ per mole, which represents the difference in energy between a Kekulé structure and the delocalized structure: the delocalized form being of lower energy is therefore more stable. The stabilization energy can be determined by comparing the experimental value for the heat of hydrogenation of benzene with that calculated for Kekulé benzene from bond-energy data.

stabilizer A substance added to prevent chemical change (i.e. a negative catalyst).

staggered conformation *See* conformation.

standard solution A solution that contains a known weight of the reagent in a definite volume of solution. A standard flask or volumetric flask is used for this purpose. The solutions may be prepared by direct weighing for primary standards. If the reagent is not available in a pure form or is deliquescent the solution must be standardized by titration against another known standard solution. *See* primary standard.

standard state The standard conditions used as a reference system in thermodynamics: pressure is 101 325 Pa; temperature is 25°C (298.15 K); concentration is 1 mol. The substance under investigation must also be pure and in its usual state, given the above conditions.

standard temperature An internationally agreed value for which many measurements are quoted. It is the melting temperature of water, 0°C (273.15 K). *See also* STP.

starch A polysaccharide that occurs exclusively in plants. Starches are extracted commercially from maize, wheat, barley, rice, potatoes, and sorghum. The starches are storage reservoirs for plants; they can be broken down by enzymes to simple sugars and then metabolized to supply energy needs. Starch is a dietary component of animals.

Starch is not a single molecule but a mixture of two glucose polymers: *amylose* (water-soluble, blue color with iodine) and *amylopectin* (not water-soluble, violet color with iodine). The composition is amylose 10–20%, amylopectin 80–90%.

states of matter The three physical conditions in which substances occur: solid, liquid, and gas. The addition or removal of energy (usually in the form of heat) enables one state to be converted into another.

The major distinctions between the states of matter depend on the kinetic energies of their particles and the distances between them. In solids, the particles have low kinetic energy and are closely packed; in gases they have high kinetic energy and are very loosely packed; kinetic energy and separation of particles in liquids are intermediate.

Solids have fixed shapes and volumes, i.e. they do not flow, like liquids and gases, and they are difficult to compress. In solids the atoms or molecules occupy fixed positions in space. In most cases there is a regular pattern of atoms – the solid is crystalline.

Liquids have fixed volumes (i.e. low compressibility) but flow to take up the shape of the container. The atoms or molecules move about at random, but they are quite close to one another and the motion is hindered.

Gases have no fixed shape or volume. They expand spontaneously to fill the container and are easily compressed. The molecules have almost free random motion.

A plasma is sometimes considered to be a fourth state of matter.

stationary phase *See* chromatography.

Staudinger, Hermann (1881–1965) German organic chemist. Staudinger was a pioneer of polymer chemistry. It was Staudinger who introduced the word 'macromolecule' into chemistry in 1922. He proposed that polymers are very long molecules held together by ordinary chemical bonds. This correct view was not universally accepted when it was first put forward but within a decade or so decisive evidence emerged that confirmed Studinger's view. In a book entitled *Macromolecular Chemistry and Biology* (1947) he anticipated some of the key features of molecular biology. Staudinger won the 1953 Nobel Prize for chemistry for his fundamental contributions to polymer chemistry.

steam distillation A method of isolating or purifying substances by exploiting Dalton's law of partial pressures to lower the boiling point of the mixture. When two immiscible liquids are distilled, the boiling point will be lower than that of the more volatile component and consequently will be below 100°C if one component is water. The method is particularly useful for recovering materials from tarry mixtures.

steam reforming The conversion of a methane–steam mixture at 900°C with a nickel catalyst into a mixture of carbon monoxide and hydrogen. The mixture of gases (synthesis gas) provides a starting material in a number of processes, e.g. the manufacture of methanol.

stearate A salt or ester of stearic acid; an octadecanoate.

stearic acid *See* octadecanoic acid.

step An elementary stage in a chemical reaction, in which energy may be transferred from one molecule to another, bonds may be broken or formed, or electrons may be transferred.

steradian Symbol: sr The SI unit of solid angle. The surface of a sphere, for example, subtends a solid angle of 4π at its center. The solid angle of a cone is the area intercepted by the cone on the surface of a sphere of unit radius.

stereochemistry The branch of chemistry concerned with the shapes of molecules and the way these affect their chemical properties.

stereoisomerism *See* isomerism.

stereospecific Describing a chemical reaction that results in a particular arrangement of atoms in space.

stereospecific polymer *See* polymerization.

steric effect An effect in which the shape of a molecule influences its reactions. A particular example occurs in molecules containing large groups, which hinder the approach of a reactant (*steric hindrance*).

steric hindrance *See* steric effect.

steroid Any member of a group of compounds having a complex basic ring structure. Examples are corticosteroid hormones (produced by the adrenal gland), sex hormones (progesterone, androgens, and estrogens), bile acids, and sterols (such as cholesterol). *See also* anabolic steroid; sterol.

sterol A steroid with long aliphatic side chains (8–10 carbons) and at least one hydroxyl group. They are lipid-soluble and often occur in membranes (e.g. cholesterol and ergosterol).

still An apparatus for distillation.

stoichiometric coefficient *See* chemical equation.

stoichiometry The proportions in which elements form compounds. A *stoichiometric compound* is one in which the atoms have combined in small whole numbers.

STP (NTP) Standard temperature and pressure. Conditions used internationally when measuring quantities that vary with both pressure and temperature (such as the density of a gas). The values are 101 325 Pa (approximately 100 kPa) and 0°C (273.15 K).

straight chain *See* chain.

Strecker synthesis A method of synthesizing amino acids. Hydrogen cyanide forms an addition product (cyanohydrin) with aldehydes:

$$RCHO + HCN \rightarrow RCH(CN)OH$$

With ammonia further substitution occurs:

$$RCH(CN)OH + NH_3 \rightarrow RCH(CN)(NH_2) + H_2O$$

Acid hydrolysis of the cyanide group then produces the amino acid

$$RCH(CN)(NH_2) \rightarrow RCH(COOH)(NH_2)$$

The method is a general synthesis for α-amino acids (with the $-NH_2$ and $-COOH$ groups on the same carbon atom).

strong acid An ACID that is almost completely dissociated into its component ions in solution.

strong base A base that is completely or almost completely dissociated into its component ions in solution. *See* acid.

structural formula *See* formula.

structural isomerism *See* isomerism.

styrene *See* phenylethene.

sublimate A solid formed by sublimation.

sublimation The conversion of a solid into a vapor without the solid first melting. For instance (at standard pressure) iodine, solid carbon dioxide, and ammonium chloride sublime. At certain conditions of external pressure and temperature an equilibrium can be established between the solid phase and vapor phase.

subshell A subdivision of an electron shell. It is a division of the orbitals that make up a shell into sets of orbitals, which are degenerate (i.e. have the same energy) in the free atom. For example, in the second shell (the L shell) there are the 2s and 2p subshells.

substituent An atom or group substituted for another in a compound. Often the term is used for groups that have replaced hydrogen in organic compounds. For example, in chlorobenzene (C_6H_5Cl) chlorine can be regarded as a substituent.

substitution reaction A reaction in which an atom or group of atoms in an organic molecule is replaced by another atom or group. The substitution of a hydrogen atom in an alkane by a chlorine atom is an example. Substitution reactions fall into three major classes depending upon the nature of the attacking substituent.

Nucleophilic substitution: the attacking substituent is a nucleophile (i.e. a molecule or ion that can donate electrons). Such reactions are very common with alcohols and halogen compounds, in which the electron-deficient carbon atom attracts the nucleophile and the leaving group readily exists alone. Examples are the hydrolysis

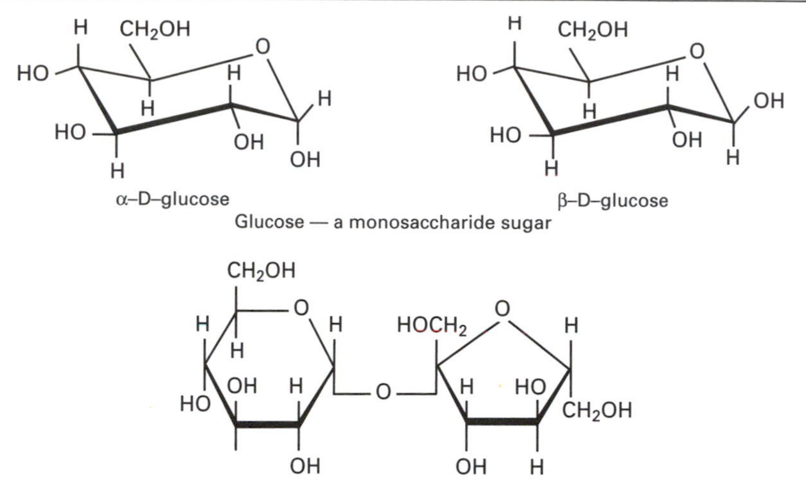

Glucose — a monosaccharide sugar

Sucrose — a disaccharide sugar

of a haloalkane and the chlorination of an alcohol:

$$C_2H_5Cl + OH^- \rightarrow C_2H_5OH + Cl^-$$

$$C_2H_5OH + HCl \rightarrow C_2H_5Cl + H_2O$$

Electrophilic substitution: the attacking substituent is an electrophile (i.e. a molecule or ion that accepts electrons). Such reactions are common in aromatic compounds, in which the electron-rich ring attracts the electrophile. The nitration of benzene in which the electrophile is NO_2^+ is an example:

$$C_6H_6 + NO_2^+ \rightarrow C_6H_5NO_2 + H^+$$

Free-radical substitution: a free radical is the attacking substituent. Such reactions can be used with compounds that are inert to either nucleophiles or electrophiles, for instance the halogenation of an alkane:

$$CH_4 + Cl_2 \rightarrow CH_3Cl + HCl$$

The term 'substitution' is very general and several reactions that can be considered as substitutions are more normally given special names (e.g. esterification, hydrolysis, and nitration). *See also* electrophilic substitution; nucleophilic substitution.

substrate A material that is acted on by a catalyst.

succinic acid *See* butanedioic acid.

sucrose (cane sugar; saccharose; $C_{12}H_{22}O_{11}$) A SUGAR that occurs in many plants. It is extracted commercially from sugar cane and sugar beet. Sucrose is a disaccharide formed from a glucose unit and a fructose unit. It is hydrolyzed to a mixture of fructose and glucose by the enzyme invertase. Since this mixture has a different optical rotation (levorotatory) than the original sucrose, the mixture is called *invert sugar*.

sugar (saccharide) One of a class of sweet-tasting simple carbohydrates. Sugars have molecules consisting of a chain of carbon atoms with –OH groups attached, and either an aldehyde or ketone group. They can exist in a chain form or in a ring formed by reaction of the ketone or aldehyde group with an OH group to form a cyclic hemiacetal (*see* acetal). Monosaccharides are simple sugars that cannot be hydrolyzed to sugars with fewer carbon atoms. Two or more monosaccharide units can be linked in *disaccharides*, *trisaccharides*, etc., by a GLYCOSIDIC LINK.

Monosaccharides are also classified according to the number of carbon atoms: a *pentose* has five carbon atoms and a *hex-*

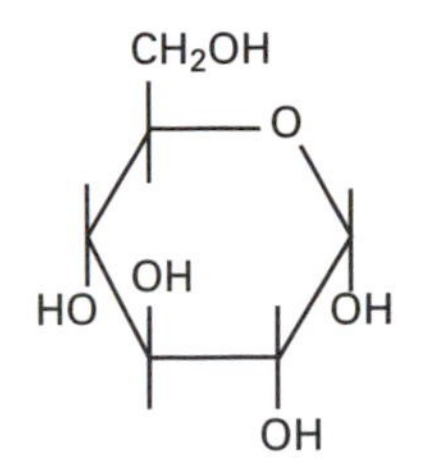

glucose – a pyranose ring

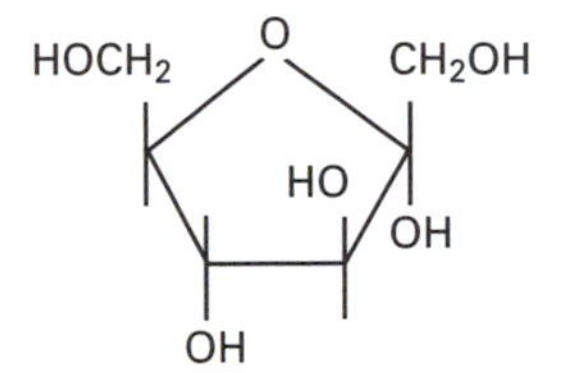

fructose – in a furanose ring form

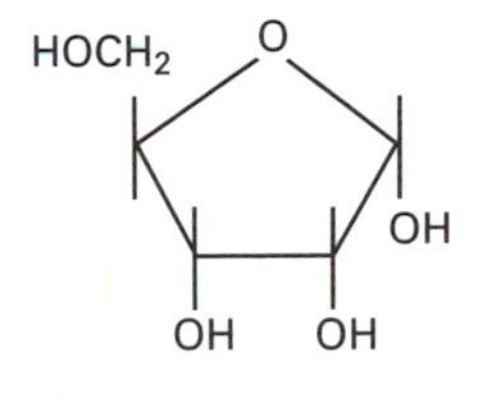

ribose – a pyranose ring

Sugars

ose six. Monosaccharides with aldehyde groups are *aldoses*; those with ketone groups are *ketoses*. Thus, an *aldohexose* is a hexose with an aldehyde group; a *ketopentose* is a pentose with a ketone group, etc.

The ring forms of monosaccharides are derived by reaction of the aldehyde or ketone group with one of the carbons at the other end of the chain. It is possible to have a six-membered (*pyranose*) ring or a five-membered (*furanose*) ring. *See also* anomer; carbohydrate; polysaccharide. *See illustration overleaf.*

sugar acid An acid formed from a monosaccharide by oxidation. Oxidation of the aldehyde group (CHO) of the aldose monosaccharides to a carboxyl group (COOH) gives an *aldonic acid*; oxidation of the primary alcohol group (CH_2OH) to COOH yields *uronic acid*; oxidation of both the primary alcohol and carboxyl groups gives an *aldaric acid*. The uronic acids are biologically important, being components of many polysaccharides, for example glucuronic acid (from glucose) is a major component of gums and cell walls, while galacturonic acid (from galactose) makes up pectin. Ascorbic acid or vitamin C is an important sugar acid found universally in plants, particularly in citrus fruits.

sugar alcohol (alditol) An alcohol derived from a monosaccharide by reduction of its carbonyl group (CO) so that each carbon atom of the sugar has an alcohol group (OH). For example, glucose yields sorbitol, common in fruits, and mannose yields mannitol.

sulfa drug *See* sulfonamide.

sulfate A salt or ester of sulfuric(VI) acid.

sulfide *See* thioether.

sulfite A salt or ester of sulfurous acid.

sulfonamide A type of organic compound with the general formula $R.SO_2.NH_2$. Sulfonamides, which are amides of sulfonic acids, are active against bacteria, and some are used in pharmaceuticals ('sulfa drugs').

sulfonate A salt or ester of a SULFONIC ACID.

sulfonation A reaction introducing the $-SO_2OH$ (sulfonic acid) group into an organic compound. Sulfonation of aromatic compounds is usually accomplished by refluxing with concentrated sulfuric acid for several hours. The attacking species is SO_3 (sulfur trioxide; sulfur(VI) oxide) and the reaction is an example of electrophilic substitution.

sulfonic acid A type of organic compound containing the $-SO_2.OH$ group. The simplest example is benzenesulfonic acid ($C_6H_5SO_2OH$). Sulfonic acids are

Glucose (an aldohexose)

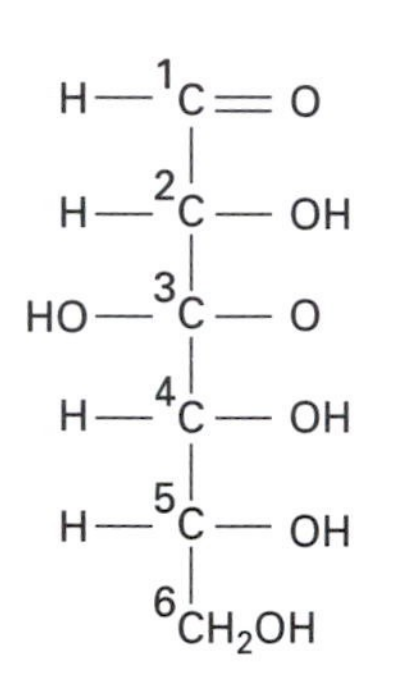

straight-chain form

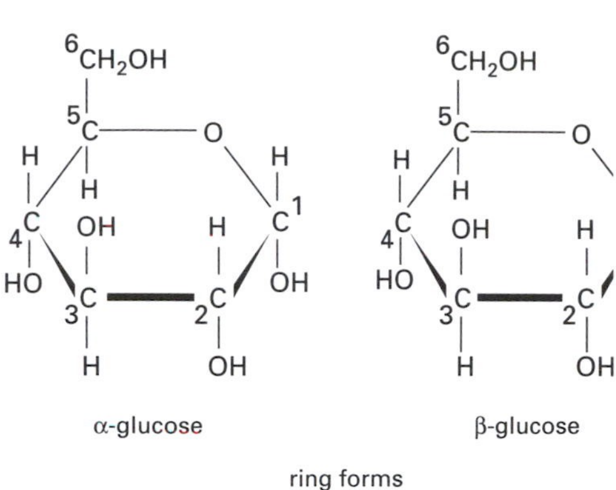

α-glucose

β-glucose

ring forms

Fructose (a ketohexose)

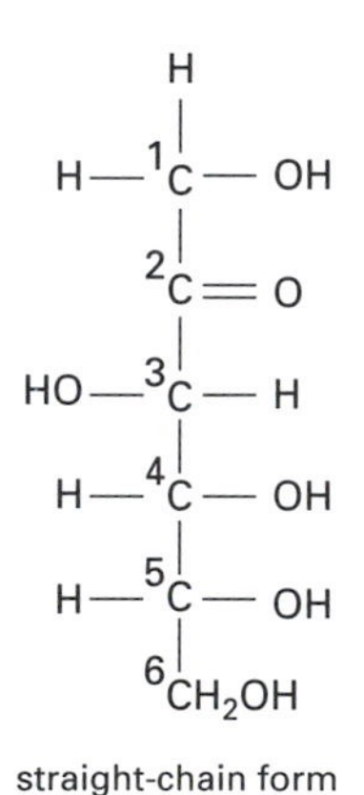

straight-chain form

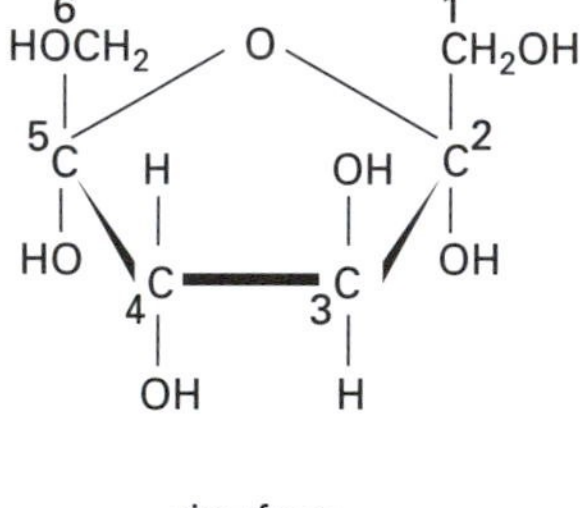

ring form

Sugar: straight-chain and ring forms

strong acids and ionize in water to form the sulfonate ion ($-SO_2O^-$). Electrophilic substitution can introduce other groups onto the benzene ring; the $-SO_2.OH$ group directs substituents into the 3-position.

sulfonium compound An organic compound of general formula R_3SX, where R is an organic radical and X is an electronegative radical or element; it contains the ion R_3S^+. An example is diethylmethylsulfonium chloride, $(C_2H_5)_2CH_3S^+Cl^-$, made by reacting diethyl sulfide with chloromethane.

sulfoxide An organic compound of general formula R^1SOR^2, where R^1 and R^2 are organic radicals. An example is dimethyl sulfoxide, $(CH_3)_2SO$, commonly used as a solvent.

sulfur A low-melting nonmetallic solid, yellow colored in its common forms; the second member of group 16 (formerly VIA) of the periodic table. It has the electronic configuration $[Ne]3s^23p^4$.

Sulfur occurs in the elemental form in Sicily and some southern states of the USA, and in large quantities in combined forms such as sulfide ores (FeS_2) and sulfate rocks ($CaSO_4$). It forms about 0.5% of the Earth's crust.

Sulfur forms a wide range of organic sulfur compounds, most of which have the typically revolting smell of H_2S.

Symbol: S; m.p. 112.8°C; b.p. 444.6°C; r.d. 2.07; p.n. 16; r.a.m. 32.066.

sulfur dichloride dioxide (sulfuryl chloride; SO_2Cl_2) A colorless fuming liquid formed by the reaction of chlorine with sulfur(IV) oxide in sunlight. It is used as a chlorinating agent.

sulfur dichloride oxide (thionyl chloride; $SOCl_2$) A colorless fuming liquid formed by passing sulfur(IV) oxide over phosphorus(V) chloride and distilling the mixture obtained. Sulfur dichloride oxide is used in organic chemistry to introduce chlorine atoms (for example, into ethanol to form monochloroethane), the organic product being easily isolated as the other products are gases.

sulfur dioxide *See* sulfur(IV) oxide.

sulfuretted hydrogen *See* hydrogen sulfide.

sulfuric(IV) acid *See* sulfurous acid.

sulfuric(VI) acid (oil of vitriol; H_2SO_4) A colorless oily liquid manufactured by the contact process. The concentrated acid is diluted by adding it slowly to water, with careful stirring. Concentrated sulfuric acid acts as an oxidizing agent, giving sulfur(IV) oxide as the main product, and also as a dehydrating agent. The diluted acid acts as a strong dibasic acid, neutralizing bases and reacting with active metals and carbonates to form sulfates.

Sulfuric acid is used in the laboratory to dry gases (except ammonia), to prepare nitric acid and ethene, and to absorb alkenes. In industry, it is used to manufacture fertilizers (e.g. ammonium sulfate), rayon, and detergents, to clean metals, and in vehicle batteries. *See* contact process.

sulfur monochloride *See* disulfur dichloride.

sulfurous acid (sulfuric(IV) acid; H_2SO_3) A weak acid found only in solution, made by passing sulfur(IV) oxide into water. The solution is unstable and smells of sulfur(IV) oxide. It is a reducing agent, converting iron(III) ions to iron(II) ions, chlorine to chloride ions, and orange dichromate(VI) ions to green chromium(III) ions.

sulfur(IV) oxide (sulfur dioxide; SO_2) A colorless choking gas prepared by burning sulfur or heating metal sulfides in air, or by treating a sulfite with an acid. It is a powerful reducing agent, used as a bleach. It dissolves in water to form sulfurous acid (sulfuric(IV) acid) and combines with oxygen, in the presence of a catalyst, to form sulfur(VI) oxide. This latter reaction is important in the manufacture of sulfuric(VI) acid.

sulfur(VI) oxide (sulfur trioxide; SO_3) A fuming volatile white solid prepared by passing sulfur(IV) oxide and oxygen over hot vanadium(V) oxide (acting as a catalyst) and cooling the product in ice. Sulfur(VI) oxide reacts vigorously with water to form sulfuric acid. *See also* contact process.

sulfur trioxide *See* sulfur(VI) oxide.

sulfuryl chloride *See* sulfur dichloride dioxide.

superacid An acid with a high proton-donating ability. They are substances such as $HF–SbF_3$ and $HSO_3–SbF_5$. Sometimes known as *magic acids*, they are able to produce carbenium ions from some saturated hydrocarbons.

superheating The raising of a liquid's temperature above its boiling temperature, by increasing the pressure.

supernatant Denoting a clear liquid that lies above a sediment or a precipitate.

supersaturated solution *See* saturated solution.

supersaturated vapor *See* saturated vapor.

supplementary units *See* dimensionless units.

supramolecular chemistry A branch of chemistry concerned with the synthesis and study of large structures consisting of molecules assembled together in a definite pattern. In a *supramolecule* the molecular units are joined by intermolecular bonds – i.e. by hydrogen bonds or by ionic attractions. A particular interest in supramolecular research is the idea of *self-assembly* – i.e. that the molecules form well-defined structures spontaneously as a result of their geometry and chemical properties. In this way, supramolecular chemistry is 'chemistry beyond molecules'.

There are various different examples of supramolecular structures. For instance, single layers of carboxylic acid molecules held by hydrogen bonds may form two-dimensional crystalline structures with novel electrical properties. Large organic polymeric structures may be formed in which a number of individual chains radiate from a central point or region. These polymers are called *dendritic polymers* or *dendrimers* and they have a number of possible applications. Another type of supramolecule is a *helicate*, which has a double helix made of two chains of bipyridyl units held by copper ions along the axis. The structure is analogous to the double helix of DNA. *See also* host–guest chemistry.

supramolecule *See* supramolecular chemistry.

surfactant A substance that lowers surface tension and has properties of wetting, foaming, detergency, dispersion, and emulsification. SOAPS and other DETERGENTS have surfactant properties.

suspension A system in which small particles of a solid or liquid are dispersed in a liquid or gas.

synclinal conformation *See* conformation.

syndiotactic polymer *See* polymerization.

Synge, Richard Laurence Millington (1914–94) British biochemist. Synge is best known for his work with Archer MARTIN on paper chromatography which led to them sharing the 1952 Nobel Prize for chemistry. Synge used this technique to determine the exact structure of a simple antibiotic peptide gramicidin-S.

synperiplanar conformation *See* conformation.

synthesis The preparation of chemical compounds from simpler compounds.

synthesis gas A mixture of carbon monoxide and hydrogen produced by steam reforming of natural gas.

$$CH_4 + H_2O \rightarrow CO + 3H_2$$

Synthesis gas is a useful starting material for the manufacture of a number of organic compounds.

synthon *See* retrosynthetic analysis.

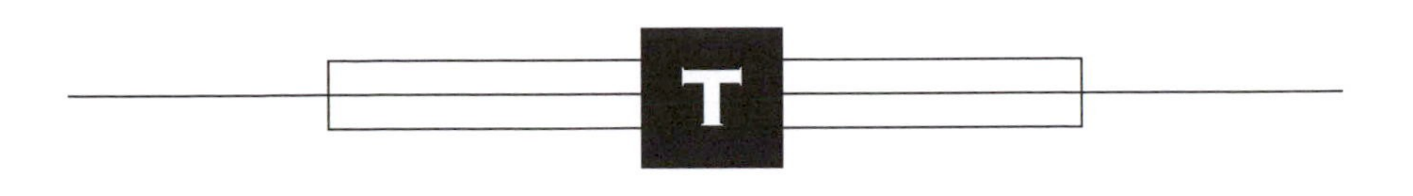

tactic polymer *See* polymerization.

tannic acid *See* tannin.

tannin Any of several yellow organic compounds found in vegetable sources such as bark of trees, oak galls, and tea. They are used in tanning animal skins to make leather and as mordants in dyeing. *Tannic acid* (a type of tannin) is a white solid heterocyclic organic acid extracted from oak galls and used for making dyes and inks.

tar (bitumen) A dark oily viscous liquid obtained by the destructive distillation of coal or the fractionation of petroleum. Tars are mixtures of mainly high-molecular weight hydrocarbons and phenols.

tartaric acid A crystalline hydroxy carboxylic acid with the formula:

$$HOOC(CHOH)_2COOH$$

Its systematic name is 2,3-dihydroxybutanedioic acid. It is used as an additive in foodstuffs. *See also* optical activity.

tartrate A salt or ester of tartaric acid.

tautomerism Isomerism in which each isomer can convert into the other, so that the two isomers are in equilibrium. The isomers are called *tautomers*. Tautomerism often results from the migration of a hydrogen atom. *See* keto–enol tautomerism.

TCA cycle (tricarboxylic acid cycle) *See* Krebs cycle.

TCCD *See* dioxin.

Teflon (*Trademark*) The synthetic polymer polytetrafluoroethane.

temperature scale A practical scale for measuring temperature. A temperature scale is determined by fixed temperatures (*fixed points*), which are reproducible systems assigned an agreed temperature. On the Celsius scale the two fixed points are the temperature of pure melting ice (the *ice temperature*) and the temperature of pure boiling water (the *steam temperature*). The difference between the fixed points is the *fundamental interval* of the scale, which is subdivided into temperature units. The International Temperature Scale has 11 fixed points that cover the range 13.81 kelvin to 1337.58 kelvin.

tera- Symbol: T A prefix denoting 10^{12}. For example, 1 terawatt (TW) = 10^{12} watts (W).

terephthalic acid *See* benzene-1,4-dicarboxylic acid.

ternary compound A chemical compound formed from three elements; e.g. Na_2SO_4 or $LiAlH_4$.

terpene Any of a class of natural unsaturated hydrocarbons with formulae $(C_5H_8)_n$, found in plants. Terpenes consist of isoprene units,

$$CH_2=C(CH_3)CH=CH_2.$$

Monoterpenes have two units ($C_{10}H_{16}$), *diterpenes* four units ($C_{20}H_{32}$), *triterpenes* six units ($C_{30}H_{48}$), etc. *Sesquiterpenes* have three isoprene units ($C_{15}H_{24}$).

tertiary alcohol *See* alcohol.

tertiary amine *See* amine.

tertiary structure *See* protein.

Terylene (*Trademark*) A polymer made by condensing benzene-1,4-dicarboxylic acid (terephthalic acid) and ethane-1,2-diol (ethylene glycol), used for making fibers for textiles.

tesla Symbol: T The SI unit of magnetic flux density, equal to a flux density of one weber of magnetic flux per square meter. 1 T = 1 Wb m^{-2}. The unit is named for the Croatian–US electrical engineer Nikola Tesla (1870–1943).

tetrachloroethene (ethylene tetrachloride; tetrachloroethylene; $CCl_2{:}CCl_2$) A colorless poisonous liquid organic compound (a haloalkene) used as a solvent in dry cleaning and as a de-greasing agent.

tetrachloroethylene *See* tetrachloroethene.

tetrachloromethane (carbon tetrachloride; CCl_4) A colorless nonflammable liquid made by the chlorination of methane. Its main use is as a solvent although it is being replaced by other compounds for safety reasons.

tetraethyl lead *See* lead tetraethyl.

tetrafluoroethene ($CF_2{:}CF_2$) A gaseous organic compound (a fluorocarbon and a haloalkene) used to make the plastic polytetrafluoroethene (PTFE). *See* polytetrafluoroethene.

tetrahydrofuran (THF; C_4H_8O) A colorless liquid widely used as a solvent and for making polymers.

tetrapyrrole A structure of four pyrrole molecules linked together, found in heme, chlorophyll, and other compounds. Usually a metal ion is coordinated to the four nitrogen atoms on the pyrrole rings. *See* porphyrin.

theobromine An alkaloid found in the cacao bean. Its action is similar to caffeine. The systematic name is *3,7-dimethylxanthine.*

theophylline An alkaloid similar in action to caffeine. Its systematic name is *1,3-dimethylxanthine.*

thermal dissociation The decomposition of a chemical compound into component atoms or molecules by the action of heat. Often it is temporary and reversible.

thermochemistry The branch of chemistry concerned with heats of reaction, solvation, etc.

thermodynamics The study of heat and other forms of energy and the various related changes in physical quantities such as temperature, pressure, density, etc.

The *first law of thermodynamics* states that the total energy in a closed system is conserved (constant). In all processes energy is simply converted from one form to another, or transferred from one system to another.

A mathematical statement of the first law is:

$$\delta Q = \delta U + \delta W$$

Here, δQ is the heat transferred to the system, δU the change in internal energy (resulting in a rise or fall of temperature), and δW is the external work done *by* the system.

The *second law of thermodynamics* can be stated in a number of ways, all of which are equivalent. One is that heat cannot pass from a cooler to a hotter body without some other process occurring. Another is the statement that heat cannot be totally converted into mechanical work, i.e. a heat engine cannot be 100% efficient.

The *third law of thermodynamics* states that the entropy of a substance tends to zero as its thermodynamic temperature approaches zero.

Often a zeroth law of thermodynamics is given: that if two bodies are each in thermal equilibrium with a third body, then they are in thermal equilibrium with each other. This is considered to be more fundamental than the other laws because they assume it.

thermodynamic temperature Symbol: *T* A temperature measured in kelvins. *See also* absolute temperature.

thermoplastic polymer *See* polymer.

thermosetting polymer *See* polymer.

THF *See* tetrahydrofuran.

thiamine (vitamin B_1) One of the water-soluble B-group of vitamins. Good sources of thiamine are unrefined cereal grains, liver, heart, and kidney. Thiamine deficiency predominantly affects the peripheral nervous system, the gastrointestinal tract, and the cardiovascular system. Thiamine has been shown to be of value in the treatment of beriberi. Thiamine, in the form of thiamine diphosphate, is the coenzyme for the decarboxylation of acids such as pyruvic acid. *See also* vitamin B complex.

thiazine (C_4H_4NS) Any of a group of heterocyclic organic compounds that have a six-membered ring containing four carbon atoms, one nitrogen atom, and one sulfur atom.

thiazole (C_3H_3NS) A colorless volatile liquid, a beterocyclic compound with a five-membered ring containing three carbon atoms, one nitrogen atom, and one sulfur atom. It resembles PYRIDINE in its reactions and is used in making dyes.

Thiele, Friedrich Karl Johannes (1865–1918) French organic chemist. Thiele is best known for his concept of partial valence which he put forward in 1899. In order to explain the lack of reactivity of the benzene molecule Thiele postulated that when single and double bonds alternate then the overall bonding is less conducive to reactivity than purely double bonds. The concept of partial valence was not properly understood until the chemical bonding and valence of molecules such as benzene were explained in terms of quantum mechanics. Thiele also worked extensively on organic compounds of nitrogen.

thin-layer chromatography A type of CHROMATOGRAPHY widely used for the analysis of mixtures. Thin-layer chromatography employs a solid stationary phase, such as alumina or silica gel, spread evenly as a thin layer on a glass plate. A base line is carefully scratched near the bottom of the plate, and a small sample of the mixture is spotted onto the base line. The plate is then stood upright in solvent, which rises up to the base line and beyond by capillary action. The components of the spot of the sample will dissolve in the solvent and tend to be carried up the plate. However, some of the components will cling more readily to the solid phase than others and will not move up the plate so rapidly. In this way, different fractions of the mixture eventually become separated. When the solvent has almost reached the top, the plate is removed and quickly dried. The plate is developed to locate the positions of colorless fractions by spraying with a suitable chemical or by exposure to ultraviolet radiation. The components are identified by comparing the distance they have moved up the plate with standard solutions that have been run simultaneously, or by computing an R_F VALUE.

thio alcohol *See* thiol.

thiocarbamide (thiourea; NH_2CSNH_2) A colorless crystalline organic compound (the sulfur analog of urea). It is converted to the inorganic compound ammonium thiocyanate on heating. It is used as a sensitizer in photography and in medicine.

thioether (sulfide) A compound of the type RSR′. They are the sulfur analogs of ethers and are generally more reactive than the corresponding oxygen compound. With halogen compounds they form sulfonium compounds:

$$CH_3SCH_3 + CH_3Cl \rightarrow (CH_3)_3\,S^+Cl^-$$

They can also be oxidized to sulfoxides:

$$(CH_3)_2S + O \rightarrow (CH_3)_2S{=}O$$

thiol (thio alcohol; mercaptan) A com-

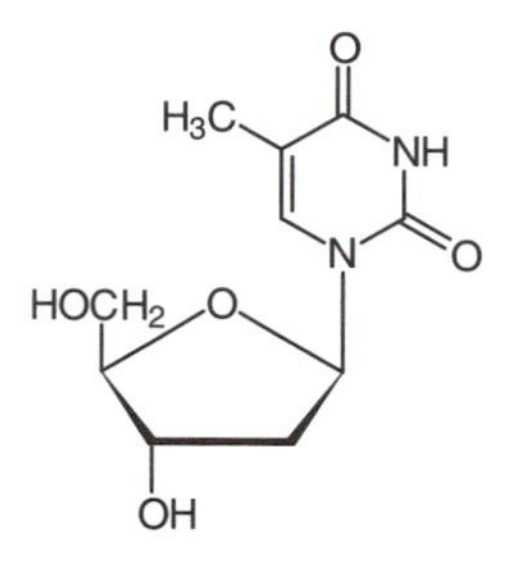

Thymidine

pound of the type RSH, similar to an alcohol with the oxygen atom replaced by sulfur. Typically they have a strong unpleasant odor. They are more reactive than the corresponding alcohols and can form salts of the type $RS^{-}M^{+}$.

thionyl chloride *See* sulfur dichloride oxide.

thiophene (C_4H_4S) A colorless liquid that smells like benzene, a heterocyclic compound with a five-membered ring containing four carbon atoms and one sulfur atom. It occurs as an impurity in commercial benzene and is used as a solvent and in organic syntheses.

thiourea ($(NH_2)_2CS$) *See* thiocarbamide.

threonine *See* amino acid.

thymidine The NUCLEOSIDE formed when thymine is linked to D-ribose by a β-glycosidic bond.

thymine A nitrogenous base found in DNA. It has a PYRIMIDINE ring structure.

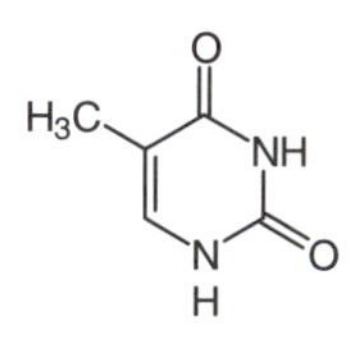

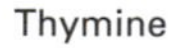

Thymine

tincture A solution in which alcohol (ethanol) is the solvent.

Tiselius, Arne Wilhelm Kaurin (1902–71) Swedish chemist. Tiselius is best known for developing electrophoresis as a technique for studying proteins. He was able to separate the proteins in horse serum using this technique and to confirm that there are four types of proteins: albumins and alpha, beta and gamma globulins. Tiselius also used other techniques such as chromatography and partition and gel filtration to separate proteins. Tiselius won the 1948 Nobel Prize for chemistry for his work using electrophoresis and other techniques in the analysis of proteins.

titrant *See* titration.

titration A procedure in volumetric analysis in which a solution of known concentration (called the *titrant*) is added to a solution of unknown concentration from a burette until the equivalence point or end point of the titration is reached. *See* volumetric analysis.

TNT *See* trinitrotoluene.

tocopherol *See* vitamin E.

Todd, Alexander Robertus, Lord (1907–97) British organic chemist. In the early part of his career, Todd was largely concerned with the vitamins B_1, B_{12} and E. In the late 1940s and 1950s he synthesized the purine and pyrimidine bases that occur in nucleic acids such as DNA and

RNA and established their structures. He also synthesized a number of other important compounds such as adenosine triphosphate (ATP) and adenosine diphosphate (ADP), which are of crucial importance in energy transfer in biological systems. Todd won the 1957 Nobel Prize for chemistry for his work on biologically important molecules. He published an autobiography entitled *A Time to Remember* in 1983.

Tollen's reagent A solution of the complex ion $Ag(NH_3)_2^+$ produced by precipitation of silver oxide from silver nitrate with a few drops of sodium hydroxide solution, and subsequent dissolution of the silver oxide in aqueous ammonia. Tollen's reagent is used in the 'silver-mirror test' for aldehydes, where the Ag^+ ion is reduced to silver metal. It is also a test for alkynes with a triple bond in the 1-position. A yellow precipitate of silver carbide is formed in this case.

$$RCCH + Ag^+ \rightarrow RCC^-Ag^+ + H^+$$

It is named for Bernhard Tollens (1841–1918). *See also* silver-mirror test.

toluene *See* methylbenzene.

toluidine (methylaniline; aminotoluene; $CH_3C_6H_4NH_2$) An aromatic amine used in making dyestuffs and drugs. There are three isomers; the 1,2- (ortho-aminotoluene) and 1,3- (meta-) forms are liquids, the 1,4- (para-) isomer is a solid.

tonne Symbol: t A unit of mass equal to 10^3 kilograms (i.e. one megagram).

torr A unit of pressure equal to a pressure of 101 325/760 pascals (133.322 Pa). One torr is equal to one mmHg. The unit is named for the Italian physicist Evangelista Torricelli (1609–47).

torsion angle *See* conformation.

tracer An isotope of an element used to investigate chemical reactions or physical processes (e.g. diffusion). *See* isotope.

trans- Designating an isomer with groups that are on opposite sides of a bond or structure. *See* isomerism.

trans fat *See* hardening.

transient species A short-lived intermediate in a chemical reaction.

transition state (activated complex) Symbol: ‡ A short-lived high-energy molecule, radical, or ion formed during a reaction between molecules possessing the necessary activation energy. The transition state decomposes at a definite rate to yield either the reactants again or the final products. The transition state can be considered to be at the top of the energy profile.

For the reaction,

$$X + YZ = X...Y...Z^{\ddagger} \rightarrow XY + Z$$

the sequence of events is as follows. X approaches YZ and when it is close enough the electrons are rearranged producing a weakening of the bond between Y and Z. A partial bond is now formed between X and Y producing the transition state. Depending on the experimental conditions, the transition state either breaks down to form the products or reverts back to the reactants. *See also* activation energy.

transition temperature A temperature at which some definite physical change occurs in a substance. Examples of such transitions are change of state, change of crystal structure, and change of magnetic behavior.

triacylglycerol *See* triglyceride.

tricarboxylic acid cycle (TCA cycle) *See* Krebs cycle.

triaminotriazine *See* melamine.

triatomic molecule A molecule consisting of three atoms, such as O_3 or H_2O.

triazine ($C_3H_3N_3$) A heterocyclic organic compound with a six-membered ring containing three carbon atoms and three nitrogen atoms. There are three isomers, used as dyestuffs and herbicides. *See also* melamine.

triazole ($C_2H_3N_3$) A heterocyclic organic compound with a five-membered ring containing two carbon atoms and three nitrogen atoms. There are two isomers.

tribasic acid An acid with three replaceable hydrogen atoms (such as phosphoric(V) acid, H_3PO_4). *See* acid.

tribromomethane (bromoform; $CHBr_3$) A colorless liquid compound. *See* haloform.

trichloroacetic acid *See* chloroethanoic acid.

trichloroethanal (chloral; CCl_3CHO) A colorless liquid aldehyde made by chlorinating ethanal. It was used to make the insecticide DDT. It can be hydrolyzed to give *2,2,2-trichloroethanediol* (*chloral hydrate*, $CCl_3CH(OH)_2$). Most compounds with two –OH groups on the same carbon atom are unstable. However, in this case the effect of the three chlorine atoms stabilizes the compound. It is used as a sedative.

trichloroethanoic acid *See* chloroethanoic acid.

trichloromethane (chloroform; $CHCl_3$) A colorless volatile liquid formerly used as an anesthetic. Now its main use is as a solvent and raw material for making other chlorinated compounds. Trichloromethane is made by reacting ethanal, ethanol, or propanone with chlorinated lime. *See also* haloform.

triglyceride A GLYCERIDE in which esters are formed with all three –OH groups of glycerol.

trihydric alcohol *See* triol.

triiodomethane (iodoform; CHI_3) A yellow crystalline compound made by warming ethanal with an alkaline solution of an iodide:

$$CH_3CHO + 3I^- + 4OH^- \rightarrow CHI_3 + HCOO^- + 3H_2O$$

The reaction also occurs with all ketones of general formula CH_3COR (R is an alkyl group) and with secondary alcohols $CH_3CH(OH)R$. Iodoform is used as a test for such reactions (the *iodoform reaction*). *See also* haloform.

trimer A molecule (or compound) formed by addition of three identical molecules. *See* ethanal; methanal.

trimethylaluminum (aluminum trimethyl; $(CH_3)_3Al$) A colorless liquid produced by the reduction of dimethyl aluminum chloride using sodium. It ignites spontaneously on contact with air and reacts violently with water, acids, halogens, alcohols, and amines. Aluminum alkyls are used in the ZIEGLER PROCESS for the manufacture of high-density polyethene.

trimolecular Describing a reaction or step that involves three molecules interacting simultaneously with the formation of a product. For example, the final step in reaction between hydrogen peroxide and acidified potassium iodide is trimolecular:

$$HOI + H^+ + I^- \rightarrow I_2 + H_2O$$

It is uncommon for reactions to take place involving trimolecular steps.

trinitroglycerine *See* nitroglycerine.

trinitrophenol *See* picric acid.

trinitrotoluene (TNT; $CH_3C_6H_2(NO_2)_3$) A yellow crystalline solid. It is a highly unstable substance, used as an explosive. The compound is made by nitrating methylbenzene and the nitro groups are in the 2, 4, and 6 positions.

triol (trihydric alcohol) An alcohol that has three hydroxyl groups (–OH) per molecule of compound.

triose A SUGAR that contains three carbon atoms.

trioxygen *See* ozone.

tripeptide *See* peptide.

triple bond A covalent bond formed between two atoms in which three pairs of electrons contribute to the bond. One pair forms a sigma bond (equivalent to a single bond) and two pairs give rise to two pi bonds. It is conventionally represented as three lines, thus H–C≡C–H. The bond occurs in ALKYNES. *See* multiple bond.

triple point The only point at which the gas, solid, and liquid phases of a substance can coexist in equilibrium. The triple point of water (273.16 K at 101 325 Pa) is used to define the kelvin.

triplet state *See* carbene.

triterpene *See* terpene.

tritiated compound A compound in which one or more ^1H atoms have been replaced by tritium (^3H) atoms.

tritium Symbol: T, ^3H A radioactive isotope of hydrogen of mass number 3. The nucleus contains 1 proton and 2 neutrons. Tritium decays with emission of low-energy beta radiation to give ^3He. The half-life is 12.3 years. It is useful as a tracer in studies of chemical reactions. Compounds in which ^3H atoms replace the usual ^1H atoms are said to be *tritiated*. A positive tritium ion, T^+, is a *triton*.

triton *See* tritium.

tropylium ion The positive ion $C_7H_7^+$, having a symmetrical seven-membered ring of carbon atoms. The tropylium ion ring shows nonbenzenoid aromatic properties. *See* aromatic compound.

tryptophan *See* amino acid.

turpentine (pine-cone oil) A yellow viscous RESIN obtained from coniferous trees. It can be distilled to produce turpentine oil (also known simply as turpentine), used in medicine and as a solvent in paints, polishes, and varnishes.

twist-boat conformation *See* cyclohexane.

tyrosine *See* amino acid.

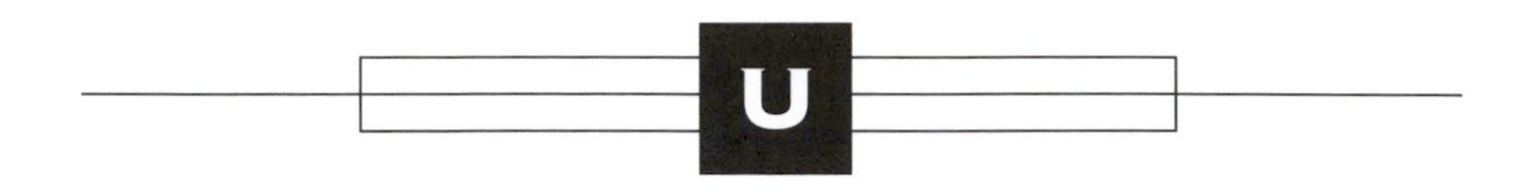

ubiquinone An electron-transporting coenzyme that is a component in the electron-transport chain. It was formerly called *coenzyme Q*.

ultracentrifuge A high-speed centrifuge used for separating out very small particles. The sedimentation rate depends on the particle size, and the ultracentrifuge can be used to measure the mass of colloidal particles and large molecules (e.g. proteins).

ultraviolet (UV) A form of electromagnetic radiation, shorter in wavelength than visible light. Ultraviolet wavelengths range between about 1 nm and 400 nm. Ordinary glass is not transparent to these waves; quartz is a much more effective material for making lenses and prisms for use with ultraviolet. Like light, ultraviolet radiation is produced by electronic transitions between the outer energy levels of atoms. However, having a higher frequency, ultraviolet photons carry more energy than those of light and can induce photolysis of compounds and photoionization. *See also* electromagnetic radiation.

unimolecular Describing a reaction (or step) in which only one molecule is involved. For example, radioactive decay is a unimolecular reaction:

$$Ra \rightarrow Rn + \alpha$$

Only one atom is involved in each disintegration.

In a unimolecular chemical reaction, the molecule acquires the necessary energy to become activated and then decomposes. The majority of reactions involve only uni- or bimolecular steps. The following reactions are all unimolecular:

$$N_2O_4 \rightarrow 2NO_2$$
$$PCl_5 \rightarrow PCl_3 + Cl_2$$
$$CH_3CH_2Cl \rightarrow C_2H_4 + HCl$$

unit A reference value of a quantity used to express other values of the same quantity. *See also* SI units.

unit cell The smallest group of atoms, ions, or molecules that, when repeated at regular intervals in three dimensions, will produce the lattice of a crystal system. There are seven basic types of unit cells, which result in seven CRYSTAL SYSTEMS.

unit processes (chemical conversions) The recognized steps used in chemical processes, e.g. alkylation, distillation, hydrogenation, pyrolysis, nitration, etc. Industrial processing and the economics, design, and use of the equipment are based on these unit processes rather than consideration of each reaction separately.

univalent *See* monovalent.

universal gas constant *See* gas constant.

universal indicator (multiple-range indicator) A mixture of indicator dyestuffs that shows a gradual change in color over a wide pH range. A typical formulation contains methyl orange, methyl red, bromothymol blue, and phenolphthalein and changes through a red, orange, yellow, green, blue, and violet sequence between pH 3 and pH 10. Several commercial preparations are available as both solutions and test papers.

unsaturated compound An organic

compound that contains at least one double or triple bond between two of its carbon atoms. The ALKENES and ALKYNES are examples of unsaturated compounds. Unsaturated compounds typically undergo addition reactions to form single bonds. *Compare* saturated compound.

unsaturated solution *See* saturated solution.

unsaturated vapor *See* saturated vapor.

UPVC Unplasticized PVC; a hard-wearing form of PVC used in building work (e.g. for window frames).

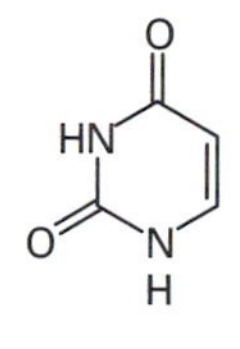

Uracil

uracil A nitrogenous base that is found in RNA, replacing the thymine of DNA. It has a pyrimidine ring structure.

urea (carbamide; $CO(NH_2)_2$) A white crystalline compound made from ammonia and carbon dioxide. It is used in the manufacture of urea–formaldehyde (methanal) resins. Urea is the end product of metabolism in many animals and is present in urine.

urea cycle *See* ornithine cycle.

urea–formaldehyde resin A synthetic POLYMER made by copolymerizing urea with formaldehyde (methanal, HCHO).

urethane (ethyl carbamate; $CO(NH_2)$-OC_2H_5) A poisonous flammable organic compound, used in medicine, as a solvent, and as an intermediate in the manufacture of POLYURETHANE resins.

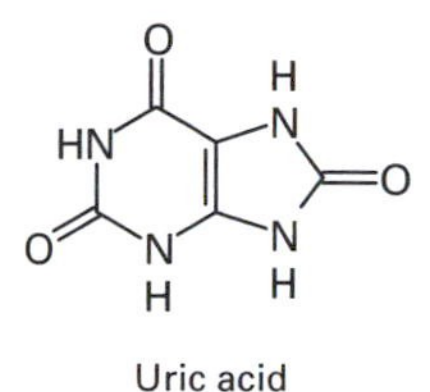

Uric acid

uric acid A nitrogen compound produced from purines. In certain animals (uricotetic animals), it is the main excretory product resulting from breakdown of amino acids. In humans, uric acid crystals in the joints are the cause of gout.

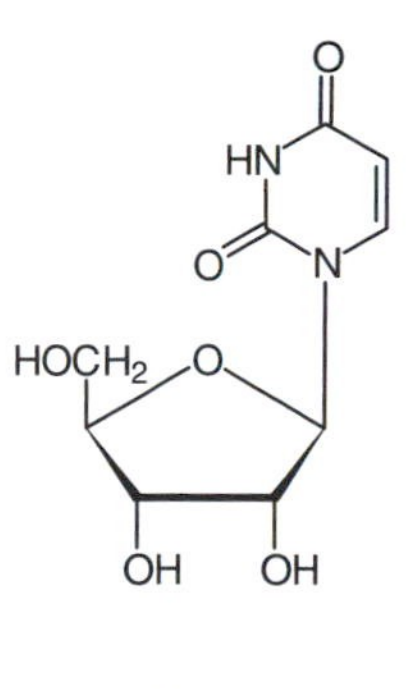

Uridine

uridine The nucleoside formed when uracil is linked to D-ribose by a β-glycosidic bond.

uronic acid *See* sugar acid.

UV *See* ultraviolet.

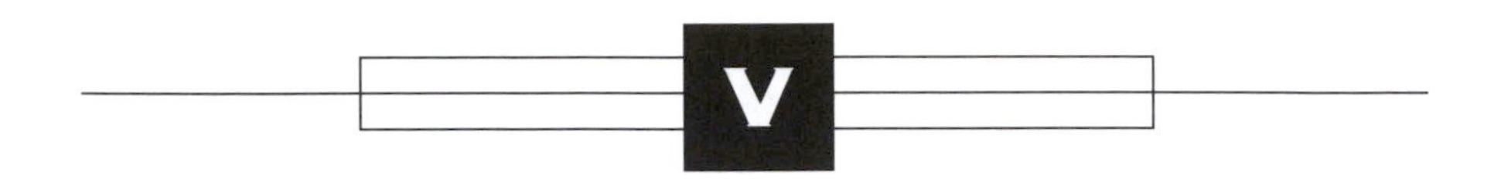

vacuum distillation The distillation of liquids under a reduced pressure, so that the boiling point is lowered. Vacuum distillation is a common laboratory technique for purifying or separating compounds that would decompose at their 'normal' boiling point.

vacuum flask *See* Dewar flask.

valence (valency) The combining power of an element or radical, equal to the number of hydrogen atoms that will combine with or displace one atom of the element. For simple covalent molecules the valence is obtained directly, for example C in CH_4 is tetravalent; N in NH_3 is trivalent. For ions the valence is regarded as equivalent to the magnitude of the charge; for example Ca^{2+} is divalent, CO_3^{2-} is a divalent radical. The rare gases are zero-valent because they do not form compounds under normal conditions. As the valence for many elements is constant, the valence of some elements can be deduced without reference to compounds formed with hydrogen. Thus, as the valence of chlorine in HCl is 1, the valence of aluminum in $AlCl_3$ is 3; as oxygen is divalent (H_2O) silicon in SiO_2 is tetravalent. The product of the valence and the number of atoms of each element in a neutral compound must be equal. For example, in Al_2O_3 for the two aluminum atoms (valence 3) the product is 6 and for the three oxygen atoms (valence 2) the product is also 6.

The valence of an element is generally equal to either the number of valence electrons or eight minus the number of valence electrons. Transition metal ions display variable valence.

valence electron An outer electron in an atom that can participate in forming chemical bonds.

valency *See* valence.

valeric acid *See* pentanoic acid.

valine *See* amino acid.

van der Waals equation An equation of state for real gases. For n moles of gas the equation is

$$(p + n^2a/V^2)(V - nb) = nRT$$

where p is the pressure, V the volume, and T the thermodynamic temperature. a and b are constants for a given substance and R is the gas constant. The equation gives a better description of the behavior of real gases than the perfect gas equation ($pV = nRT$).

The equation contains two corrections: b is a correction for the nonnegligible size of the molecules; a/V^2 corrects for the fact that there are attractive forces between the molecules, thus slightly reducing the pressure from that of an ideal gas. It is named for the Dutch physicist Johannes Diderik van der Waals (1837–1923). *See also* gas laws.

van der Waals force An intermolecular force of attraction, considerably weaker than chemical bonds and arising from weak electrostatic interactions between molecules (the energies are often less than 1 J mol^{-1}).

The van der Waals interaction contains contributions from three effects; permanent *dipole–dipole interactions* found for any polar molecule; *dipole-induced dipole interactions*, where one dipole causes a slight charge separation in bonds that have a high polarizability; and *dispersion forces*, which result from temporary polarity arising from an asymmetrical distribution of electrons around the nucleus. Even atoms of the rare gases exhibit dispersion forces.

Van't Hoff, Jacobus Henricus (1852–1911) Dutch theoretical chemist. Van't Hoff made important contributions to stereochemistry, thermodynamics, the kinetics of chemical reactions and the theory of chemical solutions. In 1874 Van't Hoff initiated the subject of stereochemistry when he postulated that the four chemical bonds which a carbon atom can form are directed toward the corners of a regular tetrahedron. This enabled the phenomenon of optical activity to be understood in terms of the structures of optical isomers. Van't Hoff introduced this idea independently of Joseph le Bel. Many of the contributions of Van't Hoff to thermodynamics, kinetics and solutions were expounded in his book *Studies of Chemical Dynamics* (1884). This included the application of thermodynamics to chemical equilibrium. He was awarded the first Nobel Prize for chemistry in 1901.

van't Hoff factor Symbol: i The ratio of the number of particles present in a solution to the number of undissociated molecules added. It is used in studies of colligative properties, which depend on the number of entities present. For example, if n moles of a compound are dissolved and dissociation into ions occurs, then the number of particles present will be in. Osmotic pressure (π), for instance, will be given by the equation

$$\pi V = inRT$$

It is named for the Dutch theoretical chemist Jacobus Henricus van't Hoff.

van't Hoff isochore The equation:

$$\mathrm{d}(\log_e K)/\mathrm{d}T = \Delta H/RT^2$$

showing how the equilibrium constant, K, of a reaction varies with thermodynamic temperature, T. ΔH is the enthalpy of reaction and R is the gas constant.

vapor A gas formed by the VAPORIZATION of a solid or liquid. Some particles near the surface of a liquid acquire sufficient energy in collisions with other particles to escape from the liquid and enter the vapor; some particles in the vapor lose energy in collisions and re-enter the liquid. At a given temperature an equilibrium is established, which determines the vapor pressure of the liquid at that temperature.

vapor density The ratio of the mass of a certain volume of a vapor to the mass of an equal volume of hydrogen (measured at the same temperature and pressure). Determination of vapor densities is one method of finding the relative molecular mass of a compound (equal to twice the vapor density). VICTOR MEYER'S METHOD, DUMAS' METHOD, or HOFMANN'S METHOD can be used.

vaporization The process by which a liquid or solid is converted into a gas or vapor by heat. Unlike boiling, which occurs at a fixed temperature, vaporization can occur at any temperature. Its rate increases as the temperature rises.

vapor pressure The pressure exerted by a vapor. The *saturated vapor pressure* is the pressure of a vapor in equilibrium with its liquid or solid. It depends on the nature of the liquid or solid and the temperature.

vat dyes A class of insoluble dyes applied by first reducing them to derivatives that are soluble in dilute alkali. In this condition they have a great attraction for certain fibers, such as cotton. The solution is applied to the material and the insoluble dye is regenerated in the fibers by atmospheric oxidation. Indigo and indanthrene are examples of vat dyes.

velocity constant *See* rate constant.

vesicant A substance that causes blistering of the skin. Mustard gas is an example.

vicinal positions Positions in a molecule at adjacent atoms. For example, in 1,2-dichloroethane the chlorine atoms are in vicinal positions, and this compound can thus be named *vic*-dichloroethane.

Viktor Meyer's method A method for determining VAPOR DENSITIES in which a given weight of sample is vaporized and the volume of air displaced by it is measured. In practice, a bulb in a heating bath is connected via a fairly long tube to a water-bath gas-collection arrangement. The system is brought to equilibrium and the sample is then added (without opening the apparatus to the atmosphere). As the air displaced by gas is collected over water a correction for the vapor pressure of water is necessary and the method may fail if the vapor is soluble in water. It is named for the German chemist Viktor Meyer (1848–97).

violaxanthin A xanthophyll pigment found in the brown algae. *See* photosynthetic pigments.

vinegar A dilute solution (about 4% by volume) of ethanoic acid (acetic acid), often with added coloring and flavoring such as caramel. Natural vinegar is produced by the bacterial fermentation of cider or wine; it can also be made synthetically. *See* ethanoic acid.

vinylation A catalytic reaction in which a compound adds across the triple bond of ethyne (acetylene) to form an ethenyl (vinyl) compound. For example, an alcohol can add as follows:

$$ROH + HC{\equiv}CH \rightarrow RHC{=}CH(OH)$$

vinyl benzene *See* phenylethene.

vinyl chloride *See* chloroethene.

vinyl group The group CH_2:CH–.

viscose rayon *See* rayon.

viscosity Symbol: η The resistance to flow of a fluid.

visible radiation *See* light.

visual purple *See* rhodopsin.

vitalism *See* Wöhler's synthesis.

vitamin One of a number of organic compounds that are essential in small quantities for metabolism. The vitamins have no energy value; most of them seem to act as catalysts for essential chemical changes in the body, each one influencing a number of vital processes. Vitamins A, D, E, and K are the fat-soluble vitamins, occurring mainly in animal fats and oils. Vitamins B and C are the water-soluble vitamins. If a diet lacks vitamins, this results in the breakdown of normal bodily activities and produces disease symptoms. Such deficiency diseases can usually be remedied by including the necessary vitamins in the diet. Plants can synthesize vitamins from simple substances, but animals generally require them in their diet, though there are exceptions to this. These include vitamins synthesized by bacteria in the gut, and some that can be manufactured by the animal itself. A precursor of vitamin D_2 (ergosterol), for example, can be converted in the skin by ultraviolet radiation.

vitamin A (vitamin A_1; retinol) A fat-soluble vitamin (a derivative of the yellow pigment, carotene) occurring in milk, butter, cheese, liver, and cod-liver oil. It can also be formed in the body by oxidation of carotene, which is present in fresh green vegetables and carrots. Deficiency in vitamin A can result in a reduced resistance to disease and in night blindness.

vitamin B complex A group of ten or more water-soluble vitamins, which tend to occur together. They can be obtained from whole grains of cereals and from meat and liver. Since the B vitamins are present in most unprocessed food, deficiency diseases only occur in populations living on restricted diets. Many of the B vitamins act as coenzymes involved in the normal oxidation of carbohydrates during respiration.

The vitamins of the B complex include thiamine (vitamin B_1), riboflavin (vitamin B_2), nicotinic acid (niacin), pantothenic acid (vitamin B_5), pyridoxine (vitamin B_6), cyanocobalamin (vitamin B_{12}), biotin, lipoic acid, and folic acid.

vitamin C (ascorbic acid) A water-soluble vitamin, which is widely required in metabolism. The major sources of vitamin C are fresh fruit and vegetables and severe deficiency results in scurvy.

vitamin D A fat-soluble vitamin found in fish-liver oil, butter, milk, cheese, egg yolk, and liver. Its principal action is to increase the absorption of calcium and phosphorus from the intestine. The vitamin also has a direct effect on the calcification process in bone. Deficiency results in inadequate deposition of calcium in the bones, causing rickets in young children and osteomalacia in adults.

The term vitamin D refers, in fact, to a group of compounds, all sterols, of very similar properties. The most important are vitamin D_2 (*calciferol*) and vitamin D_3 (cholecalciferol). Precursors of these are converted to the vitamins in the body by the action of ultraviolet radiation.

vitamin E (tocopherol) A fat-soluble vitamin found in wheat germ, dairy products, and in meat. Severe deficiency in infants may lead to high rates of red-blood cell destruction and hence to anemia. However, there are very few deficiency effects apparent in adults.

vitamin K (phylloquinone; menaquinone) A fat-soluble vitamin that is required to catalyze the synthesis of prothrombin, a blood-clotting factor, in the liver. Intestinal microorganisms are capable of synthesizing considerable amounts of vitamin K in the intestine and this, together with dietary supply, insures that deficiency is unlikely to occur in any but the newborn. A newborn child may be deficient because the intestine is sterile at birth and the level supplied by the mother during gestation is limited. Thus during the first few days of life blood-clotting deficiency may be observed, but this is readily rectified by a small injection of the vitamin.

VLDL (very low-density lipoprotein) *See* lipoprotein.

volatile Easily converted into a vapor.

volt Symbol: V The SI unit of electrical potential, potential difference, and e.m.f., defined as the potential difference between two points in a circuit between which a constant current of one ampere flows when the power dissipated is one watt. $1 V = 1 J C^{-1}$. The unit is named for the Italian physicist Alessandro Volta (1745–1827).

volumetric analysis One of the classical wet methods of quantitative analysis. It involves measuring the volume of a solution of accurately known concentration that is required to react with a solution of the substance being determined. The solution of known concentration (the standard solution) is added in small portions from a burette. The process is called a titration and the equivalence point is called the *end point*. End points are observed with the aid of indicators or by instrumental methods, such as conduction or light absorption. Volumetric analysis can also be applied to gases. The gas is typically held over mercury in a graduated tube, and volume changes are measured on reaction or after absorption of components of a mixture.

vulcanite (ebonite) A hard black insulator made by vulcanizing rubber with a large amount of sulfur.

vulcanization A process of improving the quality of rubber (hardness and resistance to temperature changes) by heating it with sulfur (about 150°C). Accelerators are used to speed up the reaction. Certain sulfur compounds can also be used for vulcanization.

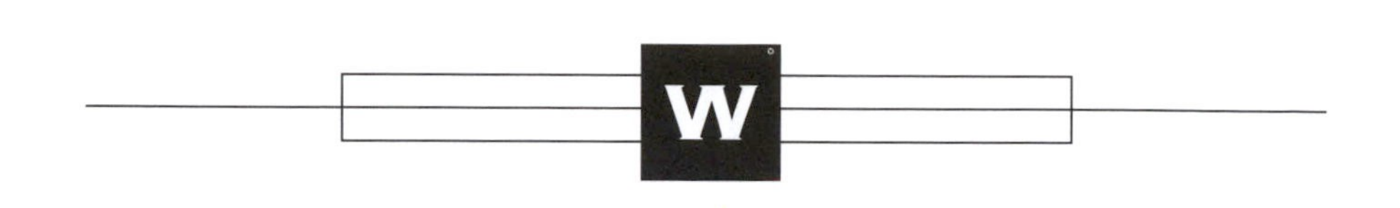

Wacker process An industrial process for making ethanal (and other carbonyl compounds). To produce ethanal, ethene and air are bubbled through an acid solution of palladium(II) chloride and copper(II) chloride (20–60°C and moderate pressure):

$$C_2H_4 + Pd^{2+} + O_2 \rightarrow CH_3CHO + Pd + 2H^+$$

The reaction involves an intermediate complex between palladium(II) ions and ethane in which the palladium bonds to the pi electrons. The purpose of the copper(II) chloride is to oxidize the palladium back to Pd^{2+} ions:

$$Pd + 2Cu^{2+} \rightarrow Pd^{2+} + 2Cu^+$$

The copper(I) ions spontaneously oxidize to copper(II) ions in air. The process provides a cheap source of ethanal (and, by oxidation, ethanoic acid) from the readily available ethene. It is named for Alexander von Wacker (1846–1922).

Walden inversion A reaction in which an optically active compound reacts to give an optically active product in which the configuration has been inverted. This happens in the S_N2 mechanism. *See* nucleophilic substitution. It is named for the German chemist Paul Walden (1863–1957), who discovered it in 1896.

Walker, John E. (1941–) British biochemist who worked on the enzyme mechanism underlying the synthesis of ATP. He was awarded the 1997 Nobel Prize for chemistry jointly with P. D. Boyer. The prize was shared with J. C. Skou.

water (H_2O) A colorless liquid that freezes at 0°C and, at atmospheric pressure, boils at 100°C. In the gaseous state water consists of single H_2O molecules. Due to the presence of two lone pairs the atoms do not lie in a straight line, the angle between the central oxygen atom and the two hydrogen atoms being 105°; the distance between each hydrogen atom and the oxygen atom is 0.099 nm. When ice forms, hydrogen bonds some 0.177 nm long develop between the hydrogen atom and oxygen atoms in adjacent molecules, giving ice its tetrahedral crystalline structure with a density of 916.8 kg m^{-3} at STP. Different ice structures develop under higher pressures. When ice melts to form liquid water, the tetrahedral structure breaks down, but some hydrogen bonds continue to exist; liquid water consists of groups of associated water molecules, $(H_2O)_n$, mixed with some monomers and some dimers. This mixture of molecular species has a higher density than the open-structured crystals. The maximum density of water, 999.97 kg m^{-3}, occurs at 3.98°C. This accounts for the ability of ice to float on water and for the fact that water pipes burst as ice expands on freezing.

Although water is predominantly a covalent compound, a very small amount of ionic dissociation occurs ($H_2O \rightleftharpoons H^+ + OH^-$). In every liter of water at STP there is approximately 10^{-7} mole of each ionic species. It is for this reason that, on the pH scale, a neutral solution has a value of 7.

As a polar liquid, water is the most powerful solvent known. This is partly a result of its high dielectric constant and partly its ability to hydrate ions. This latter property also accounts for the incorporation of water molecules into some ionic crystals as water of crystallization.

Water is decomposed by reactive metals (e.g. sodium) when cold and by less active metals (e.g. iron) when steam is passed over the hot metal. It is also decomposed by electrolysis.

water gas A mixture of carbon monoxide and hydrogen produced when steam is passed over red-hot coke or made to combine with hydrocarbons, e.g.

$$C(s) + H_2O(g) \rightarrow CO(g) + H_2(g)$$

The amount of hydrogen can be increased by combining it with the *water gas shift reaction*:

$$CO + H_2O \rightleftharpoons CO_2 + H_2$$

Water gas was once an important source of hydrogen for the production of ammonia. Most hydrogen is now obtained from methane by steam REFORMING.

$$CH_4(g) + H_2O(g) \rightarrow CO(g) + 3H_2(g)$$

The production of water gas using methane is an important step in the preparation of hydrogen for ammonia synthesis. *Compare* producer gas.

Watson, James Dewey (1928–) American molecular biologist. James Watson is famous for his work in 1953 with Francis CRICK on the structure of DNA. Watson gave a controversial account of this work in his book *The Double Helix* (1968). He also wrote several other autobiographical and popular books as well as co-authoring textbooks on molecular biology. He shared the 1962 Nobel Prize for medicine with Crick and Maurice WILKINS for their discovery of the structure of DNA.

watt Symbol: W The SI unit of power, defined as a power of one joule per second. 1 W = 1 J s^{-1}. The unit is named for the British inventor James Watt (1739–1819).

wave function Symbol: Ψ A function that describes the quantum state of a system in WAVE MECHANICS. The physical significance of the wave function for a particle is that the square of its absolute value at a point is proportional to the probability of finding the particle in a small element of volume, $dxdydz$, at that point. *See also* orbital.

wavelength Symbol: λ The distance between the ends of one complete cycle of a wave. Wavelength is related to the speed (c) and frequency (ν) thus:

$$c = \nu\lambda$$

wave mechanics A formulation of quantum mechanics put forward by the German physicist Erwin Schrödinger (1887-1961) in 1926, following the suggestion of the French physicist Louis de broglie (1892-1987) that particles such as electrons might also have wavelike properties. The basic equation of wave mechanics is the *Schrödinger equation*, which is a wave equation describing the system. Solutions of the equation give rise to WAVE FUNCTIONS. *See also* quantum theory; orbital.

wave number Symbol: σ The reciprocal of the wavelength of a wave. It is the number of wave cycles in unit distance, and is often used in spectroscopy. The unit is the $meter^{-1}$ (m^{-1}). The circular wave number (symbol: k) is given by:

$$k = 2\pi\sigma$$

wax One of a group of water-insoluble substances with a very high molecular weight; they are esters of long-chain alcohols with fatty acids. Waxes form protective coverings to leaves, stems, fruits, seeds, animal fur, and the cuticles of insects, serving principally as waterproofing. For example, waxy deposits on some plant organs add to the efficiency of the cuticle in reducing transpiration, as well as cutting down airflow over the surface and forming a highly reflective surface, thus reducing energy available for evaporation. They may also occur in plant cell walls, e.g. leaf mesophyll. They are used in varnishes, polishes, and candles.

weak acid An ACID that is not fully dissociated in solution.

weak base A base that is not fully dissociated in solution. *See* acid.

weber Symbol: Wb The SI unit of magnetic flux, equal to the magnetic flux that, linking a circuit of one turn, produces an e.m.f. of one volt when reduced to zero at uniform rate in one second. 1 Wb = 1 V s. The unit is named for the German physicist Wilhelm Weber (1804–91).

white spirit A liquid hydrocarbon resembling kerosene obtained from petroleum, used as a solvent and in the manufacture of paints and varnishes.

Wieland, Heinrich Otto (1877–1957) German organic chemist. Much of Wieland's early work was devoted to the organic compounds of nitrogen. He started to study the bile acids in 1912. He found that three of these acids: cholic acid, deoxycholic acid, and lithocholic acid, are all steroids and are all related to cholesterol. This led him to propose a structure of cholesterol. He won the 1927 Nobel Prize for chemistry for his work on steroids. However, subsequent work by Wieland and his colleagues in 1932 showed that the structure he proposed was incorrect, leading to the now generally accepted modified structure being proposed by Wieland *et al.*

Wilkins, Maurice Hugh Frederick (1916–) New Zealand-born British molecular biologist. Maurice Wilkins was one of the key figures in the determination of the structure of DNA. He was originally a physicist but turned to biophysics after the end of World War II. He began to study DNA by x-ray diffraction. Some of the x-ray diffraction pictures produced by his colleague Rosalind FRANKLIN provided essential clues to Francis CRICK and James WATSON in their search for the structure of DNA. Wilkins shared the 1962 Nobel Prize for Medicine with Crick and Watson. Wilkins also determined the structure of ribonucleic acid (RNA) using x-ray diffraction. In 2003 Wilkins published his autobiography.

Wilkinson, Sir Geoffrey (1921–96) British chemist. Wilkinson did a lot of notable work on 'sandwich compounds', i.e. molecules such as ferrocene in which an iron atom is sandwiched between two carbon rings which each have five sides. Wilkinson studied many organo-metallic compounds including $RhCl(P(C_6H_5)_3)$. This compound is known as *Wilkinson's catalyst*. It is used as a catalyst in the hydrogenation of alkenes. Wilkinson shared the 1973 Nobel Prize for chemistry with Ernst Otto Fischer for their work on sandwich compounds.

Williamson's synthesis **1.** A method of preparing simple ethers by dehydration of alcohols with concentrated sulfuric acid. The reaction is carried out at 140°C under reflux with an excess of the alcohol:

$$2ROH \rightarrow ROR + H_2O$$

The concentrated sulfuric acid both catalyzes the reaction and displaces the equilibrium to the right. Also the ether may be distilled off during the reaction (in which case it is called *Wilkinson's continuous process*). The product, ether, is termed 'simple', because the R groups are identical. There are two possible mechanisms for the process, depending on the nature of the alcohol. In the case of primary alcohols, there is a hydrogensulfate formed. For example, with ethanol:

$$C_2H_5OH + H_2SO_4 \rightleftharpoons C_2H_5O.SO_2.OH + H_2O$$

With another alcohol molecule, an oxonium ion is formed:

$$C_2H_5OH + C_2H_5O.SO_2.OH \rightarrow (C_2H_5)_2OH^+ + HSO_4^-$$

The oxonium ion loses a proton to give the ether:

$$(C_2H_5)_2OH^+ + HSO_4^- \rightarrow C_2H_5O + H_2SO_4$$

In the case of tertiary alcohols the first step is production of a carbocation. For example, with isobutanol (2-methylpronan-2-ol; $(CH_3)_3COH$)):

$$(CH_3)_3COH + H^+ \rightarrow H_2O + (CH_3)_3C^+$$

In such cases the ion is stabilized by the alkyl groups. The ion is attached by the lone pair on the oxygen of another alcohol molecule to form an oxonium ion:

$$(CH_3)_3C^+ + (CH_3)_3COH \rightarrow ((CH_3)_3C)_2OH^+$$

As in the above mechanism, this loses a proton to give the ether:

$$((CH_3)_3C)_2OH^+ \rightarrow (CH_3)_3C.O.C(CH_3)_3 + H^+$$

2. A method for the preparation of mixed ethers by nucleophilic substitution. A haloalkane is refluxed with an alcoholic solution of sodium alkoxide (from sodium dissolved in alcohol):

$$R^1Cl + R^2O^-Na^+ \rightarrow R^1OR^2 + NaCl$$

The product ether is termed 'mixed' if the alkyl groups R^1 and R^2 are different. This synthesis can be used to produce both simple and mixed ethers.

Both reactions are named for the British chemist Alexander Williamson (1824–1904).

will-o'-the-wisp *See* ignis fatuus.

Willstätter, Richard (1872–1942) German organic chemist. Willstätter's early work was concerned with the structure of alkaloids such as atropine and cocaine. In 1905 he started the work on plant pigments such as chlorophyll for which he is best known. Using the technique of chromatography he was able to establish that chlorophyll is not a single compound. He was able to work out the chemical formulae of these compounds. He won the 1915 Nobel Prize for chemistry for his work on plant pigments.

Windaus, Adolf Otto Reinhold (1876–1959) German chemist. He was awarded the Nobel Prize for chemistry in 1928 for his research into the constitution of the sterols and their connection with the vitamins.

Wöhler, Friedrich (1800–82) German chemist. Wöhler is most famous for his discovery in 1828 that urea can be made by heating ammonium thiocyanate. The significance of this discovery was that it was the first time an organic substance had been synthesized in the laboratory. Previously there was a widespread view, called vitalism, that organic substances could be synthesized only by living organisms. Together with Justus von LIEBIG, he was responsible for a number of important discoveries in organic chemistry such as isomerism. Wöhler also made significant discoveries in inorganic chemistry. This included the isolation of aluminium (1827) and beryllium (1828).

Wöhler's synthesis A synthesis of urea by evaporating the inorganic compound ammonium cyanate ($NH_4^+NCO^-$), performed in 1828 by Friedrich WÖHLER. At the time it was believed that there was a distinction between organic and inorganic compounds in that organic compounds could be made only be living organisms (an idea known as *vitalism*). Urea is present in the urine of mammals and was regarded as definitely 'organic' in this sense. Ammonium cyanate is a definite inorganic compound. The discovery is sometimes said to mark the death of vitalism, although, in fact, it was many years before the idea was finally abandoned.

wood alcohol *See* methanol.

Woodward, Robert Burns (1917–79) American organic chemist. Robert Woodward and his colleagues synthesized many complicated molecules. In 1944 together with William von Eggers Doering, Woodward succeeded in synthesizing quinine. In the late 1940s he determined the structures of penicilin and strychnine. Among the molecules he synthesized were cholesterol and cortisone, lysergic acid (1954), chlorophyll (1960), and vitamin B_{12} (1971). His work on vitamin B_{12} led Woodward and Roald HOFFMANN to put forward the Woodward–Hoffmann rules governing the conservation of orbital symmetry in the mid 1960s. Woodward and Hoffmann gave an account of this work in the book *Conservation of Orbital Symmetry* (1970). Woodward won the 1965 Nobel Prize for chemistry for his work in synthesizing complex organic molecules. He would have shared another Nobel Prize with Hoffmann if he had not died in 1979.

Woodward–Hoffmann rules A set of rules used in analyzing the progress of PERICYCLIC REACTIONS. They are based on a fundamental analysis of the symmetry of the p orbitals in the reactants and how it correlates with the symmetry of the products. It is an alternative to the related FRONTIER ORBITAL theory. This approach was put forward in the 1960s by Woodward and the Polish–American theoretical chemist Roald HOFFMANN.

Wurtz–Fittig reaction *See* Wurtz reaction.

Wurtz reaction A reaction for preparing alkanes by refluxing a haloalkane (RX) with sodium metal in dry ether:

$$2RX + 2Na \rightarrow RR + 2NaX$$

The reaction involves the coupling of two alkyl radicals. The *Fittig reaction* is a similar process for preparing alkyl-benzene hydrocarbons by using a mixture of halogen compounds. For example, to obtain methylbenzene:

$$C_6H_5Cl + CH_3Cl + 2Na \rightarrow C_6H_5CH_3 + 2NaCl$$

In this mixed reaction phenylbenzene ($C_6H_5C_6H_5$) and ethane (CH_3CH_3) are also produced by side reactions.

The Wurtz reaction is named for the French chemist Charles Adolphe Wurtz (1817–84), who developed the method in 1855. The Fittig reaction is named for Rudolph Fittig (1835–1910), a German chemist who worked with Wurtz. It is often called the *Wurtz–Fittig reaction*.

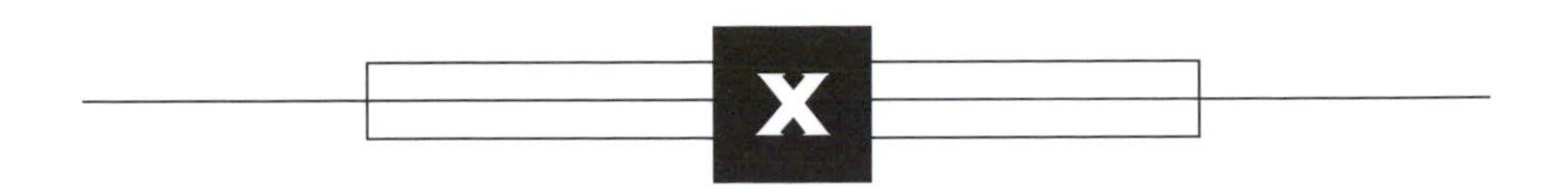

xanthate A salt or ester of XANTHIC ACID containing the ion $^{-}SCS(OR)$ or the group $-SCS(OR)$ (where R is an organic group). Cellulose xanthate is used to make RAYON.

xanthene ($CH_2(C_6H_4)_2O$) A yellow crystalline organic compound, used in making dyestuffs and fungicides.

xanthic acid ($HSCS(OR)$) Any of several unstable organic acids (where R is an organic group). The esters and salts of xanthic acid have various industrial applications. *See* xanthate.

xanthine (2,6-dioxypurine; $C_5H_4N_2O_2$) A poisonous colorless crystalline organic compound that occurs in blood, coffee beans, potatoes, and urine. It is used as a chemical intermediate.

xanthone (dibenzo-4-pyrone; $CO(C_6H_4)_2O$) A colorless crystalline organic compound found as a pigment in gentians and other flowers. It is used as an insecticide and in making dyestuffs.

xanthophyll One of a class of yellow to orange pigments derived from carotene, the commonest being lutein. *See* carotenoids; photosynthetic pigments.

xanthoproteic test A standard test for proteins. Concentrated nitric acid is added to the test solution. A yellow precipitate produced either immediately or on gentle heating indicates a positive result.

x-radiation An energetic form of electromagnetic radiation. The wavelength range is 10^{-11} m to 10^{-8} m. X-rays are normally produced by absorbing high-energy electrons in matter. The radiation can pass through matter to some extent (hence its use in medicine and industry for investigating internal structures). It can be detected with photographic emulsions and devices like the Geiger-Müller tube.

X-ray photons result from electronic transitions between the inner energy levels of atoms. When high-energy electrons are absorbed by matter, an x-ray line spectrum results. The structure depends on the substance and is thus used in x-ray spectroscopy. The line spectrum is always formed in conjunction with a continuous background spectrum. The minimum (cutoff) wavelength λ_0 corresponds to the maximum x-ray energy, W_{max}. This equals the maximum energy of electrons in the beam producing the x-rays. Wavelengths in the continuous spectrum above λ_0 are caused by more gradual energy loss by the electrons, in the process called *bremsstrahlung* (braking radiation).

x-ray crystallography The study of the internal structure of crystals using the technique of x-ray diffraction.

x-ray diffraction A technique used to determine crystal structure by directing x-rays at the crystals and examining the diffraction patterns produced. At certain angles of incidence a series of spots are produced on a photographic plate; these spots are caused by interaction between the x-rays and the planes of the atoms, ions, or molecules in the crystal lattice.

x-rays *See* x-radiation.

xylene *See* dimethylbenzene.

ylid (ylide) A type of ZWITTERION in which the two charges are on adjacent atoms.

ylide *See* ylid.

yocto- Symbol: y A prefix denoting 10^{-24}. For example, 1 yoctometer (ym) = 10^{-24} meter (m).

yotta- Symbol: Y A prefix denoting 10^{24}. For example, 1 yottameter (Ym) = 10^{24} meter (m).

Z

Zeisel reaction The reaction of an ether with excess concentrated hydroiodic acid. On refluxing, a mixture of iodoalkanes is formed:

$$ROR' + 2HI \rightarrow H_2O + RI + R'I$$

Analysis to identify the iodoalkanes gives information about the composition of the original ether. It was developed by S. Zeisel in 1886.

zeolite A member of a group of hydrated aluminosilicate minerals, which occur in nature and are also manufactured for their ion-exchange and selective-absorption properties. They are used for water softening and for sugar refining. The zeolites have an open crystal structure and can be used as molecular sieves.

See also ion exchange; molecular sieve.

zepto- Symbol: z A prefix denoting 10^{-21}. For example, 1 zeptometer (zm) = 10^{-21} meter (m).

zero order Describing a chemical reaction in which the rate of reaction is independent of the concentration of a reactant; i.e.

$$\text{rate} = k[X]^0$$

The concentration of the reactant remains constant for a period of time although other reactants are being consumed. The hydrolysis of 2-bromo-2-methylpropane using aqueous alkali has a rate expression,

$$\text{rate} = k[\text{2-bromo-2-methylpropane}]$$

i.e. the reaction is zero order with respect to the concentration of the alkali. The rate constant for a zero reaction has the units mol dm^{-3} s^{-1}.

zero point energy The energy possessed by the atoms and molecules of a substance at absolute zero (0 K).

zetta- Symbol: Z A prefix denoting 10^{21}. For example, 1 zettameter (Zm) = 10^{21} meter (m).

Ziegler, Karl (1898–1973) German organic chemist. Karl Ziegler is best known for his research into polymers, particularly *Ziegler–Natta catalysts*. In 1953 Ziegler found that catalysts consisting of organometallic compounds mixed with metals such as titanium polymerize ethene into a long-chain polymer with useful properties such as a high melting point. This method of producing polymers does not need high temperatures or pressures. This type of catalysis was developed further by Giulio NATTA. Ziegler and Natta shared the 1963 Nobel Prize for chemistry for their work on Ziegler–Natta catalysts.

Ziegler process A method for the manufacture of high-density polyethene using a catalyst of titanium(IV]] chloride and triethyl aluminum ($Al(C_2H_5)_3$) under slight pressure. The mechanism involves formation of titanium alkyls

$$TiCl_3(C_2H_5)$$

which coordinate the sigma orbitals of titanium with the π bond of ethene. The chain-length, and henfce the density, of the polymer can be controlled. It is named for the German chemist Karl Ziegler (1896–1973), who introduced it in 1953. *See also* Natta process.

zwitterion (ampholyte ion) An ion that has both a positive and negative charge on the same species. Zwitterions occur when a molecule contains both a basic group and an acidic group; formation of the ion can be regarded as an internal acid-base reaction. For example, amino-ethanoic acid (glycine) has the formula $H_2N.CH_2.$-$COOH$. Under neutral conditions it exists as the zwitterion $H_3N.CH_2.COO^-$, which can be formed by transfer of a proton from the carboxyl group to the amine group. At low pH (acidic conditions) the ion $^+H_3N.$-$CH_2.COOH$ is formed; at high pH (basic conditions) $H_2N.CH_2.COO^-$ is formed. *See also* amino acid.

APPENDIXES

Appendix I

Carboxylic Acids

In the examples below, the systematic name is given first, followed by the trivial (common) name.

Simple saturated monocarboxylic acids:

methanoic	formic	$HCOOH$
ethanoic	acetic	CH_3COOH
propanoic	proprionic	C_2H_5COOH
butanoic	butyric	C_3H_7COOH
pentanoic	valeric	C_4H_9COOH
hexanoic	caproic	$C_5H_{11}COOH$
heptanoic	enathic	$C_6H_{13}COOH$
octanoic	caprylic	$C_7H_{15}COOH$
nonanoic	pelargonic	$C_8H_{17}COOH$
decanoic	capric	$C_9H_{19}COOH$

Other simple saturated acids are found in naturally occurring glycerides. They all contain even numbers of carbon atoms:

dodecanoic	lauric	$C_{11}H_{23}COOH$
tetradecanoic	myristic	$C_{13}H_{27}COOH$
hexadecanoic	palmitic	$C_{15}H_{31}COOH$
octadecanoic	stearic	$C_{17}H_{35}COOH$
eicosanoic	arachidic	$C_{19}H_{39}COOH$
docosanoic	behenic	$C_{21}H_{43}COOH$
tetracosanoic	lignoceric	$C_{23}H_{47}COOH$
hexacosanoic	cerotic	$C_{25}H_{51}COOH$

Certain important unsaturated monocarboxylic acids occur naturally:

octadec-9-enoic	oleic	$C_8H_{17}CH=CHC_7H_{14}COOH$
octadeca-9,12- dienoic	linoleic	$C_5H_{11}CH=CHCH_2CH=CHC_7H_{14}COOH$
octadeca-9,12,15- trienoic	linolenic	$C_2H_5CH=CHCH_2CH=CHCH_2CH=CH-C_7H_{14}COOH$

There are a number of common saturated dicarboxylic acids:

ethanedioic	oxalic	$HOOCCOOH$
propanedioic	malonic	$HOOCCH_2COOH$
butanedioic	succinic	$HOOCC_2H_4COOH$
pentanedioic	glutaric	$HOOCC_3H_6COOH$
hexanedioic	adipic	$HOOCC_4H_8COOH$

Examples of unsaturated dicarboxylic acids are:

cis-butenedioic	maleic	HOOCCH=CHCOOH
trans-butenedioic	fumaric	HOOCCH=CHCOOH

Certain hydroxy acids occur naturally:

hydroxyethanoic	glycolic	$CH_2(OH)COOH$
2-hydroxypropanoic	lactic	$CH_3CH(OH)COOH$
hydroxybutanedioic	malic	$CH(OH)CH_2(COOH)_2$
2-hydroxy-propane-1,2,3-tricarboxylic	citric	$(CH_2)_2C(OH)(COOH)_3$

Naturally occurring amino carboxylic acids are shown in Appendix II.

Appendix II

Amino Acids

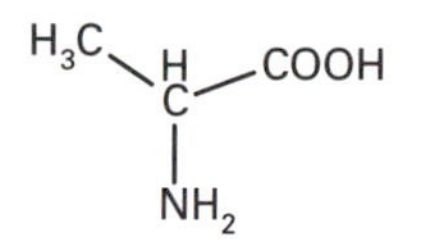

alanine

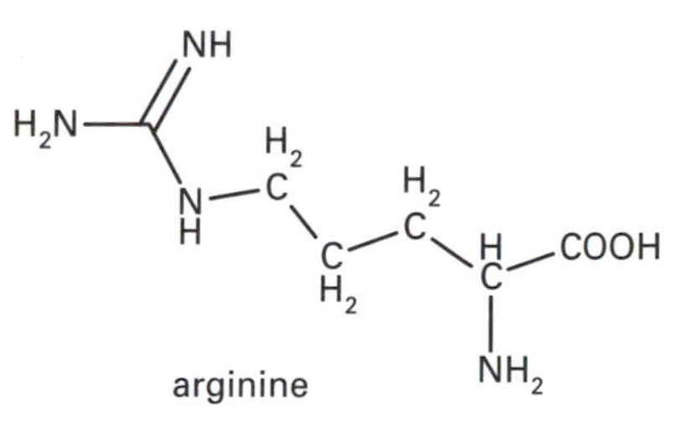

arginine

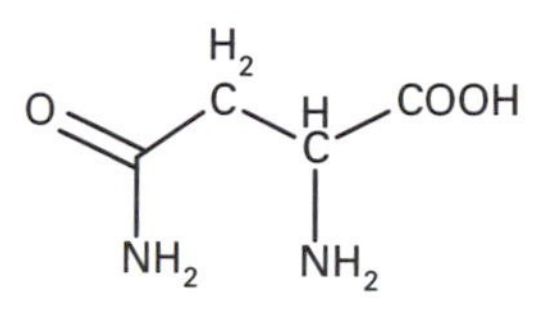

asparagine

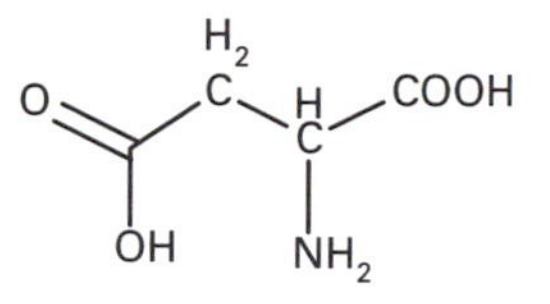

aspartine

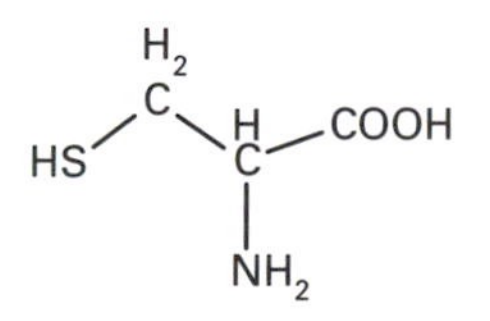

cysteine

HOOC

H_2

C

C

H_2

H

C

COOH

NH_2

glutamic acid

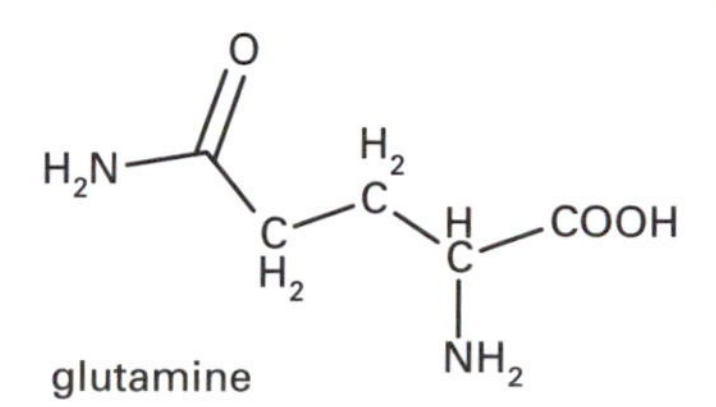

glutamine

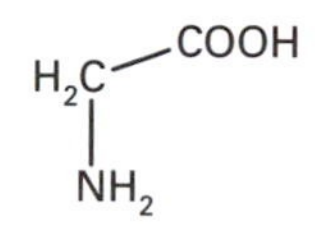

glycine

Amino Acids

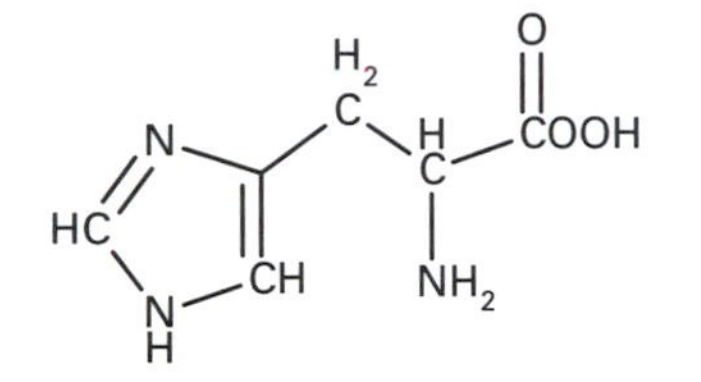

histidine

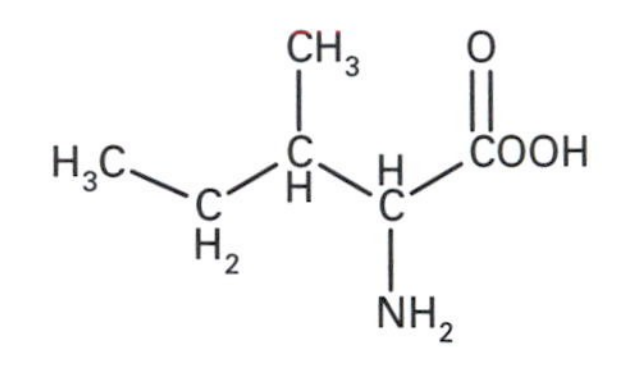

isoleucine

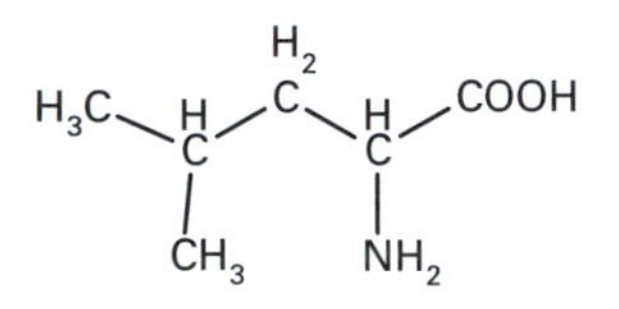

leucine

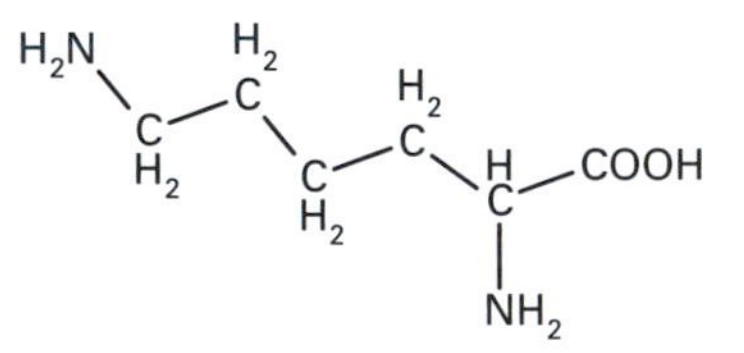

lysine

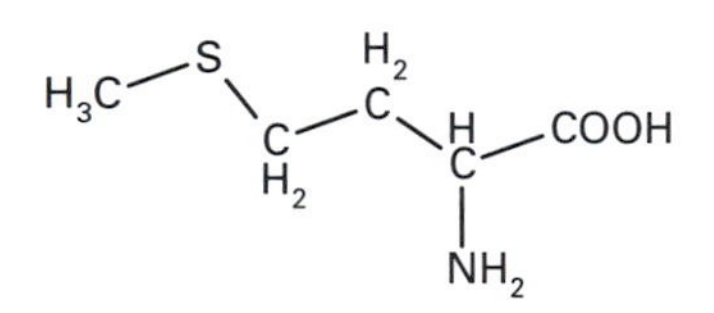

methionine

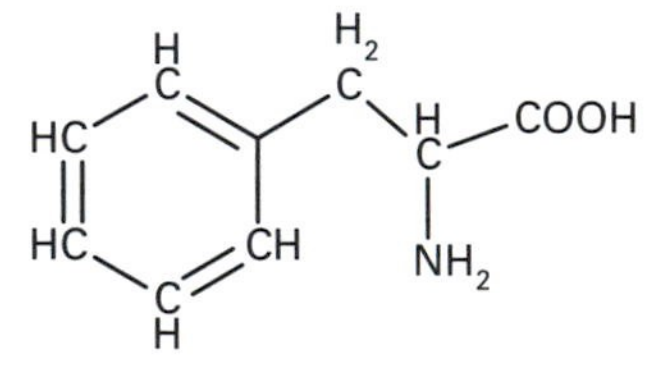

phenylalanine

proline

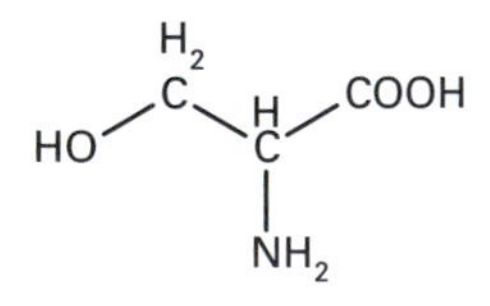

serine

Amino Acids

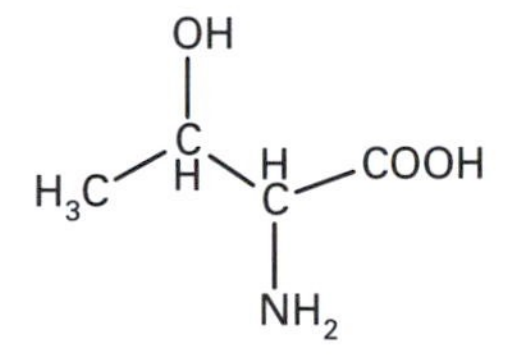

threonine

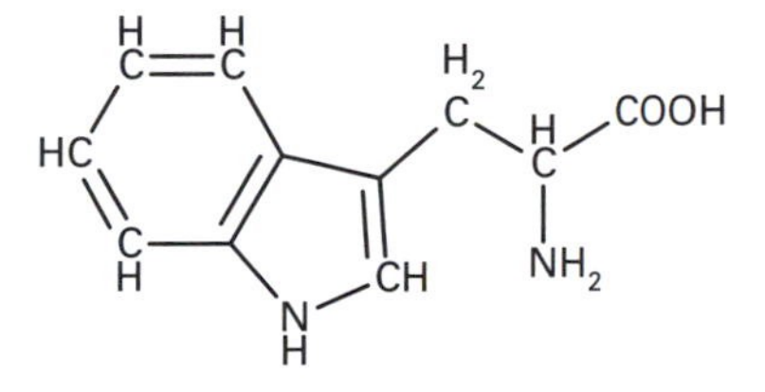

tryptophan

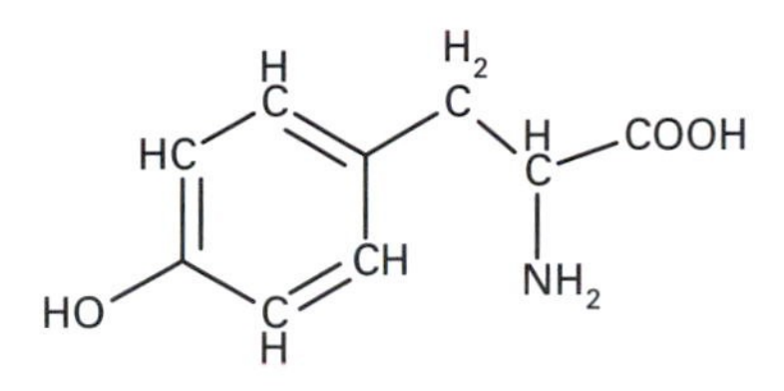

tyrosine

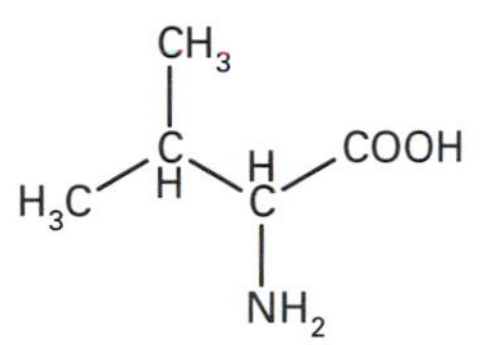

valine

Appendix III

Sugars

Some simple monosaccharides. The β-D-form is shown in each case

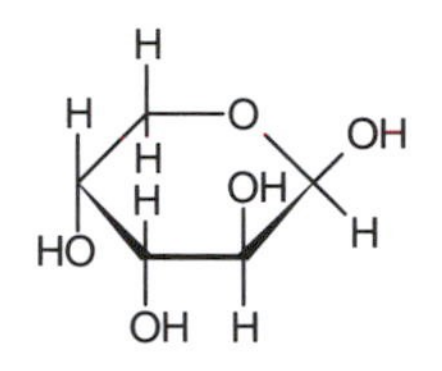

arabinose

fructose

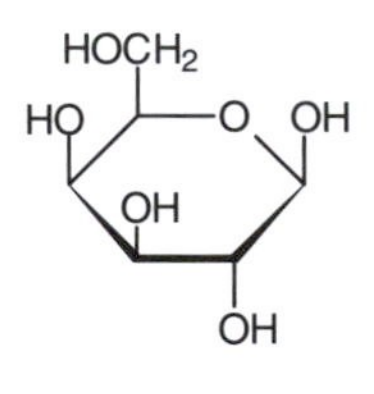

galactose

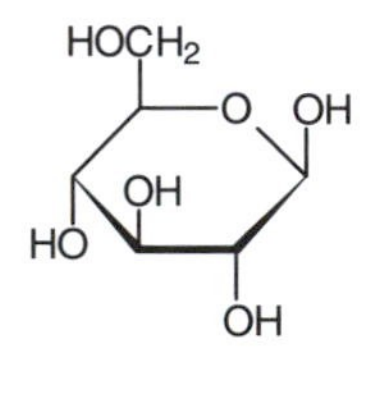

glucose

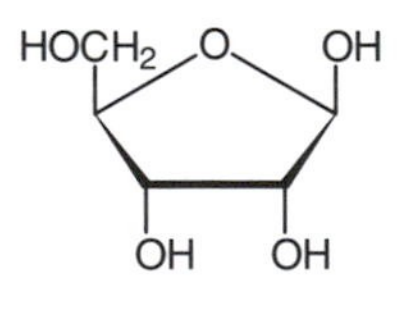

ribose

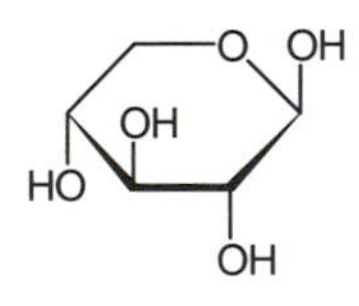

xylose

Appendix IV

Nitrogenous Bases and Nucleosides

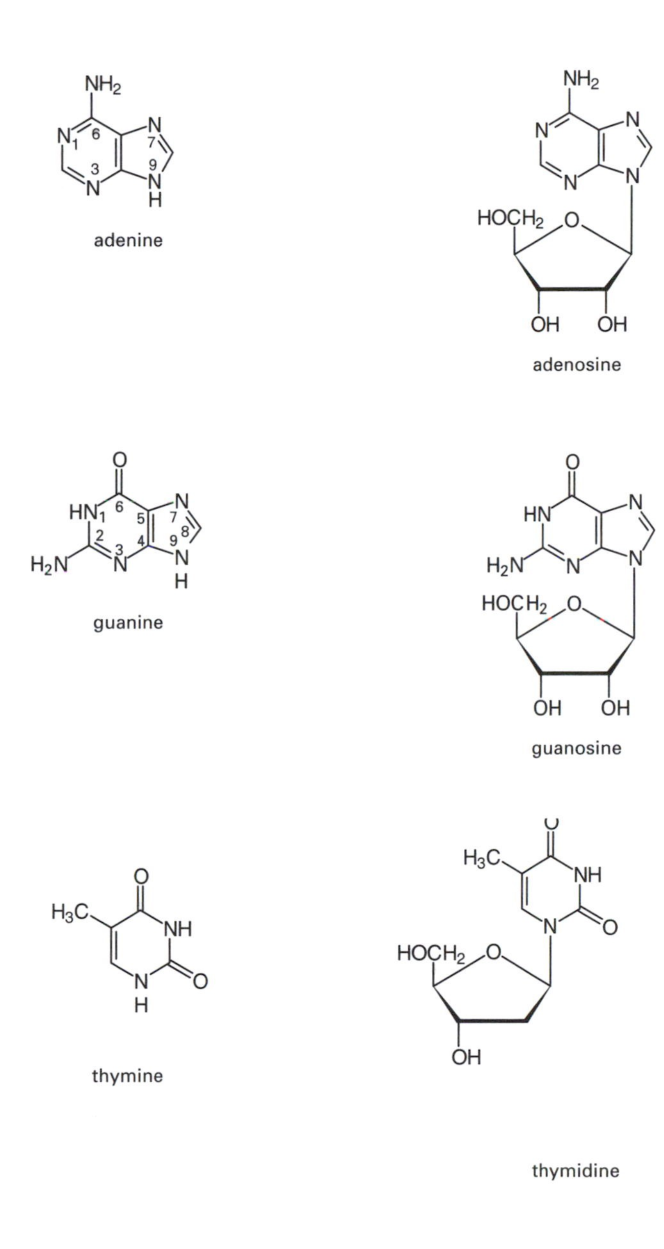

Nitrogenous Bases and Nucleosides

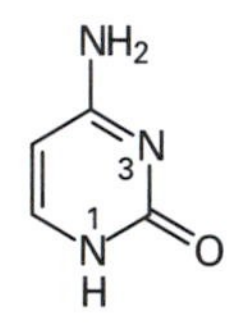

cytosine

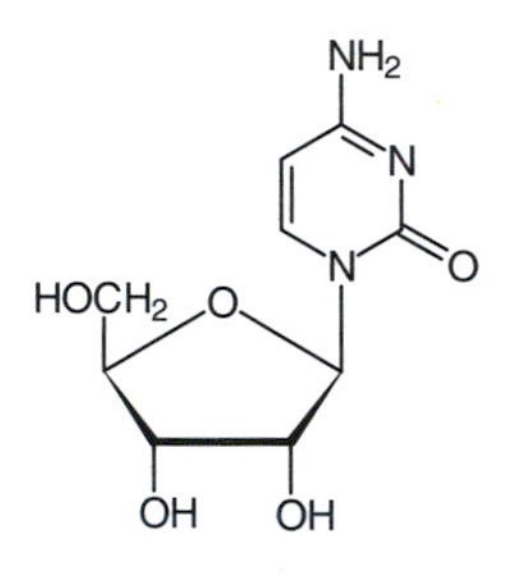

cytidine

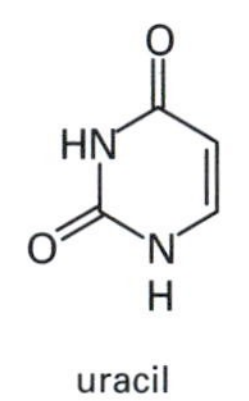

uracil

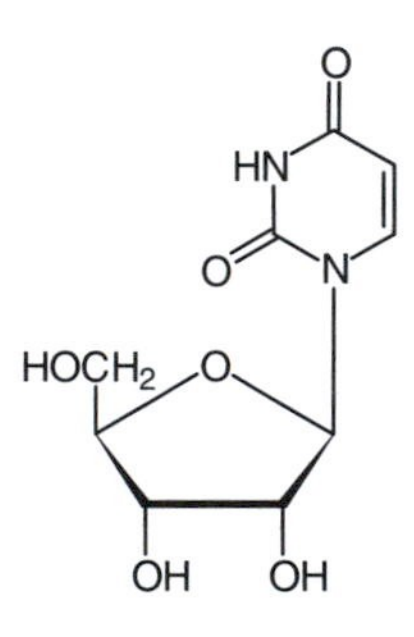

uridine

Appendix V

The Chemical Elements

(* *indicates the nucleon number of the most stable isotope*)

Element	*Symbol*	*p.n.*	*r.a.m*	*Element*	*Symbol*	*p.n.*	*r.a.m*
actinium	Ac	89	227*	europium	Eu	63	151.965
aluminum	Al	13	26.982	fermium	Fm	100	257*
americium	Am	95	243*	fluorine	F	9	18.9984
antimony	Sb	51	112.76	francium	Fr	87	223*
argon	Ar	18	39.948	gadolinium	Gd	64	157.25
arsenic	As	33	74.92	gallium	Ga	31	69.723
astatine	At	85	210	germanium	Ge	32	72.61
barium	Ba	56	137.327	gold	Au	79	196.967
berkelium	Bk	97	247*	hafnium	Hf	72	178.49
beryllium	Be	4	9.012	hassium	Hs	108	265*
bismuth	Bi	83	208.98	helium	He	2	4.0026
bohrium	Bh	107	262*	holmium	Ho	67	164.93
boron	B	5	10.811	hydrogen	H	1	1.008
bromine	Br	35	79.904	indium	In	49	114.82
cadmium	Cd	48	112.411	iodine	I	53	126.904
calcium	Ca	20	40.078	iridium	Ir	77	192.217
californium	Cf	98	251*	iron	Fe	26	55.845
carbon	C	6	12.011	krypton	Kr	36	83.80
cerium	Ce	58	140.115	lanthanum	La	57	138.91
cesium	Cs	55	132.905	lawrencium	Lr	103	262*
chlorine	Cl	17	35.453	lead	Pb	82	207.19
chromium	Cr	24	51.996	lithium	Li	3	6.941
cobalt	Co	27	58.933	lutetium	Lu	71	174.967
copper	Cu	29	63.546	magnesium	Mg	12	24.305
curium	Cm	96	247*	manganese	Mn	25	54.938
darmstadtium	Ds	110	269*	meitnerium	Mt	109	266*
dubnium	Db	105	262*	mendelevium	Md	101	258*
dysprosium	Dy	66	162.50	mercury	Hg	80	200.59
einsteinium	Es	99	252*	molybdenum	Mo	42	95.94
erbium	Er	68	167.26	neodymium	Nd	60	144.24

The Chemical Elements

Element	*Symbol*	*p.n.*	*r.a.m*	*Element*	*Symbol*	*p.n.*	*r.a.m*
neon	Ne	10	20.179	scandium	Sc	21	44.956
neptunium	Np	93	237.048	seaborgium	Sg	106	263*
nickel	Ni	28	58.69	selenium	Se	34	78.96
niobium	Nb	41	92.91	silicon	Si	14	28.086
nitrogen	N	7	14.0067	silver	Ag	47	107.868
nobelium	No	102	259*	sodium	Na	11	22.9898
osmium	Os	76	190.23	strontium	Sr	38	87.62
oxygen	O	8	15.9994	sulfur	S	16	32.066
palladium	Pd	46	106.42	tantalum	Ta	73	180.948
phosphorus	P	15	30.9738	technetium	Tc	43	99*
platinum	Pt	78	195.08	tellurium	Te	52	127.60
plutonium	Pu	94	244*	terbium	Tb	65	158.925
polonium	Po	84	209*	thallium	Tl	81	204.38
potassium	K	19	39.098	thorium	Th	90	232.038
praseodymium	Pr	59	140.91	thulium	Tm	69	168.934
promethium	Pm	61	145*	tin	Sn	50	118.71
protactinium	Pa	91	231.036	titanium	Ti	22	47.867
radium	Ra	88	226.025	tungsten	W	74	183.84
radon	Rn	86	222*	uranium	U	92	238.03
rhenium	Re	75	186.21	vanadium	V	23	50.94
rhodium	Rh	45	102.91	xenon	Xe	54	131.29
rubidium	Rb	37	85.47	ytterbium	Yb	70	173.04
ruthenium	Ru	44	101.07	yttrium	Y	39	88.906
rutherfordium	Rf	104	261*	zinc	Zn	30	65.39
samarium	Sm	62	150.36	zirconium	Zr	40	91.22

Appendix VI

Periodic Table of the Elements - giving group, atomic number, and chemical symbol

Period	1	2	3	4	5	6	7	8	9	10	11	12	13	14	15	16	17	18
1	1 **H**																	2 **He**
2	3 **Li**	4 **Be**											5 **B**	6 **C**	7 **N**	8 **O**	9 **F**	10 **Ne**
3	11 **Na**	12 **Mg**											13 **Al**	14 **Si**	15 **P**	16 **S**	17 **Cl**	18 **Ar**
4	19 **K**	20 **Ca**	21 **Sc**	22 **Ti**	23 **V**	24 **Cr**	25 **Mn**	26 **Fe**	27 **Co**	28 **Ni**	29 **Cu**	30 **Zn**	31 **Ga**	32 **Ge**	33 **As**	34 **Se**	35 **Br**	36 **Kr**
5	37 **Rb**	38 **Sr**	39 **Y**	40 **Zr**	41 **Nb**	42 **Mo**	43 **Tc**	44 **Ru**	45 **Rh**	46 **Pd**	47 **Ag**	48 **Cd**	49 **In**	50 **Sn**	51 **Sb**	52 **Te**	53 **I**	54 **Xe**
6	55 **Cs**	56 **Ba**	57-71 **La-Lu**	72 **Hf**	73 **Ta**	74 **W**	75 **Re**	76 **Os**	77 **Ir**	78 **Pt**	79 **Au**	80 **Hg**	81 **Tl**	82 **Pb**	83 **Bi**	84 **Po**	85 **At**	86 **Rn**
7	87 **Fr**	88 **Ra**	89-103 **Ac-Lr**	104 **Rf**	105 **Db**	106 **Sg**	107 **Bh**	108 **Hs**	109 **Mt**	110 **Ds**	111 **Uuu**	112 **Uub**	113	114 **Uuq**	115	116 **Uuh**		

Period																
6	Lanthanides	57 **La**	58 **Ce**	59 **Pr**	60 **Nd**	61 **Pm**	62 **Sm**	63 **Eu**	64 **Gd**	65 **Tb**	66 **Dy**	67 **Ho**	68 **Er**	69 **Tm**	70 **Yb**	71 **Lu**
7	Actinides	89 **Ac**	90 **Th**	91 **Pa**	92 **U**	93 **Np**	94 **Pu**	95 **Am**	96 **Cm**	97 **Bk**	98 **Cf**	99 **Es**	100 **Fm**	101 **Md**	102 **No**	103 **Lr**

The above is the modern recommended form of the table using 18 groups. Older group designations are shown below.

Modern form	1	2	3	4	5	6	7	8	9	10	11	12	13	14	15	16	17	18
European convention	IA	IIA	IIIA	IVA	VA	VIA	VIIA	VIII (or VIIIA)			IB	IIB	IIIB	IVB	VB	VIB	VIIB	0 (or VIIIB)
N. American convention	IA	IIA	IIIB	IVB	VB	VIB	VIIB	VIII (or VIIIB)			IB	IIB	IIIA	IVA	VA	VIA	VIIA	VIIIA (or 0)

Appendix VII

The Greek Alphabet

Α	α	alpha	Ν	ν	nu
Β	β	beta	Ξ	ξ	xi
Γ	γ	gamma	Ο	ο	omikron
Δ	δ	delta	Π	π	pi
Ε	ε	epsilon	Ρ	ρ	rho
Ζ	ζ	zeta	Σ	σ	sigma
Η	η	eta	Τ	τ	tau
Θ	θ	theta	Υ	υ	upsilon
Ι	ι	iota	Φ	φ	phi
Κ	κ	kappa	Χ	χ	chi
Λ	λ	lambda	Ψ	ψ	psi
Μ	μ	mu	Ω	ω	omega

Appendix VIII

Fundamental Constants

speed of light	c	$2.997\ 924\ 58 \times 10^{8}$ m s^{-1}
permeability of free space	μ_o	$4\pi \times 10^{-7}$
		$= 1.256\ 637\ 0614 \times 10^{-6}$ H m^{-1}
permittivity of free space	$\varepsilon_0 = \mu_0^{-1} c^{-2}$	$8.854\ 187\ 817 \times 10^{-12}$ F m^{-1}
charge of electron or proton	e	$\pm 1.602\ 177\ 33 \times 10^{-19}$ C
rest mass of electron	m_e	$9.109\ 39 \times 10^{-31}$ kg
rest mass of proton	m_p	$1.672\ 62 \times 10^{-27}$ kg
rest mass of neutron	m_n	$1.674\ 92 \times 10^{-27}$ kg
electron charge-to-mass ratio	e/m	$1.758\ 820 \times 10^{11}$ C kg^{-1}
electron radius	r_e	$2.817\ 939 \times 10^{-15}$ m
Planck constant	h	$6.626\ 075 \times 10^{-34}$ J s
Boltzmann constant	k	$1.380\ 658 \times 10^{-23}$ J K^{-1}
Faraday constant	F	$9.648\ 531 \times 10^{4}$ C mol^{-1}

Appendix IX

Webpages

Chemical society webpages include:

American Chemical Society	www.chemistry.org/
Royal Society of Chemistry	www.rsc.org/
The International Union of Pure and Applied Chemistry	www.iupac.org

Information on nomenclature is available at:

Queen Mary College, London	www.chem.qmul.ac.uk/iupac/
Advanced Chemistry Development, Inc	www.acdlabs.com/iupac/nomenclature/

An extensive set of organic chemistry links can be found at:

WWW Virtual Library, Chemistry Section	www.liv.ac.uk/Chemistry/Links/

Bibliography

There are a number of comprehensive texts covering organic chemistry. These include:

Carey, Francis & Sundberg, Richard J. *Advanced Organic Chemistry: Structure and Mechanism*. 4th ed. New York: Plenum, 2000

Carey, Francis & Sundberg, Richard J. *Advanced Organic Chemistry: Reactions*. 4th ed. New York: Plenum, 2000

Clayden, Jonathon; Greeves, Nick; Warren, Stuart; & Wothers, Peter *Organic Chemistry*. Oxford, U.K.: Oxford University Press, 2000

McMurray, John *Organic Chemistry*. 6th ed. Pacific Grove, Calif.: Brooks/Cole, 2004

March, Jerry *Advanced Organic Chemistry: Reactions, Mechanisms, and Structure*. 4th ed. New York: Wiley, 1992

Volhardt, Peter K. & Schore, Neil E. *Organic Chemistry: Structure and Function*. 4th ed. New York: W. H. Freeman, 2003

More specialized books are:

Eliel, Ernest L.; Wilen, Samuel H.; & Doyle, Michael P. *Basic Organic Stereochemistry*. New York: Wiley, 2001

Gilchrist, Thomas L. *Heterocyclic Chemistry*. 3rd ed. Harlow, U.K.: Longman, 1997

Isaacs, Neil S. *Physical Organic Chemistry*. Harlow, U.K.: Longman, 1987

Sykes, Peter *Guidebook to Mechanism in Organic Chemistry*. 6th ed. Harlow, U.K.: Longman, 1996

An advanced text on biochemistry is:

Nelson, David L. & Cox, Michael M. *Lehninger Principles of Biochemistry*. 3rd ed. New York: Worth Publishers, 2000